INTERNATIONAL SERIES OF MONOGRAPHS ON PHYSICS

INTERNATIONAL SERIES OF MONOGRAPHS ON PHYSICS

Nonlocal Gravity

Bahram Mashhoon

*Professor of Physics (retired), Department of Physics and Astronomy,
University of Missouri*

OXFORD
UNIVERSITY PRESS

OXFORD
UNIVERSITY PRESS

Great Clarendon Street, Oxford, OX2 6DP,
United Kingdom

Oxford University Press is a department of the University of Oxford.
It furthers the University's objective of excellence in research, scholarship,
and education by publishing worldwide. Oxford is a registered trade mark of
Oxford University Press in the UK and in certain other countries

© Bahram Mashhoon 2017

The moral rights of the author have been asserted

First Edition published in 2017

Impression: 1

Published in the United States of America by Oxford University Press
198 Madison Avenue, New York, NY 10016, United States of America

British Library Cataloguing in Publication Data

Data available

Library of Congress Control Number: 2017935192

ISBN 978–0–19–880380–5

Printed and bound by
CPI Group (UK) Ltd, Croydon, CR0 4YY

Preface

Gravitation is the only basic physical interaction that does not yet have a basis in microphysics. Moreover, no relativistic quantum gravity phenomenon has been experimentally identified up to now. It is indeed not clear at this time how to bring Einstein's macrophysical description of gravitation into harmony with the quantum theory. Therefore, this book deals with Einstein's classical theory of general relativity.

The theory of relativity is based on a postulate of *locality*. In this context, locality means that the measured value of a physical quantity at an event in space-time is directly influenced only by quantities in its immediate neighborhood. In fact, in relativity theory, observables are spacetime scalars that result from the projection of physical quantities onto the local (tetrad) frames of the observers. In this process, locality is maintained if the result is not directly affected by variables from the past; that is, relativistic physics becomes nonlocal if the past history of the observer has to be taken into account. Nonlocality is thus associated with spacetime memory.

In relativity theory, locality is first invoked in the special theory of relativity when Lorentz invariance is extended to accelerated observers. Lorentz invariance is a basic symmetry of nature and Lorentz transformations relate the measurements of uniformly moving ideal inertial observers. An inertial observer moves on a straight world line from minus infinity to plus infinity with constant speed. Thus inertial observers have little to do with actual observers that are all more or less accelerated. What does an accelerated observer measure? According to special relativity theory, Lorentz transformations must be applied event by event along the world line of the accelerated observer in order to interpret its measurements. This makes physical sense, of course, if during the measurement of a physical quantity the velocity of the accelerated observer can be considered effectively constant; that is, the locality postulate is valid if the accelerated observer can be considered locally inertial during an elementary act of measurement.

Classically, motion occurs via particles as well as electromagnetic waves. The interaction of classical particles and rays of radiation, in the geometric optics limit, can indeed be reduced to pointlike coincidences. Clearly, the pointwise application of Lorentz transformations would make physical sense for classical point particles and rays. However, electromagnetic waves are in general intrinsically nonlocal. Imagine, for instance, the measurement of the properties of an incident electromagnetic wave packet by an accelerated observer. To ascertain the frequency and wavelength content of the wave packet, instantaneous Lorentz transformations must be employed over an extended period of time, during which the classical state of the accelerated observer necessarily undergoes various changes. Furthermore, Bohr and Rosenfeld pointed out in 1933 that, as a matter of principle, classical electric and magnetic fields cannot be measured instantaneously and only their spacetime averages have immediate physical significance. Thus in the process of field measurement, the past history of the observer must be taken into account. This issue acquires added urgency when one recognizes that accelerated observers are endowed with intrinsic invariant length- and timescales associated with their motions. The acceleration scales can be constructed from the speed of light and the appropriate magnitudes of the observers' translational and rotational accelerations.

In 1993, I published a nonlocal theory of accelerated systems in Minkowski space-time. In formulating a theory that depends upon the past history of an observer, a *kernel* function must be introduced that can in general depend upon both the past and the present. The acceleration kernel acts as a weight function for the averaging process that takes the past history of the observer into account. To determine the kernel, I used the phenomenon of spin–rotation coupling to formulate the physical principle that no observer can stand completely still with respect to a fundamental radiation field. Implementing this principle, I obtained an integral equation for the kernel function that can have many possible solutions. Furthermore, to obtain a unique solution, I chose a simplifying assumption regarding the functional form of the kernel that is usually adopted in the dynamics of continuous media to describe memory-dependent phenomena. Further work revealed that my simplifying choice for the kernel could give unphysical results if the observer's acceleration is not uniform. The correct answer turned out to be a kernel function that only depends upon the past. The complete resolution of the problem was achieved, thanks mainly to significant contributions by Carmen Chicone, in two papers that Chicone and I published in 2002. The path was now clear for the development of a *nonlocal special relativity theory*, which was accomplished by 2008 in close collaboration with Carmen Chicone and Friedrich W. Hehl. I devote the first three chapters of this book to the explanation of the main tenets of this nonlocal extension of special relativity theory in Minkowski spacetime.

Once history dependence has been incorporated into the physics of accelerated observers in Minkowski spacetime, it would seem natural to extend the theory to the gravitational domain. The basis for such an extension is Einstein's fundamental insight that the principle of equivalence of inertial and gravitational masses implies a deep connection between inertia and gravitation. The universality of free fall is a crucial and observationally well-established property of gravitation within the classical domain of physics. In the context of Newtonian gravity, this experimental result implies the principle of equivalence of inertial and gravitational masses. Einstein interpreted this principle to mean that there exists in nature a profound relationship between acceleration and gravity. Einstein's heuristic principle of equivalence, which is the cornerstone of his general theory of relativity, establishes a definite *local* relationship between an observer in a gravitational field and an accelerated observer in Minkowski spacetime. The accelerated observer is locally (i.e. pointwise) inertial by the postulate of locality; therefore, Einstein's principle of equivalence renders observers pointwise inertial in a gravitational field and opens the path toward a geometric interpretation of gravitation as the Riemannian curvature of the spacetime manifold.

It is important to observe that Einstein's principle of equivalence loses its basic operational significance if one does not know a priori what accelerated observers in Minkowski spacetime actually measure; that is, the locality postulate plays a crucial role in Einstein's extension of relativity theory to the gravitational domain. In nonlocal special relativity theory in Minkowski spacetime, an accelerated observer in general carries the memory of its past acceleration. Invoking Einstein's original insight in this more general circumstance, one expects that gravity should be nonlocal as well and in nonlocal gravity, the gravitational memory of past events must be taken into account. A gravitational *kernel* is thus needed to incorporate history dependence into gravitation theory. However, it is not possible to deduce the kernel of nonlocal gravity theory from the kernel of accelerated observers in Minkowski spacetime; indeed, one cannot invoke Einstein's principle of equivalence in this case due to its inherent extreme locality. To develop a nonlocal general relativity via a direct extension of nonlocal special relativity, one must go beyond Einstein's strictly local principle of equivalence. For instance, suppose that we start with Einstein's linear approximation to general relativity theory as a classical spin-2 field in Minkowski spacetime. Nonlocal theory of accelerated systems can then be applied in this case to generate a linear nonlocal spin-2 field theory as determined by accelerated observers in Minkowski spacetime. It is however not clear how to generalize in this case Einstein's local identification of an observer in a gravitational field with a certain accelerated observer in Minkowski spacetime. Thus, as described in detail in Chapter 4, a *direct* attempt at a nonlocal generalization of Einstein's theory of gravitation appears to be futile. It seems that an indirect approach may be necessary.

Einstein's general relativity is a field theory of gravitation patterned after Maxwell's field theory of electromagnetism. It is interesting to recall that the electrodynamics of media is inherently *nonlocal*; that is, in general, nonlocal constitutive relations naturally arise in the treatment of electrodynamics of bulk matter (Jackson 1999). This circumstance directly leads to nonlocal Maxwell's equations for the electrodynamics of continuous media. Nonlocal characterization of the properties of continua has a long history (Poisson 1823; Liouville 1837; Hopkinson 1877); indeed, the corresponding memory-dependent phenomena, such as hysteresis, have been the subject of many investigations—see, for example, Bertotti (1998). Along this line of thought, one wonders whether a similar constitutive approach can be adopted for gravitation. Can general relativity be rendered nonlocal in analogy with the electrodynamics of media?

It is possible to arrive at general relativity (GR) from the standpoint of the gauge theories of gravitation. Indeed, the gauge approach to gravity naturally leads to spacetime theories with curvature and torsion. There is a spectrum of such theories such that at one end of the spectrum, one has GR based upon a pseudo-Riemannian spacetime manifold with only curvature and no torsion, while at the other end of the spectrum are spacetime theories with torsion and no curvature. Of the latter, there is a unique one that is essentially equivalent to Einstein's general relativity: this is the teleparallel equivalent to general relativity ($GR_{||}$), where gravity is described in terms of local frames in Weitzenböck spacetime. Teleparallelism has a long history; its application to gravitational physics has been considered by many authors starting with Einstein in 1928. $GR_{||}$ is the gauge theory of the Abelian group of spacetime translations. As such, it bears a certain formal resemblance to electrodynamics, which is the gauge theory of the Abelian U(1) group. The analogy with electrodynamics led Friedrich W. Hehl to suggest that one should attempt a nonlocal $GR_{||}$ through a nonlocal constitutive relation as an indirect way of constructing a nonlocal generalization of GR. This fruitful suggestion was then developed in two papers that Hehl and I published in 2009. Within our formal framework for a nonlocal theory of gravity, the kernel must satisfy certain requirements but is otherwise undetermined. It is possible that the kernel could be derived from a more comprehensive future theory.

In nonlocal gravity, the gravitational field is local but satisfies partial integro-differential field equations. Using simple assumptions regarding the constitutive kernel, our preliminary studies revealed that gravity can be nonlocal even in the Newtonian limit; that is, in the Newtonian regime, we find an integro-differential equation for the gravitational potential. This equation can be expressed as the Poisson equation of Newtonian gravitation, except that the source term now includes, in addition to matter density, a term induced by nonlocality that is reminiscent of the density of *dark matter*. Surprised by this development, we sent a preliminary version of our paper to the late Jacob Bekenstein for his comments. He kindly pointed out to us that our modified Poisson equation had already been proposed by Jeffrey R. Kuhn in the 1980s. From Bekenstein we learned of the Tohline–Kuhn modified gravity approach to the explanation of the "flat" rotation curves of spiral galaxies. The nonlocally modified Newtonian gravitation appears to provide a natural explanation for the dark matter problem; that is, nonlocality appears to mimic dark matter. In the absence of a deeper understanding of the gravitational interaction, we therefore adopt the view that the kernel of nonlocal gravity must be determined from observational data regarding dark matter. In other words, there is no dark matter in nonlocal gravity; therefore, what appears as dark matter in astrophysics and cosmology must be the nonlocal aspect of the gravitational interaction.

Among the basic interactions in nature, gravitation has the unique feature of *universality*. We assume that it is also history-dependent. That is, the gravitational interaction has an additional feature of nonlocality in the sense of an influence ("memory") from the past that endures. Is there any compelling evidence that Einstein's theory of gravitation should be modified? The theory is in good agreement with observational data from submillimeter scales to the scale of the Solar System and binary star systems. However, on galactic scales and beyond the theory fails unless one invokes the existence of the hypothetical dark matter. Indeed, on such large scales, gravity is dominated by the attraction of dark matter.

Can nonlocal gravity explain away what appears as dark matter in astronomy? It is important to note that the persistent negative result of experiments that have searched for the particles of dark matter naturally leads to the possibility that what appears as dark matter in astrophysics and cosmology is in fact an aspect of the gravitational interaction. The nonlocal character of gravity, however, cannot yet replace dark matter on all physical scales. Indeed, dark matter is currently needed for explaining: (i) gravitational dynamics of galaxies and clusters of galaxies, (ii) gravitational lensing observations in general and the Bullet Cluster in particular and (iii) the formation of structure in cosmology and the large-scale structure of the universe. We emphasize that nonlocal gravity theory is so far in the early stages of development and only some of its implications have been confronted with observation, thanks to the work of Sohrab Rahvar on the rotation curves of a sample of spiral galaxies as well as on the internal dynamics of a sample of Chandra X-ray clusters of galaxies. Indeed, the establishment of nonlocal gravity theory on both theoretical and experimental fronts is certainly work in progress and much remains to be done.

Nonlocal gravity is presented in this book within an extended general relativistic framework that includes the Weitzenböck connection. This framework is described in Chapter 5 and the field equation of the nonlocal generalization of Einstein's theory of gravitation is developed in Chapter 6. I assume throughout that the reader is familiar with the basic tenets of general relativity; in fact, the required background material can be found in standard introductory textbooks such as Ryder (2009), Misner, Thorne and Wheeler (1973), Weinberg (1972) and Landau and Lifshitz (1971). No exact solution of the field equation of nonlocal gravity beyond Minkowski spacetime is known. The absence of any exact nontrivial solution of the theory implies that the nonlinear regime of the theory has yet to be studied. Thus exact cosmological models or issues involving the formation and evolution of black holes are beyond the scope of the present work. Therefore, Chapters 7, 8 and 9 are essentially concerned with the implications of the general linear approximation of nonlocal gravity. Chapter 10 deals with the difficult question of whether nonlocal gravity can potentially solve the problem of structure formation in cosmology without invoking any dark matter. The work reported in this book would not have been possible without decades of invaluable collaboration with Carmen Chicone and Friedrich W. Hehl. More recently, I have benefitted greatly from my collaborations with Donato Bini and Sohrab Rahvar on the theoretical and observational aspects of nonlocal gravity, respectively.

The constitutive relations in the electrodynamics of media naturally depend upon the medium under consideration. Similarly, there are many possible constitutive relations in the case of nonlocal gravity and they can lead to different theories of nonlocal gravity. For instance, as discussed in Chapter 7, it is possible to choose the constitutive relation such that nonlocality enters gravitation theory only in the nonlinear regime of the theory, in which case nonlocality cannot provide an explanation for dark matter. On the other hand, we have tentatively adopted a constitutive relation in this book such that nonlocality survives even in the Newtonian regime and therefore has the potential to do away with the hypothetical dark matter. The constitutive kernel in the Newtonian regime is a universal function that depends upon three constant parameters. These should account for all of the astrophysical data regarding dark matter. However, if the present approach to dark matter fails or the elusive particles of dark matter are eventually discovered, nonlocal gravity theory could still survive with a different constitutive relation.

In this book, I attempt to present a coherent account of the work that has been done on a particular approach to the nonlocal generalization of Einstein's theory of gravitation. I am deeply grateful to the many colleagues, at the University of Missouri and elsewhere, who have supported me in this endeavor. Most of the material covered in this work is based on collaborative efforts, which I gratefully acknowledge. Particular thanks are due to Donato Bini, Carmen Chicone, Friedrich W. Hehl, Lorenzo Iorio, Jeffrey R. Kuhn, Roy Maartens, José W. Maluf, Yuri Obukhov, Sohrab Rahvar and Haojing Yan for their generous help and advice. I am solely responsible for any errors or deficiencies in this book.

Finally, it is important to mention here that there are other approaches to nonlocal gravitation theory; see, for instance, Soussa and Woodard (2003), Barvinsky (2003), Biswas, Mazumdar and Siegel (2006), Biswas *et al.* (2012), Briscese *et al.* (2013), Tsamis and Woodard (2014), Conroy *et al.* (2015) and the references cited therein. The inspiration for such theories often has its origin in developments in quantum field theory. The consideration of such theories is beyond the scope of the present book, which is primarily based on a critical analysis of the fundamental assumption of locality that underlies the standard theory of relativity. It is possible that Einstein's classical theory of gravitation needs modification on galactic scales along the lines indicated in this book in order to explain observational phenomena associated with what has come to be known in astronomy and cosmology as "dark matter". I can only hope that progress in astronomy will support the approach to nonlocal gravity theory adopted in this book, thereby leading to a deeper understanding of the gravitational interaction and possibly providing a clue towards its eventual quantization (Becker and Reuter 2014).

Columbia, Missouri, 15 August 2016

Contents

1
Introduction

A basic locality postulate permeates through the standard special and general theories of relativity. The purpose of this initial chapter is to identify the locality assumption and briefly study its physical origin as well as the significant role that it plays in relativity theory.

1.1 Lorentz Invariance

The principle of relativity—namely, the assertion that the laws of microphysics are the same in all inertial frames of reference—refers to the measurements of ideal inertial observers. The transition from Galilean invariance of Newtonian physics to Lorentz invariance marks the beginning of modern relativity theory. Lorentz invariance is the invariance of the fundamental laws of microphysics under the group of passive inhomogeneous Lorentz transformations. Lorentz invariance has firm observational support; therefore, we assume throughout that Lorentz invariance is a fundamental symmetry of nature. The basic laws of microphysics have been formulated with respect to ideal inertial observers, since these are conceived to be free of the various limitations associated with actual observers. Each ideal inertial observer is forever at rest in a global inertial frame of reference, namely, a Cartesian coordinate system that is homogeneous and isotropic in space and time and in which Newton's fundamental laws of motion are valid.

The global inertial frames of reference are all related to each other by passive inhomogeneous Lorentz transformations of the form

$$x^\mu = \Lambda^\mu{}_\nu \, x'^\nu + b^\mu, \tag{1.1}$$

where an event is characterized by inertial coordinates $x^\mu = (ct, \mathbf{x})$, $(\Lambda^\mu{}_\nu)$ is a Lorentz matrix and b^μ is a constant vector of spacetime translation. The set of all such events constitutes flat Minkowski spacetime. In our convention, Greek indices run from 0 to 3, while Latin indices run from 1 to 3; moreover, the spacetime metric has signature $+2$.

The inhomogeneous Lorentz transformations form the Poincaré group, which is the ten-parameter group of isometries of Minkowski spacetime. That is, the Minkowski spacetime interval ds given by

$$ds^2 = \eta_{\alpha\beta} \, dx^\alpha \, dx^\beta \tag{1.2}$$

is preserved under the Poincaré group. Here, $\eta_{\alpha\beta}$ is the Minkowski metric tensor given by $\mathrm{diag}(-1, 1, 1, 1)$, in accordance with our convention regarding metric signature.

Nonlocal Gravity. Bahram Mashhoon. © Bahram Mashhoon 2017. Published 2017 by Oxford University Press.

The four-parameter Abelian group of spacetime translations and the six-parameter Lorentz group, which consists of boosts and rotations, are subgroups of the Poincaré group. It follows from eqns (1.1) and (1.2) that

$$\eta_{\alpha\beta} \Lambda^{\alpha}{}_{\mu} \Lambda^{\beta}{}_{\nu} = \eta_{\mu\nu}. \tag{1.3}$$

For $\mu = \nu = 0$, eqn (1.3) can be expressed as

$$(\Lambda^{0}{}_{0})^2 = 1 + \sum_{i=1}^{3} (\Lambda^{i}{}_{0})^2. \tag{1.4}$$

Moreover, taking the determinant of both sides of eqn (1.3), we find

$$\det(\Lambda^{\alpha}{}_{\mu}) = \pm 1. \tag{1.5}$$

Henceforth, we work with the proper orthochronous Lorentz group (Streater and Wightman 1964), whose elements satisfy $\det(\Lambda^{\alpha}{}_{\mu}) = 1$ as well as $\Lambda^{0}{}_{0} \geq 1$. Clearly, the Lorentz group contains the identity element $\Lambda^{\alpha}{}_{\mu} = \delta^{\alpha}_{\mu}$.

The determination of temporal and spatial intervals constitutes the most basic measurements of a physical observer. We assume that each inertial observer has access to an ideal clock as well as infinitesimal measuring rods, and carries along its world line an orthonormal tetrad frame (or vierbein), that is, a set of four unit axes that are orthogonal to each other and characterize the observer's local temporal and spatial axes. For instance, the local axes of the class of *fundamental observers* at rest in a global inertial frame are parallel to the corresponding global axes and are given by

$$\lambda^{\mu}{}_{\hat{0}} = (1, 0, 0, 0), \qquad \lambda^{\mu}{}_{\hat{1}} = (0, 1, 0, 0), \tag{1.6}$$

$$\lambda^{\mu}{}_{\hat{2}} = (0, 0, 1, 0), \qquad \lambda^{\mu}{}_{\hat{3}} = (0, 0, 0, 1). \tag{1.7}$$

The hatted tetrad indices at an event enumerate the tetrad axes in the tangent space at that event. In particular, $\lambda^{\mu}{}_{\hat{0}}$ is the temporal axis of the observer, while $\lambda^{\mu}{}_{\hat{i}}$, for $i = 1, 2, 3$, constitutes the observer's spatial frame. Each inertial observer in Minkowski spacetime belongs to a class of fundamental observers.

For a tetrad frame $\lambda^{\mu}{}_{\hat{\alpha}}$ carried by an arbitrary inertial observer in a global inertial frame, the orthonormality condition takes the form

$$\eta_{\mu\nu} \lambda^{\mu}{}_{\hat{\alpha}} \lambda^{\nu}{}_{\hat{\beta}} = \eta_{\hat{\alpha}\hat{\beta}}. \tag{1.8}$$

The tetrads that we consider throughout are adapted to the observers under consideration, which have future-directed timelike world lines and employ right-handed spatial frames in conformity with the right-handed Cartesian coordinates of the background space. Therefore, we limit our considerations to tetrads for which $\det(\lambda^{\mu}{}_{\hat{\alpha}}) = 1$. Spacetime indices are in general raised and lowered via the spacetime metric tensor $g_{\mu\nu}$, which is equal to $\eta_{\mu\nu}$ in the present case, while the hatted tetrad indices—that is, the local Lorentz indices—are raised and lowered via the Minkowski metric tensor $\eta_{\hat{\mu}\hat{\nu}}$.

In connection with spacetime measurements, we imagine static inertial observers in a global inertial frame and assume that their clocks are all synchronized; that is, adjacent clocks can be synchronized. Moreover, the adiabatic transport of a clock to another location can be so slow as to have negligible practical impact on synchronization. In a similar way, lengths can be measured in general by placing infinitesimal rods together. Furthermore, it is assumed in general that for physical measurements, inertial observers have access to ideal measuring devices. These are free from the specific practical limitations of laboratory devices that are usually due to the nature of their construction and modes of operation. The measurements of moving inertial observers are related to those at rest via Lorentz invariance, which preserves the causal sequence of events.

An equivalent ("radar") approach to spacetime measurements relies on the transmission and reception of light signals. In this procedure, a static inertial observer $\mathcal{O}_1$ sends out a light signal at time t_1 to static inertial observer $\mathcal{O}_2$. The signal is immediately transponded without delay back to $\mathcal{O}_1$ and is received at time t_2. If the clocks at $\mathcal{O}_1$ and $\mathcal{O}_2$ are synchronized, they would both register time $t = (t_1 + t_2)/2$ at the instant the signal is received at $\mathcal{O}_2$. Moreover, the distance between $\mathcal{O}_1$ and $\mathcal{O}_2$ is $D_{12} = c\,(t_2 - t_1)/2$. Thus $t_2 - t = t - t_1 = D_{12}/c$.

The inertial physics that is based on the ideal inertial observers and their tetrad frames has played a significant role in the development of theoretical physics. Inertial physics was originally established by Newton (Cohen 1960).

1.1.1 Inertial observers

Imagine a background global inertial frame in Minkowski spacetime. The ideal inertial observers in this arena are either at rest with local spatial reference frames that are related to the global axes by a constant rotation or move with constant speeds on straight lines from minus infinity to plus infinity and carry constant local reference frames. The *fundamental* inertial observers are all at rest and carry orthonormal tetrad frames with axes that coincide with the global Cartesian spacetime axes of the background inertial frame of reference.

The translational motion of the observer in spacetime fixes its local temporal axis as well as its spatial frame but only up to an element of the rotation group. Consider, for illustration, a background inertial frame with coordinates $(ct, \mathbf{x})$ in Minkowski spacetime. An inertial observer moves with constant velocity $\mathbf{v}$ relative to the background frame. The Lorentz transformation to the rest frame of the moving observer $(ct', \mathbf{x}')$ involving a pure boost with *no rotation* is given by

$$t = \gamma \left(t' + \frac{\mathbf{v} \cdot \mathbf{x}'}{c^2} \right), \tag{1.9}$$

$$\mathbf{x} = \mathbf{x}' + \frac{1}{v^2}(\gamma - 1)(\mathbf{x}' \cdot \mathbf{v})\,\mathbf{v} + \gamma\,\mathbf{v}\,t', \tag{1.10}$$

where $v = |\mathbf{v}|$ and γ is the corresponding Lorentz factor, namely, $\gamma = 1/\sqrt{1 - v^2/c^2}$. In the $(ct', \mathbf{x}')$ frame, the tetrad of the fundamental inertial observers at rest is given by $h'^{\mu}{}_{\hat{\alpha}} = \delta^{\mu}_{\hat{\alpha}}$. Transforming the local tetrad frame of the moving observer to the $(ct, \mathbf{x})$

system via the Lorentz boost matrix that can be simply deduced from eqns (1.9) and (1.10), we find

$$h^\mu{}_{\hat{0}} = \gamma\left(1, \frac{\mathbf{v}}{c}\right), \tag{1.11}$$

$$h^\mu{}_{\hat{i}} = \delta^\mu_i + v_i\left(\frac{\gamma}{c}, \frac{(\gamma-1)}{v^2}\mathbf{v}\right). \tag{1.12}$$

Up to a constant rotation, this is the orthonormal tetrad frame of the boosted inertial observer with respect to the background $(ct, \mathbf{x})$ system. For the generalization of this result to curved spacetimes, see Mashhoon and Obukhov (2014).

The temporal axis of the moving inertial observer's tetrad, $h^\mu{}_{\hat{0}}$, is equal to the unit timelike vector that is the 4-velocity of the observer, $h^\mu{}_{\hat{0}} = dx^\mu/d\tau$, where the temporal parameter τ is simply related to the invariant spacetime interval along the observer's world line. That is, $ds^2 = -d\tau^2$, where $d\tau = c\,dt/\gamma$. We can clearly identify τ with ct', the *proper* time of the moving inertial observer. The observer's spatial frame consists, up to a constant rotation, of the three orthogonal unit spacelike axes given by $h^\mu{}_{\hat{i}}$ for $i = 1, 2, 3$.

Ideal inertial observers all have *straight world lines*. Imagine, for instance, an inertial observer $\mathcal{O}_0$ moving along the positive z axis with constant speed v. Let us introduce the rapidity parameter Θ_0 such that $v/c = \tanh\Theta_0$; then, eqns (1.11) and (1.12) imply that the orthonormal tetrad frame of $\mathcal{O}_0$ is given by

$$h^\mu{}_{\hat{0}} = (\cosh\Theta_0, 0, 0, \sinh\Theta_0), \qquad h^\mu{}_{\hat{1}} = (0, 1, 0, 0), \tag{1.13}$$

$$h^\mu{}_{\hat{2}} = (0, 0, 1, 0), \qquad h^\mu{}_{\hat{3}} = (\sinh\Theta_0, 0, 0, \cosh\Theta_0). \tag{1.14}$$

The path of the observer in spacetime is then rectilinear; that is, it follows from the integration of the 4-velocity vector of $\mathcal{O}_0$, $u^\mu = dx^\mu/d\tau = h^\mu{}_{\hat{0}}$, that

$$ct = \tau\cosh\Theta_0, \qquad x = y = 0, \qquad z = \tau\sinh\Theta_0, \tag{1.15}$$

where the integration constants have been chosen such that at $t = 0$, $\tau = 0$ and the observer is at the origin of spatial coordinates.

1.1.2 Examples of uniformly accelerated observers

Realistic observers in a global inertial frame in Minkowski spacetime would all be more or less accelerated. We consider here some examples of uniformly accelerated observers.

Let us first imagine a noninertial observer $\hat{\mathcal{O}}$ that has the same history as $\mathcal{O}_0$ for $\tau < 0$, but $\hat{\mathcal{O}}$ is forced to accelerate uniformly along the positive z direction starting at $\tau = 0$, when $\hat{\mathcal{O}}$ is at the origin of spacetime coordinates. For $\tau > 0$, the observer's orthonormal tetrad frame is then $\lambda^\mu{}_{\hat{\alpha}}$ such that

$$\lambda^\mu{}_{\hat{0}} = (\cosh\Theta, 0, 0, \sinh\Theta) \tag{1.16}$$

and

$$\lambda^\mu{}_{\hat{1}} = (0, 1, 0, 0), \qquad \lambda^\mu{}_{\hat{2}} = (0, 0, 1, 0), \qquad \lambda^\mu{}_{\hat{3}} = (\sinh\Theta, 0, 0, \cosh\Theta). \tag{1.17}$$

Here,

$$\Theta = \Theta_0 + \frac{g_0\,\tau}{c^2} \tag{1.18}$$

and g_0 is the *constant* invariant acceleration of $\hat{O}$. The 4-acceleration $\mathfrak{a}^\mu$ of $\hat{O}$ is

$$\mathfrak{a}^\mu = \frac{d\lambda^\mu{}_{\hat{0}}}{d\tau} = \frac{g_0}{c^2}\lambda^\mu{}_{\hat{3}}. \tag{1.19}$$

It follows in general from $u_\mu u^\mu = -1$ that $\mathfrak{a}_\mu u^\mu = 0$, so that the acceleration 4-vector is spacelike with $\mathfrak{a}_\mu \mathfrak{a}^\mu = \tilde{g}^2$, where $\tilde{g}(\tau) > 0$ is the magnitude of the translational acceleration with dimensions of $(\text{length})^{-1}$. For constant linear acceleration, for instance, $\tilde{g} = |g_0|/c^2$.

In connection with the propagation of the tetrad frame along the world line of $\hat{O}$, let us briefly digress here and discuss a more general situation that involves an observer following a timelike path in an inertial frame in Minkowski spacetime. The observer has 4-velocity $u^\mu = dx^\mu/d\tau$ and 4-acceleration $\mathfrak{a}^\mu = du^\mu/d\tau$ and carries a vector V^μ along its path. One can decompose this vector into its parallel and perpendicular components relative to the path, namely $V^\mu = V^\mu_{\|} + V^\mu_{\perp}$, where the component parallel to the curve is $V^\mu_{\|} = -(u \cdot V)u^\mu$ and the corresponding perpendicular component is $V^\mu_{\perp} = (\eta^{\mu\nu} + u^\mu u^\nu)V_\nu$. Here, $u \cdot V = u_\mu V^\mu$ and $|u \cdot V|$ is the length of the parallel component of V. It follows that $V \cdot V = V_{\|} \cdot V_{\|} + V_{\perp} \cdot V_{\perp}$, which is reminiscent of the Pythagorean theorem. Let us now suppose that V^μ is so transported that it does not rotate and its length remains constant; that is, along the path we have

$$\frac{d\,(u \cdot V)}{d\tau} = 0, \qquad \left(\frac{d\,V^\mu_{\perp}}{d\tau}\right)_{\perp} = 0. \tag{1.20}$$

These relations mean that while the magnitude of the parallel component of vector V^μ remains constant along the path, the perpendicular component cannot change in the perpendicular direction; otherwise, vector V^μ would rotate. That is, the net variation of the perpendicular component along the path can only be in the direction parallel to the path. In this way, we find from (1.20) that

$$\frac{d\,V^\mu}{d\tau} = (V \cdot \mathfrak{a})\,u^\mu - (V \cdot u)\,\mathfrak{a}^\mu, \tag{1.21}$$

which is the Fermi–Walker transport law. We note, in particular, that the 4-velocity of the observer u^μ satisfies eqn (1.21).

Returning to the case of observer $\hat{O}$, it is straightforward to check that its tetrad frame is Fermi–Walker transported along its world line. The path of the observer for $\tau > 0$ is given by

$$ct = \frac{c^2}{g_0}\,(\sinh\Theta - \sinh\Theta_0), \qquad x = y = 0, \qquad z = \frac{c^2}{g_0}\,(\cosh\Theta - \cosh\Theta_0), \tag{1.22}$$

which is a *hyperbola* in the (ct, z) plane. That is, it follows from (1.22) that

$$\left(z + \frac{c^2}{g_0}\cosh\Theta_0\right)^2 - \left(ct + \frac{c^2}{g_0}\sinh\Theta_0\right)^2 = \frac{c^4}{g_0^2}, \tag{1.23}$$

where $c^2/|g_0|$ is the invariant acceleration length associated with $\hat{O}$ for $\tau > 0$.

We note that at any instant of time $\tau > 0$, the tetrad frame (1.16)–(1.17) of the hyperbolic observer $\hat{\mathcal{O}}$ is of the general form of the frame (1.13)–(1.14) with an instantaneous speed $c \tanh \Theta$ such that $|\tanh \Theta| \to 1$ for $\tau \to \infty$. This corresponds to the fact that the asymptotes of the hyperbola in (1.23) are null lines. Over the period of proper time τ from $0 \to \infty$, an external source causing the constant acceleration of $\hat{\mathcal{O}}$ must provide an infinite amount of energy to sustain the complete hyperbolic motion. To avoid such unphysical situations, we assume that in general the observer's acceleration is turned on only over time intervals such that the net amount of energy transfer is always finite.

Observers in a laboratory fixed on the Earth in general rotate in space as the Earth rotates about its axis. Moreover, such observers generally refer their measurements to spatial axes that are rigidly attached to the Earth. Thus their spatial frames rotate with the Earth as well. It is therefore interesting to study tetrad frames adapted to such rotating observers.

To describe uniformly rotating observers in Minkowski spacetime, let us first imagine observers $\tilde{\mathcal{O}}$ at rest in a global inertial frame with spacetime coordinates (ct, x, y, z). However, these static observers are not inertial since instead of the (x, y, z) axes of the background frame, they refer their measurements to axes (x', y', z') that rotate with angular speed Ω about the z axis. That is, they employ

$$x' = x \cos \varphi + y \sin \varphi, \qquad y' = -x \sin \varphi + y \cos \varphi \qquad (1.24)$$

and $z' = z$, where $\varphi = \Omega t$. Thus the orthonormal tetrad frame $\tilde{\lambda}^\mu{}_{\hat{\alpha}}$ of such static noninertial observers is given in (ct, x, y, z) coordinates by

$$\tilde{\lambda}^\mu{}_{\hat{0}} = (1, 0, 0, 0), \qquad (1.25)$$
$$\tilde{\lambda}^\mu{}_{\hat{1}} = (0, \cos \varphi, \sin \varphi, 0), \qquad (1.26)$$
$$\tilde{\lambda}^\mu{}_{\hat{2}} = (0, -\sin \varphi, \cos \varphi, 0), \qquad (1.27)$$
$$\tilde{\lambda}^\mu{}_{\hat{3}} = (0, 0, 0, 1), \qquad (1.28)$$

so that at $t = 0$, $\tilde{\lambda}^\mu{}_{\hat{\alpha}}$ coincides with the tetrad frame of the fundamental static inertial observers.

Let us next imagine an observer $\mathcal{O}$ moving uniformly for $t < 0$ on a plane parallel to the (x, y) plane with $x = r$, $y = vt$ and $z = z_0$, where r and z_0 are constant lengths and $v = r\Omega$. At $t = 0$, $\mathcal{O}$ is forced to rotate uniformly on a circle of radius r about the z axis on the plane that is at fixed $z = z_0$ and is parallel to the (x, y) plane as in Fig. 1.1.

For any instant of time $t \geq 0$, the natural orthonormal tetrad frame $\lambda^\mu{}_{\hat{\alpha}}$ of $\mathcal{O}$, adapted to the rotating system, can be simply obtained from that of the corresponding static observer $\tilde{\mathcal{O}}$ by a pure boost with speed $v = r\Omega$ along the tangential direction to the circular trajectory (see Fig. 1.1); that is,

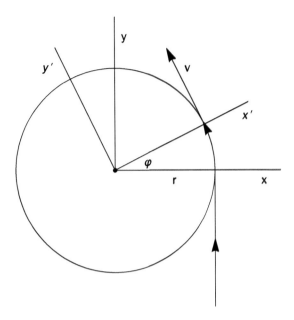

Fig. 1.1 Schematic plot depicting the path of accelerated observer $\mathcal{O}$ that moves on the plane $z = z_0$ with uniform speed $v = r\,\Omega$ along a straight line parallel to the y axis for $t < 0$ but for $t \geq 0$ undergoes uniform rotation with angular speed Ω about the z axis such that $\varphi = \Omega t$.

$$\lambda^{\mu}{}_{\hat{0}} = \gamma \left[\tilde{\lambda}^{\mu}{}_{\hat{0}} + \beta\,\tilde{\lambda}^{\mu}{}_{\hat{2}} \right], \tag{1.29}$$

$$\lambda^{\mu}{}_{\hat{1}} = \tilde{\lambda}^{\mu}{}_{\hat{1}}, \tag{1.30}$$

$$\lambda^{\mu}{}_{\hat{2}} = \gamma \left[\tilde{\lambda}^{\mu}{}_{\hat{2}} + \beta\,\tilde{\lambda}^{\mu}{}_{\hat{0}} \right], \tag{1.31}$$

$$\lambda^{\mu}{}_{\hat{3}} = \tilde{\lambda}^{\mu}{}_{\hat{3}}, \tag{1.32}$$

where $\beta = v/c$ and $\gamma = 1/\sqrt{1 - \beta^2}$ is the corresponding Lorentz factor. Thus, with respect to (ct, x, y, z) coordinates, $\lambda^{\mu}{}_{\hat{\alpha}}$ is given by

$$\lambda^{\mu}{}_{\hat{0}} = \gamma\,(1,\ -\beta\,\sin\varphi,\ \beta\,\cos\varphi,\ 0), \tag{1.33}$$

$$\lambda^{\mu}{}_{\hat{1}} = (0,\ \cos\varphi,\ \sin\varphi,\ 0), \tag{1.34}$$

$$\lambda^{\mu}{}_{\hat{2}} = \gamma\,(\beta,\ -\sin\varphi,\ \cos\varphi,\ 0), \tag{1.35}$$

$$\lambda^{\mu}{}_{\hat{3}} = (0,\ 0,\ 0,\ 1). \tag{1.36}$$

This is the tetrad frame of the uniformly rotating observer $\mathcal{O}$ for $t \geq 0$. At each instant $t \geq 0$ along its circular trajectory in the orbital plane that is parallel to the (x, y) plane, the spatial frame of $\mathcal{O}$, $\lambda^{\mu}{}_{\hat{i}}$, for $i = 1, 2, 3$, consists of the radial, tangential and normal directions, respectively, with respect to the orbital plane.

The uniform speed of observer $\mathcal{O}$ is given by $r\,\Omega < c$ and its proper time can be written as $\tau = ct/\gamma$ if we assume that $\tau = 0$ at $t = 0$; therefore, $\varphi = \gamma\,\Omega\,\tau/c$. Moreover, the observer's acceleration, which was turned on at $t = \tau = 0$, can be turned off after a finite period of time. Such observers exist for $0 < r < c/\Omega$, the boundary of this open cylindrical region in Minkowski spacetime is the light cylinder of radius $\mathcal{L} = c/\Omega$. The light cylinder is a timelike hypersurface; therefore, observers inside this cylinder can in principle communicate with the outside world without any difficulty. In the $r = 0$ limit, eqns (1.33)–(1.36) with $\beta = 0$ and $\gamma = 1$ reduce to the tetrad frame of the noninertial observer $\tilde{\mathcal{O}}$ that is at rest along the axis of rotation at $z = z_0$.

We have thus far discussed the measurements of inertial observers. We are also interested in the measurements of accelerated observers. What do accelerated observers measure? What are the laws of physics according to accelerated observers? What is the generalization of Lorentz invariance that applies to accelerated observers? We now turn to a discussion of these issues.

1.1.3   Nonexistence of ideal inertial observers

The special theory of relativity is about the standard relativistic physics of Minkowski spacetime, where gravity has been turned off. Physical phenomena in each global inertial frame of reference involve ideal inertial observers as well as accelerated observers. Indeed, all actual observers are accelerated; that is, inertial observers, though of deep theoretical significance, do not in fact exist. There is a basic dichotomy here involving theory and experiment that is noteworthy: The basic laws of non-gravitational physics have all been formulated with respect to ideal inertial observers, yet the experimental basis of these laws—namely, the foundation of physical science—has been established via actual observers that are all accelerated. To set the foundation of physical science on a firm basis, a connection must be established between inertial and accelerated observers. Simply stated, the fundamental microphysical laws, such as the principles of quantum mechanics, have been formulated for nonexistent ideal inertial observers, while all actual observers are accelerated. The resolution of this dichotomy requires an a priori axiom that relates inertial and accelerated observers. The observational consequences of such an axiom should then be compared with experimental results.

Ideal inertial observers are supposed to move on straight lines with constant speeds from minus infinity to plus infinity in a global inertial frame and carry constant local reference frames. It is important to note that these theoretical assumptions regarding ideal inertial observers cannot be directly verified by experiment. For instance, distant past and future states of the universe are not directly accessible to experimentation. Furthermore, repeated observational attempts to determine that an object is indeed at rest or moves uniformly on a rectilinear path will produce disturbances that cause deviations from the state of rest or uniform rectilinear motion. The ideal inertial observers are thus hypothetical and have been introduced to embody the principle of inertia perfectly. Real observers in this global inertial frame are all accelerated and we need to determine what accelerated observers actually measure. In this treatment, observers can be sentient beings or measuring devices. In either case, observers are classical macrophysical systems that are extended in space. Any real measuring device is subject to various limitations; for example, it may not operate properly under certain conditions. Moreover, an *accelerated* device is under the influence of various internal inertial effects that could, over time, affect its constitution and mode of operation. In practice, all such issues require careful consideration; however, for the purposes of this theoretical discussion, we generally follow the standard practice in the theory of relativity and represent an observer by a single timelike world line for the sake of simplicity. This is *not* considered to be a fundamental limitation; rather, it helps simplify the analysis. In fact, this notion of an elementary observer can then be extended to a reference system by considering a congruence of elementary observers that occupy a finite spacetime domain in a global inertial frame in Minkowski spacetime. That this construction is indeed possible has been demonstrated in various ways by explicit examples for simple accelerated systems (see Mashhoon 2008 and the references cited therein). A general method based on fiber bundles for the construction of such reference systems involving nonintegrable anholonomic observers has been discussed by Auchmann and Kurz (2014).

By employing pointlike observers in our theoretical treatment, we avoid the problem of determination of the integrated influence of inertial effects on measuring devices that are employed during the measurement process. All observers under consideration are thus essentially ideal pointlike systems subject to the laws of classical (i.e. non-quantum) physics. We can therefore concentrate on the theoretical distinction between pointlike inertial and accelerated observers.

Observational data, collected over time by actual observers that are all more or less accelerated, have helped establish microphysical laws and have indicated that Lorentz invariance is a fundamental symmetry of nature. Therefore, a connection must exist between inertial and accelerated observers. What is this connection?

1.2 Hypothesis of Locality

To extend Lorentz invariance to accelerated observers in Minkowski spacetime it is necessary to relate accelerated observers to inertial observers. The standard theory of relativity is based on the postulate that an accelerated observer is pointwise equivalent to an otherwise identical momentarily comoving inertial observer. The latter follows the straight world line that is instantaneously tangent to the world line of the accelerated observer. The locality postulate in effect replaces the world line of the accelerated observer at each instant by its tangent line at that event. Geometrically, the tangent line is the first Frenet approximation to the curve. The Frenet approach involves the mathematics of turning and twisting of a curve in space (O'Neill 1966). A discussion of the Frenet–Serret method of moving frames for world lines is contained in Synge (1971). The world line of the accelerated observer is the envelope of the set of straight tangent world lines; therefore, the accelerated observer may be replaced in effect by an infinite sequence of hypothetical momentarily comoving inertial observers. Thus the association between actual accelerated observers and ideal inertial observers is purely *local*, since an accelerated observer is pointwise inertial according to the standard theory of relativity.

The hypothesis of locality originates from the Newtonian mechanics of point particles, where the state of a point particle is determined at each instant of time t by its position $\mathbf{x}$ and velocity $\mathbf{v}$. The arbitrary point particle and the corresponding hypothetical momentarily comoving inertial particle of the same mass m share the same state $(\mathbf{x}, \mathbf{v})$ and are thus physically equivalent. The motion of the point particle of mass m under an external force $\mathbf{f}(t, \mathbf{x}, \mathbf{v})$ in the background global inertial frame is given by

$$\frac{d\mathbf{x}}{dt} = \mathbf{v}, \qquad \frac{d\mathbf{v}}{dt} = \frac{1}{m} \mathbf{f}(t, \mathbf{x}, \mathbf{v}). \tag{1.37}$$

The state of the particle $(\mathbf{x}, \mathbf{v})$ at time t uniquely determines the motion for all time. Moreover, if $\mathbf{f}$ is turned off at any time t, the motion continues uniformly on a straight line tangent to the path at t. The inertial tendency of the particle is thus continually interrupted by the presence of the external force, which changes the state of the particle. This is the physical explanation for the fact that the accelerated path of the particle under the external force is the envelope of the straight tangent lines. It follows that the postulate of locality is automatically satisfied in the Newtonian mechanics of point particles, as it is ingrained in the Newtonian laws of motion. Hence no new physical assumption is needed to deal with accelerated systems in Newtonian physics. The hypothesis of locality should hold equally well in the *relativistic* mechanics of classical point particles (Minkowski 1952). Moreover, it is clear that the hypothesis of locality would be exactly valid if all physical phenomena in Minkowski spacetime could be reduced to pointlike *coincidences* of classical point particles and rays of radiation (Einstein 1950). The hypothesis of locality is schematically illustrated in Fig. 1.2.

It is through the locality postulate that the consequences of Lorentz invariance can be verified by experiment within the framework of the special theory of relativity. That is, it follows from the hypothesis of locality that in a global inertial frame in Minkowski spacetime the measurements of an accelerated observer can be determined by applying Lorentz transformation point by point along its path. Consider, for instance, the measurement of time by an ideal clock following an accelerated path. At each instant of time t, the clock has velocity $\mathbf{v}(t)$ and is instantaneously inertial by the locality hypothesis and hence at rest in an inertial frame with coordinates $x'^{\mu} = (ct', \mathbf{x}')$. An instantaneous Lorentz transformation from the background inertial frame to the momentary rest frame of the moving clock leads to the formula for time dilation, namely,

$$dt' = \left(1 - \frac{v^2}{c^2}\right)^{1/2} dt, \tag{1.38}$$

which can also be obtained from the invariance of the Minkowski spacetime interval (1.2) under Lorentz transformations. Here, $v = |\mathbf{v}|$ and only positive square roots are considered throughout. The accelerated clock passes through an infinite sequence of such momentarily inertial states; therefore, its local *proper* time τ/c is a sum of infinitesimal time intervals each of the form of eqn (1.38). Hence, the proper time of the clock is given by

$$\frac{\tau}{c} = \int_0^t \left[1 - \frac{v^2(T')}{c^2}\right]^{1/2} dT', \tag{1.39}$$

where $\tau = 0$ at $t = 0$ by assumption.

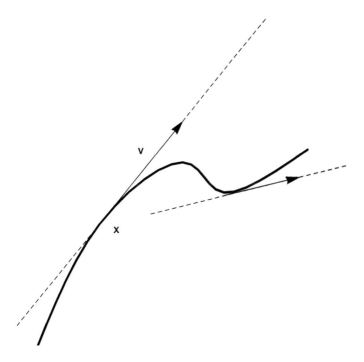

Fig. 1.2 Schematic illustration of the locality postulate that an accelerated observer is point-wise inertial. The accelerated observer at each instant of time measures what an otherwise identical hypothetical momentarily comoving inertial observer would measure at that instant. They share the same state $(\mathbf{x}, \mathbf{v})$. The accelerated observer thus passes through a continuous infinity of tangent inertial observers; indeed, its path is the envelope of the corresponding family of straight lines.

We note that $\tau/c \leq t$ in eqn (1.39). In this connection, imagine two identical ideal clocks at rest at some point P in space in the background inertial frame and synchronized to register time $t = 0$. One clock remains at rest at P, while the other moves about and eventually returns to P at time t, the proper time of the clock at rest. At t, the clock that was in motion registers proper time τ/c given by eqn (1.39), which is shorter than t. For a discussion of clock experiments using the Global Positioning System (GPS), see Ashby (2003).

The influence of acceleration on clock performance has been discussed by a number of authors: see Mainwaring and Stedman (1993), Dahia and Felix da Silva (2015) and the references cited therein. Possible deviations from locality tend to be rather small and well below the level of accuracy ($\sim 10^{-9}$) of the recent experimental verification of time dilation by Botermann *et al.* (2014).

Consider next an example of the application of the locality postulate involving length measurement. Imagine two small identical blocks A and B at rest and a distance L apart along the x axis in a global inertial frame. At $t = 0$, A and B are accelerated from rest and forced to move *in exactly the same way* along the positive x direction.

The speeds of A and B are the same by assumption for all $t \geq 0$; therefore, the distance between A and B as measured in the inertial frame is always L. At time $t > 0$, blocks A and B have the same speed $v(t)$ and are momentarily inertial by the locality postulate; hence, an instantaneous Lorentz transformation connects the background inertial frame to the instantaneous inertial frame in which both A and B are at rest. It follows from this Lorentz transformation that at time t

$$x'_B - x'_A = \gamma(t)(x_B - x_A), \qquad \gamma(t) = \left[1 - \frac{v^2(t)}{c^2}\right]^{-1/2}, \qquad (1.40)$$

so that the distance between the two blocks as measured in their momentary rest frame is generally larger than L and is given by $\gamma(t)L$, where $\gamma \geq 1$ is the Lorentz factor corresponding to speed $v(t)$. If, for instance, there is a taut string that is initially attached to A and B and these are then forced to undergo hyperbolic motion in exactly the same way, the distance between A and B will continue to increase monotonically in their momentary rest frames and the string should eventually break (Dewan and Beran 1959; Bell 1987). It is assumed here that the string is always in tension but exerts negligible force on A and B. For a detailed analysis of the problem of length measurement in accelerated systems, see Mashhoon (1990b) and Mashhoon and Muench (2002).

1.2.1 Physics of locality

The postulate of locality states that acceleration can be locally ignored. This means, in terms of realistic measurements, that the integrated influence of inertial effects over the length- and timescales characteristic of the measurement process can be neglected. Hence, the observer's tetrad frame should in effect remain constant during the process of measurement.

In retrospect, the hypothesis of locality appears rather simple and natural. For instance, Maxwell's (1880) considerations regarding optical phenomena in moving systems implicitly contained the hypothesis of locality. Its approximate validity was later assumed by Lorentz (1952) in his theory of electrons in order to ensure that an electron, conceived as a small ball of charge, would always be Lorentz contracted along its direction of motion; see Section 183 of Lorentz's book (1952). It was clearly recognized by Lorentz that this is simply an approximation whose validity rests on the supposition that the electron velocity would change over a timescale that is much longer than the period of internal oscillations of the electron, see p. 216 of Section 183 of Lorentz (1952).

A similar assumption was simply adopted by Einstein for rods and clocks as a useful *approximation*, see the footnote on p. 60 of Einstein (1950). In the early days of relativity theory, the locality assumption was discussed in terms of the "clock hypothesis", as it led to the so-called twin paradox; in this connection, Sommerfeld's notes on Minkowski's 1908 paper are quite informative, see p. 94 of Minkowski (1952). Indeed, the hypothesis of locality underlies Einstein's development of the theory of relativity. For instance, the locality assumption fits perfectly together with Einstein's local principle of equivalence to ensure that every observer in a gravitational field is pointwise inertial. In fact, to preserve the operational significance of Einstein's heuristic principle of equivalence—namely, the presumed local equivalence of an observer in a gravitational field with an accelerated observer in Minkowski spacetime—it must be coupled with a statement regarding what accelerated observers actually measure. When coupled with the hypothesis of locality, Einstein's principle of equivalence provides a physical basis for a field theory of gravitation that is consistent with (local) Lorentz invariance.

Following Einstein's development of the general theory of relativity, Weyl discussed the physical basis for the hypothesis of locality, see pp. 176–177 of Weyl (1952). In particular, Weyl noted that the locality hypothesis was an adiabaticity assumption analogous to the one for sufficiently slow processes in thermodynamics and would therefore be expected to be a good approximation only up to some acceleration.

If the hypothesis of locality is an approximation, what is the *exact* result? For instance, if the locality postulate is valid at sufficiently low accelerations, what happens at high accelerations? Can one devise an approximation scheme in which the locality postulate would be the first approximation? It appears that following the great success of general relativity, relativistic physics was generally considered to be simply *local* (Robertson 1949). The investigation of the difference between actual accelerated observers and the hypothetical inertial observers and the related problem of the domain of validity of the locality postulate received little or no attention until about thirty years ago (Mashhoon 1986, 1988, 1990a, 1990b).

1.2.2   Standard measuring devices

Ideal measuring devices that are so robust as to be essentially unaffected by acceleration are called "standard". Thus a standard clock measures proper time along its world line ("clock hypothesis"). From a modern perspective, all ideal measuring devices that are pointwise inertial are standard. That is, an ideal measuring device is practically standard if we can suppose that over the length- and timescales characteristic of typical measurements, the net impact of the internal inertial effects over the operation of the device can be neglected. Though cognizant of the possible limitations of these ideas, we will adhere to the traditional approach to spacetime measurements in relativity theory and assume that all measuring devices are standard. In this way, we will concentrate on the intrinsic nonlocality of the measurement of phenomena associated with electromagnetic fields. That is, as explained in the next chapter, even when an accelerated observer employs only ideal standard devices for measurement purposes, there are intrinsically nonlocal measurements involving electromagnetic fields that extend over the past world line of the observer and hence go beyond the postulate of locality.

1.2.3   Acceleration tensor

Inertial observers are endowed with local reference frames; therefore, it follows from the hypothesis of locality that an accelerated observer in Minkowski spacetime carries an orthonormal tetrad frame $\lambda^\mu{}_{\hat\alpha}(\tau)$ such that at each instant of proper time τ, this frame coincides with the tetrad frame carried by the momentarily comoving inertial observer. Here, $\lambda^\mu{}_{\hat0} = dx^\mu/d\tau$ is the observer's 4-velocity vector u^μ, which is the unit timelike vector that is tangent to the path, and $\lambda^\mu{}_{\hat\imath}$, $i = 1, 2, 3$, are three unit spacelike vectors that constitute the local spatial frame of the accelerated observer moving in a background inertial frame with inertial coordinates $x^\mu = (ct, \mathbf{x})$. The operational establishment of the local tetrad frame of the accelerated observer is ultimately based on the standard rods and clocks that the accelerated observer may use for local spacetime determinations. It follows from the method of moving frames that

$$\frac{d\lambda^\mu{}_{\hat{\alpha}}}{d\tau} = \Phi_{\hat{\alpha}}{}^{\hat{\beta}}(\tau)\, \lambda^\mu{}_{\hat{\beta}}, \tag{1.41}$$

where $\Phi_{\hat{\alpha}\hat{\beta}}$ is the *acceleration tensor* of the observer. The orthonormality of the frame field implies that $\Phi_{\hat{\alpha}\hat{\beta}} = -\Phi_{\hat{\beta}\hat{\alpha}}$. This invariant antisymmetric acceleration tensor can be decomposed into spacetime scalars as $\Phi_{\hat{\alpha}\hat{\beta}} \mapsto (-\tilde{\mathbf{g}}, \tilde{\boldsymbol{\Omega}})$ in close analogy with the standard decomposition of the electromagnetic field tensor $F_{\mu\nu} \mapsto (\mathbf{E}, \mathbf{B})$ into electric $(\mathbf{E})$ and magnetic $(\mathbf{B})$ components, where $F_{0i} = -E_i$, $F_{ij} = \epsilon_{ijk} B^k$ and $\epsilon_{123} = 1$ in our convention. Thus the *translational acceleration* of the observer is given by the "electric" part of the acceleration tensor, $\Phi_{\hat{0}\hat{i}} = \tilde{g}_{\hat{i}}(\tau)$, while the *angular velocity of the rotation* of the observer's spatial frame with respect to a locally non-rotating (i.e. Fermi–Walker transported) frame is given by the "magnetic" part of the acceleration tensor, namely, $\Phi_{\hat{i}\hat{j}} = \epsilon_{\hat{i}\hat{j}\hat{k}}\, \tilde{\Omega}^{\hat{k}}(\tau)$.

To clarify the interpretation of $\tilde{\boldsymbol{\Omega}}$, imagine that the accelerated observer carries a *non-rotating* orthonormal frame $n^\mu{}_{\hat{\alpha}}(\tau)$ along its world line as well. That is, $n^\mu{}_{\hat{0}} = \lambda^\mu{}_{\hat{0}} = u^\mu$ and the spatial frame is Fermi–Walker transported in accordance with eqn (1.21), namely,

$$\frac{dn^\mu{}_{\hat{i}}}{d\tau} = (n^\nu{}_{\hat{i}}\, \mathfrak{a}_\nu)\, n^\mu{}_{\hat{0}}. \tag{1.42}$$

The spatial frame of the observer can be obtained from the non-rotating frame by a time-dependent rotation,

$$\lambda^\mu{}_{\hat{i}} = M_{\hat{i}\hat{j}}\, n^{\mu\hat{j}}, \tag{1.43}$$

where $M(\tau)$ is an orthogonal matrix. The instantaneous angular velocity of the rotation is a pseudovector $\tilde{\boldsymbol{\Omega}}(\tau)$ with respect to the local frame of the observer given by

$$\frac{dM_{\hat{i}\hat{j}}}{d\tau} = \epsilon_{\hat{i}\hat{k}\hat{l}}\, \tilde{\Omega}^{\hat{l}}\, M^{\hat{k}}{}_{\hat{j}}, \tag{1.44}$$

which is consistent with the orthogonality of $M(\tau)$. It follows from eqns (1.42)–(1.44) that the rotational motion of the spatial frame is properly contained in the definition of the acceleration tensor in eqn (1.41). That is, differentiating eqn (1.43) and evaluating $d\lambda^\mu{}_{\hat{i}}/d\tau$ via eqns (1.42) and (1.44), we recover the same equation for the variation of the spatial frame that is contained in eqn (1.41).

It is possible to define a spacelike 4-vector of angular velocity $\tilde{\Omega}^\mu$ that is orthogonal to u^μ via

$$\tilde{\Omega}^\mu = \lambda^\mu{}_{\hat{k}}\, \tilde{\Omega}^{\hat{k}} \tag{1.45}$$

and then it is straightforward to show that the acceleration tensor $\Phi^{\mu\nu}$,

$$\Phi^{\mu\nu} = \lambda^\mu{}_{\hat{\alpha}}\, \lambda^\nu{}_{\hat{\beta}}\, \Phi^{\hat{\alpha}\hat{\beta}}, \tag{1.46}$$

can be expressed as

$$\Phi^{\mu\nu} = \mathfrak{a}^\mu\, u^\nu - \mathfrak{a}^\nu\, u^\mu + \epsilon^{\mu\nu\rho\sigma}\, u_\rho\, \tilde{\Omega}_\sigma, \tag{1.47}$$

where $\epsilon_{0123} = 1$ in our convention.

In Newtonian mechanics, the state of a pointlike observer is given at any instant of time by its position and velocity in space; however, the situation is clearly different for a pointlike observer in relativity theory. The state of such an observer is given by its spacetime position along a future-directed world line and its adapted orthonormal tetrad frame. The four coordinates of the event together with the six independent components of the frame (i.e. three boosts and three rotations or, stated otherwise, sixteen tetrad components subject to ten orthonormality conditions) render the state space of the elementary observer a ten-dimensional manifold. The tetrad frame moves along the timelike world line in accordance with eqn (1.41). In this connection, it is interesting to note that historically the Frenet–Serret method of moving frames for curves in space was later extended to surfaces in space by Darboux; however, the recognition of the full power of this method and its complete generalization was accomplished by E. Cartan.

It seems intuitively clear that an accelerated observer in a global inertial frame of reference may be considered practically inertial during an experiment if the observer's acceleration is such that its motion is in effect uniform and rectilinear and its spatial frame is non-rotating for the duration of the physical process under study. Lorentz invariance can then be employed to predict the result of the experiment. More generally, let λ/c be the intrinsic timescale for the process under consideration, and let $\mathcal{L}/c$ be the relevant acceleration timescale over which the tetrad frame of the observer changes appreciably; then, the condition for the validity of the hypothesis of locality is

$$\lambda \ll \mathcal{L}. \tag{1.48}$$

The time and length scales over which the state of an accelerated observer changes are given by invariants that can be constructed out of $\tilde{\mathbf{g}}(\tau)$, $\tilde{\boldsymbol{\Omega}}(\tau)$ and the speed of light c. For instance, an observer may have a translational acceleration length $1/|\tilde{\mathbf{g}}|$ and a rotational acceleration length $1/|\tilde{\boldsymbol{\Omega}}|$. For observers at rest on the surface of the Earth, $c^2/|\mathbf{g}_\oplus| \approx 1$ light year and $c/|\boldsymbol{\Omega}_\oplus| \approx 28$ astronomical units. These astronomical lengths are very large compared to laboratory dimensions of interest on Earth; hence, the hypothesis of locality is ordinarily a rather good approximation. For instance, in an optics experiment in the laboratory involving $\lambda \sim 10^{-5}$ cm, we have $\lambda/\mathcal{L} \lesssim 10^{-20}$. This means that for most physical situations the acceleration of the observer can be neglected for the duration of the physical process under consideration, which explains why locality is so effective in practice.

1.2.4 Local geodesic coordinates for accelerated observers

Imagine an accelerated observer in a *global inertial frame* in Minkowski spacetime. The observer follows the reference world line $\bar{x}^\mu(\tau)$, where τ is its proper time; moreover, it carries along this path an orthonormal tetrad frame $\lambda^\mu{}_{\hat{\alpha}}(\tau)$, where $\lambda^\mu{}_{\hat{0}} = d\bar{x}^\mu/d\tau$ is its unit temporal axis and $\lambda^\mu{}_{\hat{i}}$, $i = 1, 2, 3$, constitute its local spatial frame. The observer's acceleration tensor is given by $\Phi_{\hat{\alpha}\hat{\beta}} \mapsto (-\tilde{\mathbf{g}}, \tilde{\boldsymbol{\Omega}})$, where $\tilde{\mathbf{g}}(\tau)$ and $\tilde{\boldsymbol{\Omega}}(\tau)$ are respectively the tetrad components of the reference observer's translational acceleration of its world line and the rotational angular velocity of its spatial frame with respect to the local non-rotating (i.e. Fermi–Walker transported) frame. Let us now consider a geodesic

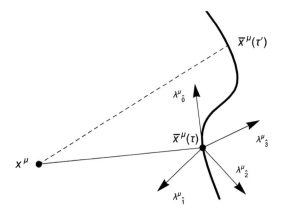

Fig. 1.3 Schematic construction of a geodesic system of coordinates $X^\mu = (cT, \mathbf{X})$ about an accelerated observer. The geodesic (Fermi) coordinates of an event x^μ in the background inertial frame are determined by the way in which x^μ can be connected orthogonally to the observer's path via a geodesic (i.e. straight) line. If such a connection occurs at $\bar{x}^\mu(\tau)$, then $cT = \tau$ and $x^\mu - \bar{x}^\mu(\tau)$ has components $\mathbf{X}$ along the axes of the spatial frame of the observer at τ. A second possibility is indicated in the plot at τ'; however, a coordinate system must *uniquely* identify events. Therefore, to avoid such a possibility, the spatial extent of the geodesic coordinate system is in general limited to a sufficiently narrow world tube along the timelike world line of the observer.

system of coordinates $X^{\hat\mu} = (cT, X^{\hat i})$ established in a certain spacetime domain in the neighborhood of the world line of the reference observer. Given any event τ along the fiducial path $\bar{x}^\mu(\tau)$, the straight spacelike geodesic lines orthogonal to the reference observer's world line at τ span a simultaneity hyperplane that is in fact the three-dimensional Euclidean space. An event on this hyperplane with inertial coordinates x^μ in the background global frame will be assigned geodesic ("Fermi") coordinates $X^{\hat\mu}$. The proper distance away from the reference world line is given by the length of the *unique* spacelike geodesic on the simultaneity hyperplane connecting $\bar{x}^\mu(\tau)$ with x^μ, see Fig. 1.3. The coordinate transformation $x \mapsto X$ is given by

$$\tau = X^{\hat 0}, \qquad x^\mu = \bar{x}^\mu(\tau) + X^{\hat i}\lambda^\mu{}_{\hat i}(\tau). \tag{1.49}$$

It follows from eqns (1.41) and (1.49) that

$$dx^\mu = \left(\mathcal{P}\lambda^\mu{}_{\hat 0} + \mathcal{Q}^{\hat j}\lambda^\mu{}_{\hat j}\right)dX^{\hat 0} + \lambda^\mu{}_{\hat i}\,dX^{\hat i}, \tag{1.50}$$

where

$$\mathcal{P} = 1 + \tilde{\mathbf{g}} \cdot \mathbf{X}, \qquad \mathcal{Q}_{\hat i} = (\tilde{\mathbf{\Omega}} \times \mathbf{X})_{\hat i}. \tag{1.51}$$

It is then simple to show that the Minkowski metric tensor $\eta_{\mu\nu}\,dx^\mu \otimes dx^\nu$ with respect to the new geodesic coordinate system can be written as $g_{\hat\mu\hat\nu}\, dX^{\hat\mu} \otimes dX^{\hat\nu}$, where

$$g_{\hat0\hat0} = -\mathcal{P}^2 + \mathcal{Q}^2, \quad g_{\hat0\hat\imath} = \mathcal{Q}_{\hat\imath}, \quad g_{\hat\imath\hat\jmath} = \delta_{\hat\imath\hat\jmath}. \tag{1.52}$$

If the acceleration tensor of the observer vanishes, the geodesic coordinate system reduces to a global inertial frame that covers Minkowski spacetime and $g_{\hat\mu\hat\nu} = \eta_{\hat\mu\hat\nu}$. The geodesic coordinates are thus quasi-inertial; therefore, for the sake of notational consistency, the use of hatted indices has been extended here to the quasi-inertial geodesic (Fermi) coordinate systems constructed on the basis of the tetrad frames of the reference observers. It is important to note that the reference observer permanently resides at the spatial origin of the geodesic coordinate system, namely, $\mathbf{X} = 0$.

It is interesting to illustrate the construction of geodesic coordinate systems about the world lines of uniformly accelerated observers of Section 1.1. Let us first consider the fiducial world line $\bar x^\mu(\tau)$ given by eqn (1.22). It follows from eqn (1.49) that an event with coordinates (ct, x, y, z) in the background inertial frame has geodesic (Fermi) coordinates (cT, X, Y, Z) such that

$$ct = \left(Z + \frac{c^2}{g_0}\right)\sinh\left(\Theta_0 + \frac{g_0\,T}{c}\right) - \frac{c^2}{g_0}\sinh\Theta_0, \tag{1.53}$$

$$z = \left(Z + \frac{c^2}{g_0}\right)\cosh\left(\Theta_0 + \frac{g_0\,T}{c}\right) - \frac{c^2}{g_0}\cosh\Theta_0, \tag{1.54}$$

$x = X$ and $y = Y$. Here, the only nonzero components of the acceleration tensor are given by $\Phi_{\hat0\hat3} = -\Phi_{\hat3\hat0} = \tilde g_{\hat3} = g_0/c^2$.

Let us next consider observers $\tilde{\mathcal{O}}$ that are all at rest in the background global inertial frame, but refer their measurements to uniformly rotating axes. In this case, we have $ct = \tau$ and

$$\bar x^\mu(\tau) = (\tau,\, x_0,\, y_0,\, z_0). \tag{1.55}$$

Then, it follows from eqn (1.49) and eqns (1.25)–(1.28) that $T = t$,

$$X = (x - x_0)\cos\Omega t + (y - y_0)\sin\Omega t, \tag{1.56}$$

$$Y = -(x - x_0)\sin\Omega t + (y - y_0)\cos\Omega t \tag{1.57}$$

and $Z = z - z_0$. For such a static observer, $\Phi_{\hat1\hat2} = -\Phi_{\hat2\hat1} = \tilde\Omega_{\hat3} = \Omega/c$ are the only non-zero elements of the acceleration tensor. Let us note that the standard classical rotating coordinate system (Landau and Lifshitz 1971) is thus the geodesic (Fermi) system established in the neighborhood of a static noninertial observer $\tilde{\mathcal{O}}$.

Finally, we consider the uniformly rotating observer $\mathcal{O}$ with $\beta = r\,\Omega/c$,

$$\bar x^\mu(\tau) = \left(\gamma\tau,\, r\cos\left(\frac{\gamma\Omega\tau}{c}\right),\, r\sin\left(\frac{\gamma\Omega\tau}{c}\right),\, z_0\right) \tag{1.58}$$

and tetrad frame $\lambda^\mu{}_{\hat a}(\tau)$ given by eqns (1.33)–(1.36). The corresponding coordinate transformation in this case is given by

$$t = \gamma \left(T + \frac{1}{c} \beta Y \right),$$ (1.59)

$$x = (X + r) \cos(\gamma \Omega T) - \gamma Y \sin(\gamma \Omega T),$$ (1.60)

$$y = (X + r) \sin(\gamma \Omega T) + \gamma Y \cos(\gamma \Omega T)$$ (1.61)

and $z = Z + z_0$. In this case, the only non-zero elements of the acceleration tensor $\Phi_{\hat{\alpha}\hat{\beta}} \mapsto (-\tilde{\mathbf{g}}, \tilde{\mathbf{\Omega}})$ are given by the centripetal acceleration $\tilde{g}_{\hat{1}} = -\gamma^2 \beta \Omega / c$ and the rotational angular velocity $\tilde{\Omega}_{\hat{3}} = \gamma^2 \Omega / c$, so that $\tilde{\mathbf{g}} \cdot \tilde{\mathbf{\Omega}} = 0$. We note that the spatial frame of the uniformly rotating observer rotates with respect to a corresponding non-rotating (i.e. Fermi–Walker transported) frame $n^\mu{}_{\hat{\alpha}}$ at a rate of $\gamma \Omega$ with respect to time t of the static inertial observers. The tetrad frame $\lambda^\mu{}_{\hat{\alpha}}$ rotates relative to static inertial observers at a rate of Ω per unit time t; therefore, the non-rotating frame $n^\mu{}_{\hat{\alpha}}$ rotates with respect to the static inertial observers at a rate of $(1 - \gamma) \Omega$ per unit time t. This corresponds to the Thomas precession frequency, which can be interpreted as being ultimately due to an overcompensation as a consequence of time dilation involving proper time of static inertial observers and the proper time of the rotating observer (Mashhoon and Obukhov 2014).

In general, the spacetime manifold can be covered by an overlapping set of admissible coordinate charts. The admissibility of a system of coordinates $x^\mu = (ct, x^i)$ is related to the possibility of making proper temporal and spatial measurements by the static observers associated with the coordinate system, namely those that remain at rest in space. Let the spacetime interval take the form $ds^2 = g_{\mu\nu} \, dx^\mu \, dx^\nu$ in the system of coordinates under consideration. Then, the 4-velocity of the observers at rest is given by $u^\mu = dx^\mu / d\tau = (1/\sqrt{-g_{00}}) \, \delta^\mu_0$, since the proper time of the static observer is given by $ds^2 = -d\tau^2 = g_{00} \, (c \, dt)^2$. Thus the temporal admissibility condition requires that $g_{00} < 0$. Furthermore, consider the measurement of the length of an infinitesimal measuring rod by the static observer. The local hypersurface of simultaneity in the immediate neighborhood of the world line of a static observer is orthogonal to its world line and is given by $u_\mu dx^\mu = 0$, so that $c \, g_{00} \, dt + g_{0i} \, dx^i = 0$. From the fact that

$$ds^2 = \frac{1}{g_{00}} \left(c \, g_{00} \, dt + g_{0i} \, dx^i \right)^2 + \gamma_{ij} \, dx^i \, dx^j,$$ (1.62)

where

$$\gamma_{ij} = g_{ij} - \frac{g_{0i} \, g_{0j}}{g_{00}},$$ (1.63)

we see that the measured length of the infinitesimal rod is given by $d\ell^2 = \gamma_{ij} \, dx^i \, dx^j$, where (γ_{ij}) is a symmetric matrix. The spatial admissibility condition is then the requirement that (γ_{ij}) be a positive-definite matrix. It turns out that the temporal and spatial admissibility conditions imply that the principal minors of the metric tensor $(g_{\mu\nu})$ must all be negative in our $(-, +, +, +)$ convention. That is, $(g_{\mu\nu})$ must be a negative-definite matrix. For further discussion and extension of these ideas, see Bini, Chicone and Mashhoon (2012).

Returning now to geodesic coordinates in Minkowski spacetime, it follows from a detailed investigation (Bini, Chicone and Mashhoon 2012) that the new geodesic coordinates are admissible in a spacetime neighborhood around $\bar{x}^{\mu}(\tau)$ so long as $g_{\hat{0}\hat{0}} < 0$. Equation (1.52) implies that the boundary of the admissible region, $g_{\hat{0}\hat{0}} = 0$, is given by $\mathcal{P}^2 = \mathcal{Q}^2$, which is a quadratic equation in the spatial coordinates and represents a surface. Such surfaces have been classified under the Euclidean group into seventeen standard forms called quadric surfaces (O'Neill 1966; Birkhoff and MacLane 1953). If the reference observer is only translationally accelerated, the quadric boundary surface degenerates to *coincident planes*, given in the case of the hyperbolic observer $\hat{\mathcal{O}}$ by $Z = -c^2/g_0$. Turning next to static noninertial observers $\tilde{\mathcal{O}}$, the boundary hypersurface of the admissible region is the circular light cylinder of radius c/Ω. Finally, for the uniformly rotating observer $\mathcal{O}$, we note that the quadric surface can be expressed as

$$(X + r)^2 + \gamma^2 Y^2 = \frac{c^2}{\Omega^2}, \tag{1.64}$$

which for all Z is an *elliptic cylinder* whose axis coincides with the Z axis. For any constant Z, eqn (1.64) represents an ellipse with semimajor axis c/Ω, semiminor axis $c/(\gamma\Omega)$ and eccentricity v/c. In fact, this ellipse appears as a circle of radius c/Ω that is Lorentz–Fitzgerald contracted along the direction of motion of $\mathcal{O}$. For $Z = 0$, the center of the ellipse is at $x = y = 0$ and $z = z_0$, and the reference observer is at one of the foci of this ellipse; as $v \to c$, $r \to c/\Omega$, the reference observer approaches the light cylinder and the area of the ellipse tends to zero.

A general discussion of the boundary of admissible geodesic coordinates about the world line of an arbitrary accelerated observer is contained in the next chapter. For a detailed discussion of the properties of the boundary hypersurface, see Mashhoon (2003a); moreover, a general discussion of the inertial effects in geodesic coordinates and further references are contained in Chicone and Mashhoon (2005).

The quasi-inertial geodesic coordinate system is by no means the only possible way to establish coordinates in order to identify events uniquely in the neighborhood of an accelerated observer. Another useful method is furnished, for example, by radar coordinates (Bini, Lusanna and Mashhoon 2005). It is indeed possible to show that for observers whose world lines are infinitesimally close to each other, the results of the radar approach are identical to those based on the geodesic coordinate system. All such systems eventually break down, however, due to the existence of acceleration lengths of the observer (Bini, Lusanna and Mashhoon 2005). On the other hand, Minkowski spacetime can always be adequately covered by an overlapping set of such local coordinate charts.

1.3   General Relativity and Locality

The hypothesis of locality provides the simplest possible way to extend relativity theory to noninertial observers in Minkowski spacetime. The next fundamental step involves the extension of relativity theory to observers in a gravitational field. This is achieved via Einstein's principle of equivalence. According to this heuristic principle, an observer in a gravitational field is presumed to be *locally* equivalent to a certain accelerated observer in Minkowski spacetime. The physical basis for this idea is the universality of the gravitational interaction in the framework of the Newtonian laws of motion as well as Newtonian gravitation. The universality of gravity in turn follows from *the principle of equivalence* of inertial and gravitational masses. The principle of equivalence, which has firm observational support, originally provided Einstein with the key to the relativistic theory of gravitation. That is, Einstein interpreted the experimentally well-established principle of equivalence to mean that there is an intimate connection between *inertia* and *gravitation*. This notion eventually led to Einstein's own extremely local principle of equivalence.

Einstein's principle of equivalence and the hypothesis of locality, taken together, imply that observers in a gravitational field are all locally inertial. That is, Einstein's principle of equivalence postulates a pointwise correspondence between measurements of an observer in a gravitational field with an accelerated observer in Minkowski spacetime, while the latter observer is pointwise equivalent to an inertial observer by the hypothesis of locality; therefore, an observer in a gravitational field is pointwise inertial. Thus, at all regular events in a gravitational field observers can define local inertial frames that are then somehow connected with each other through the structure of spacetime. This circumstance suggests that the gravitational field must be inherent in the spacetime structure. The flat Minkowski spacetime has no structure capable of accommodating a gravitational field. The simplest possible way to connect all such local inertial frames of reference for observers in the presence of gravitation is through the pseudo-Riemannian (i.e. Lorentzian) geometry of curved spacetime, where the gravitational field is then identified with the curvature of the spacetime manifold. That is, a curved spacetime manifold is at each event locally (i.e. pointwise) flat. In this general relativistic (GR) framework, free test particles and null rays follow geodesics of the curved spacetime manifold. It remains to establish, within this framework, the gravitational field equation and the correspondence of the resulting relativistic theory with the Newtonian theory of gravitation.

In GR, the curved spacetime interval is given by

$$ds^2 = g_{\alpha\beta}\, dx^\alpha\, dx^\beta, \tag{1.65}$$

where the pseudo-Riemannian metric tensor $g_{\alpha\beta}$ of Lorentzian signature cannot be transformed to the Minkowski metric tensor by a global change of coordinates unless the Riemannian curvature tensor vanishes, which would indicate the absence of a gravitational field. In the presence of spacetime curvature, the ten independent components of $g_{\alpha\beta}$ correspond to the ten gravitational potentials in GR. The correspondence with Newtonian gravitation can be established when we formally let $c \to \infty$. In this way, there is a certain correspondence between the metric tensor and Newtonian gravitational potential ϕ_N. Moreover, it follows from the correspondence between the geodesic equation and Newton's second law of motion of a test particle under gravitation that the Levi-Civita connection $({}^0\Gamma^\sigma_{\mu\nu})$ is similar to the Newtonian gravitational acceleration $(\partial_i \phi_N)$. Finally, the Jacobi equation corresponds to the tidal equation of Newtonian physics; hence, the Riemann curvature tensor $({}^0R_{\mu\nu\rho\sigma})$ is similar to the Newtonian tidal matrix $(\partial_i \partial_j \phi_N)$. To complete the structure of a proper field theory of gravitation, we need the analog of Poisson's equation of Newtonian gravity, namely,

$$\nabla^2 \phi_N = 4\pi G\rho, \tag{1.66}$$

where G is Newton's constant of gravitation and ρ is the density of matter. We note that Poisson's equation connects the matter content of space with the trace of the Newtonian tidal matrix. The simplest generalization of eqn (1.66) in the context of curved spacetime is Einstein's gravitational field equation of GR, namely,

$$^{0}R_{\mu\nu} - \frac{1}{2} g_{\mu\nu} \, ^{0}R = \frac{8\pi G}{c^4} T_{\mu\nu}. \tag{1.67}$$

Here $T_{\mu\nu}$ is the energy–momentum tensor of matter, the Ricci tensor, $^{0}R_{\mu\nu} = {^{0}R^{\alpha}}_{\mu\alpha\nu}$, is the trace of the Riemann tensor and $^{0}R = {^{0}R^{\alpha}}_{\alpha}$ is the scalar curvature. Thus the mass–energy content of spacetime is connected via eqn (1.67) to the appropriate trace of the Riemann tensor in the gravitational field equation of GR. In this way, GR is a *field* theory of gravitation in close analogy with Maxwell's field theory of electromagnetism.

Einstein's gravitational field equation describes how material sources (including non-gravitational fields) produce spacetime curvature, just as Maxwell's electro-dynamics describes how the electromagnetic field is generated by electric charges and their currents. That is, in a global inertial frame in Minkowski spacetime, the electromagnetic field equations are given by

$$\partial_{[\rho} F_{\mu\nu]} = 0, \qquad \partial_{\nu} F^{\mu\nu} = \frac{4\pi}{c} \bar{J}^{\mu}, \tag{1.68}$$

where $\bar{J}^{\mu}$ is the total current 4-vector associated with electric charges. The antisymmetry of the electromagnetic field tensor, $F_{\mu\nu} = -F_{\nu\mu}$, immediately implies that in eqn (1.68) electric charge is conserved, namely, $\partial_{\mu} \bar{J}^{\mu} = 0$. Thus we would expect that the conservation law for the source of the gravitational field, namely,

$$^{0}\nabla_{\nu} T^{\mu\nu} = 0, \tag{1.69}$$

where $^{0}\nabla$ denotes covariant differentiation, be an immediate consequence of the gravitational field equation. This is indeed the case, since eqn (1.69) is a direct consequence of the reduced Bianchi identity. Furthermore, it follows from the field character of the gravitational interaction in Einstein's theory that $T_{\mu\nu} = 0$ does not necessarily mean that spacetime is flat; indeed, the Riemann curvature tensor could be non-zero in a Ricci-flat spacetime due to the existence of gravitational waves.

The Poisson equation of Newtonian gravitation is a consequence of the inverse square force law, which is ultimately based on astronomical observations in the Solar System that originally led to Kepler's laws of planetary motion. Einstein's gravitational field equation has generalized eqn (1.66) into a consistent relativistic framework that is in good agreement with present Solar System as well as binary pulsar data (Shapiro 1980; Stairs 2003). Nevertheless, on small laboratory scales, for instance, questions remain regarding the validity of the inverse square law of gravitation; at present, efforts continue on resolving such experimental problems (Adelberger *et al.* 2003; Hoyle *et al.* 2004; Adelberger *et al.* 2007; Kapner *et al.* 2007; Little and Little 2014). It will turn out later in this book that the nonlocal aspect of the gravitational interaction can lead to possible deviations from the inverse square law on galactic scales. The resulting nonlocal extension of GR may then be used to resolve difficulties in astrophysics and cosmology, such as the problem of the rotation curves of spiral galaxies.

It is in general rather difficult to solve the Einstein field equation, since it contains a coupled system of second-order nonlinear partial differential equations for the gravitational potentials $g_{\mu\nu}(x)$. Many classes of exact solutions of these equations are known in cases involving certain symmetries; however, the general solution is unknown (Stephani *et al.* 2003).

The simplest generalization of the gravitational field equation of GR involves the addition of a *cosmological constant* $^0\Lambda$, namely,

$$^0R_{\mu\nu} - \frac{1}{2} g_{\mu\nu} \, ^0R + \, ^0\Lambda \, g_{\mu\nu} = \frac{8\pi G}{c^4} \, T_{\mu\nu}. \tag{1.70}$$

However, many other generalizations of the field equation of GR are possible. The main purpose of this book is to present a nonlocal generalization of GR, where the gravitational field is local but satisfies a partial integro-differential field equation. In this way, gravitation becomes history-dependent. It seems that in this extension of GR, the nonlocal aspect of the gravitational interaction may simulate dark matter. That is, according to nonlocal GR, what appears as dark matter in astrophysics and cosmology may essentially be a manifestation of the nonlocality of the gravitational interaction.

Finally, it is important to point out that the Einstein field equation can be derived from an action principle. Indeed, eqn (1.70) follows from the variational principle of stationary action S_{GR}, namely, $\delta\, S_{GR} = 0$, where

$$S_{GR} = \int \left(\mathfrak{L}_g + \mathfrak{L}_m - \frac{c^3}{8\pi G} \, ^0\Lambda \sqrt{-g} \right) d^4x. \tag{1.71}$$

Here,

$$\mathfrak{L}_g = \frac{c^3}{16\pi G} \sqrt{-g} \, ^0R \tag{1.72}$$

is the Lagrangian density of the gravitational field, $\mathfrak{L}_m$ is the Lagrangian density of matter and non-gravitational fields and $g = \det (g_{\mu\nu})$.

1.4 Fundamental Observers

In the transition from the flat Minkowski spacetime of special relativity to the curved spacetime of general relativity, the global inertial frames are supplanted with *local* inertial frames. Thus in GR, observers no longer have access to a *global system of parallel axes* in the presence of gravitation.

It is a characteristic feature of the nonlocal generalization of GR developed in this book that the gravitational degrees of freedom are carried by the sixteen components of the tetrad frame field adapted to *fundamental observers*. These form a class of observers throughout spacetime whose tetrad frame field is rendered globally parallel by virtue of the introduction of the Weitzenböck connection in addition to the Levi-Civita connection. Thus, in this extension of GR framework, two distant vectors in spacetime are parallel if they have the same components relative to their respective local fundamental tetrad frames ("teleparallelism"). Moreover, to find a solution of the nonlocal gravitational field equation is tantamount to the determination of the tetrad frame field of the fundamental observers throughout spacetime.

In nonlocal GR, the complete absence of the gravitational field implies that the fundamental observers reduce to those defined in Section 1.1, namely, inertial observers that are all at rest in a global inertial frame in Minkowski spacetime with local tetrad frames whose axes are aligned with the global Cartesian spacetime axes of the background inertial frame of reference.

2
Acceleration-Induced Nonlocality

Modern classical relativity theory grew out of developments in electrodynamics. There-fore, in addition to the relativistic mechanics of point particles, traditional relativistic physics involves classical electromagnetic fields and radiation. In classical physics, the value of a physical quantity $Q(t)$ at time t is based on a certain measurement process that in general started *before* time t. For pointwise coincidences involving classical point particles and rays of radiation, the locality postulate holds exactly and $Q(t)$ depends only on events at time t. However, the question naturally arises whether the locality postulate can be extended to fields. Measurements of electromagnetic fields and their properties are intrinsically nonlocal; moreover, this nonlocality cannot be ignored if the observer is accelerated, since there are basic acceleration scales asso-ciated with the motion of the observer. The purpose of this chapter is to show that field measurements of ideal accelerated observers *cannot* be performed instantaneously. Therefore, the past history of the accelerated observer must be taken into account. A history-dependent ansatz for what accelerated observers measure would eventually lead to a nonlocal special relativity theory.

2.1 Accelerated Observers

All actual observers are to some extent accelerated; therefore, to compare the predic-tions of the special theory of relativity with observational data, one needs to know how to interpret the measurements of accelerated observers. What do accelerated observers measure? The standard answer in special relativity is provided through the hypoth-esis of locality, namely, the assumption that an *accelerated* observer along its world line is at each instant physically equivalent to a hypothetical *inertial* observer that is otherwise identical and instantaneously comoving with the accelerated observer. The accelerated observer therefore measures at each instant what the corresponding iner-tial observer would measure at that instant. It is important to reiterate here that the special theory of relativity is based on Lorentz invariance as well as the hypothesis of locality. To satisfy the locality postulate, physical measurements must be essentially reduced to pointlike coincidences.

Observers are classical macrophysical entities—either sentient or measuring devices. Therefore, as in Newtonian physics, one can imagine the observing system as consisting of a collection of Newtonian point particles such that the behavior of the system can be analyzed in terms of the motion of each individual point particle. We emphasize that such a pointlike observer should still be considered a macrophysical entity subject to the laws of classical (i.e. non-quantum) physics. Therefore, in the

Newtonian picture, an elementary observer is considered a point particle following a path in space. In the spacetime picture, such an observer is endowed with a local tetrad frame as well. That is, the Newtonian description can be essentially generalized by the addition of a local spatial frame of reference. Thus an elementary observer can be described via a future directed timelike world line with a suitable tetrad frame defined along the path. A laboratory would then correspond to a congruence of such world lines and their associated tetrad frames. This *minimal* description of observers goes beyond the traditional one that involves only a future directed timelike world line (see e.g. Gödel 1949).

The *theoretical* association of a moving frame with an observer actually dates back to the early days of general relativity theory; however, this connection is now no longer a figment of imagination, but is in fact indispensable due to developments in *experimental* physics. Indeed, modern measurements in spacetime generally involve the determination of tensorial or spinorial entities and this necessitates the association of a spatial frame to each elementary observer along its world line. Imagine, for example, a gravity gradiometer on a space platform in orbit about the Earth (Paik 2008). The gradiometer registers inertial and gravitational forces. To separate out the effects of the translational acceleration as well as the rotation of the gradiometer from the tidal gravitational effects of the Earth, the spatial frame of the gradiometer must be controlled with sufficient accuracy (Mashhoon and Theiss 1982; Mashhoon, Paik and Will 1989).

The observer together with its tetrad frame moves not just in spacetime, but in the ten dimensional *state space* that is part of the *frame bundle*. That is, to describe the state of an observer at each instant of its proper time τ, one needs an event on its future directed world line as well as its spatial frame at that event. The four coordinates of the event together with the six independent components of the frame (i.e. sixteen tetrad elements subject to ten orthonormality conditions or, equivalently, three boosts plus three rotations) renders the state space a ten-dimensional manifold for a general observer.

In the standard approach, an accelerated observer is thus assumed to be instantaneously inertial and at rest in an inertial frame of reference that moves with its instantaneous velocity with respect to the background global inertial frame. The inhomogeneous Lorentz transformation (1.1) that connects the background frame $x^\mu = (ct, \mathbf{x})$ to the instantaneous frame $x'^\mu = (ct', \mathbf{x}')$ can be employed to determine what the accelerated observer would measure. This method is clearly reasonable so long as the phenomena under consideration involve only pointlike coincidences involving classical point particles and electromagnetic rays that have, by definition, vanishing wavelengths.

Consider, for instance, the reception of electromagnetic radiation by an accelerated observer. The incident wave packet consists, via Fourier analysis, of a spectrum of plane monochromatic waves each with propagation vector $k^\mu = (\omega/c, \mathbf{k})$, as determined by the static inertial observers in the background global inertial frame. The phase differential $d\,(\text{phase}) = k_\mu\, dx^\mu$ associated with each component of the wave packet is a Lorentz-invariant quantity; therefore, $k'_\alpha = k_\mu \Lambda^\mu{}_\alpha$. Thus the accelerated observer measures an *instantaneous* spectrum with components $k'^\mu = (\omega'/c, \mathbf{k}')$ given by the standard formulas for the Doppler effect and aberration of light,

$$\omega' = \gamma(\omega - \mathbf{v} \cdot \mathbf{k}), \tag{2.1}$$

$$\mathbf{k}' = \mathbf{k} + \frac{\gamma - 1}{v^2}(\mathbf{v} \cdot \mathbf{k})\mathbf{v} - \frac{1}{c^2}\gamma\omega\,\mathbf{v}, \tag{2.2}$$

where $\mathbf{v}(t)$ is the instantaneous velocity of the observer and γ is the corresponding Lorentz factor, cf. eqns (1.9) and (1.10). To measure wave properties, the observer needs to register at least a few periods of the wave in order to make a reasonable determination. The observer's velocity $\mathbf{v}(t)$ in general changes from one instant to the next; therefore, as discussed in detail in the following section, one must conclude that phase invariance as well as eqns (2.1) and (2.2) can be valid only in the geometric optics or eikonal *limit* of wave motion corresponding to rays of radiation.

2.2   Frequency Measurement

Consider the measurement of the frequency of an incident plane monochromatic wave of frequency ω and wave vector $\mathbf{k}$, $\omega = c\,|\mathbf{k}|$, by an accelerated observer in a global inertial frame in Minkowski spacetime. If the observer moving with velocity $\mathbf{v}(t)$ could be considered inertial, then the local inertial frame of the observer can be related to the background global inertial frame by a Lorentz transformation. Subsequently, the Doppler effect—which is a consequence of the invariance of the phase of the wave under Lorentz transformations—may be employed to give $\omega'(t) = \gamma\,[\omega - \mathbf{v}(t) \cdot \mathbf{k}\,]$. Physically, the observer must register at least a few oscillations of the incident wave before an adequate determination of its frequency can be attempted; on the other hand, the formula for the Doppler effect is valid if during this time $\mathbf{v}(t)$ changes very little. We can express this condition from the standpoint of the fundamental observers—that is, inertial observers at rest in the background global inertial frame with local tetrad frames that are parallel to the corresponding Cartesian coordinate axes—as

$$n\,\bar{T}\left|\frac{d\mathbf{v}(t)}{dt}\right| \ll v(t). \tag{2.3}$$

Here, $\bar{T} = 2\pi/\omega$ is the period of the incident wave, $v(t)$ is the magnitude of $\mathbf{v}(t)$ and n is the number of cycles of oscillation needed for a reasonable determination of the frequency. Thus $n > 1$ is an integer. In general, the frequency can be determined more accurately if more cycles are used; for very large n, however, one finds from eqn (2.3) that the acceleration must then be very small.

Let us assume that the accelerated observer moves in a direction that is generally parallel to the direction of wave propagation. If the observer has just a constant linear acceleration, then using $\omega = 2\pi c/\lambda$ and $v(t) < c$, eqn (2.3) can be written as

$$\lambda \ll \frac{c^2}{g_0}, \tag{2.4}$$

where g_0 is the constant magnitude of the acceleration of the observer. On the other hand, if the observer is at the time rotating uniformly on a circular orbit with angular speed Ω, then it has centripetal acceleration of magnitude Ωv. It follows from eqn (2.3) that $n \bar{T} \Omega \ll 1$, so that with $\lambda = c\bar{T}$ and $n > 1$, we find

$$\lambda \ll \frac{c}{\Omega}. \tag{2.5}$$

Equations (2.4) and (2.5) illustrate the relation $\lambda \ll \mathcal{L}$, which ensures that locality in these specific instances is a good approximation. For sufficiently low accelerations, the relevant acceleration scales can be large enough to ensure that $\mathcal{L} \gg \lambda$, in which case the standard local theory is quite adequate. Otherwise, the local theory breaks down. To have a complete theory, it is necessary to go beyond the locality postulate, since the standard local theory of relativity does not appear to be capable of dealing with the measurement of electromagnetic wave phenomena involving $\lambda \gtrsim \mathcal{L}$.

2.3 Radiating Charged Particle

To illustrate a situation where $\lambda \sim \mathcal{L}$, consider the path of an accelerated *charged* particle. On dimensional grounds, one would expect that an inertial charged particle would not radiate, since v/c is dimensionless and the inertial particle's acceleration lengths are all infinite; hence, no finite length scale (that could result in a wavelength) can be associated with its motion. On the other hand, the radiation emitted by an accelerating charged particle has dominant wavelengths $\lambda \sim \mathcal{L}$; therefore, locality is violated for the particle. This circumstance is reflected in its equation of motion. That is, the accelerated charged particle's measurements cannot be theoretically predicted by pointwise Lorentz transformations from the background inertial frame to the momentarily comoving inertial frame at the position of the particle on its path. Indeed, the particle radiates away its energy via electromagnetic waves and this loss of energy must show up in its equation of motion through a radiation reaction force. In the nonrelativistic approximation, the Abraham–Lorentz equation of motion of the particle of mass m and charge $\bar{q}$ is given by

$$\frac{d\mathbf{v}}{dt} - \frac{2}{3} \frac{\bar{q}^2}{mc^3} \frac{d^2\mathbf{v}}{dt^2} + \cdots = \frac{1}{m} \mathbf{f}(t, \mathbf{x}, \mathbf{v}). \tag{2.6}$$

The Newtonian postulate of locality no longer holds here as $\mathbf{x}(t)$ and $\mathbf{v}(t)$ are not sufficient to specify the state of the particle at time t; for this purpose, at least the acceleration of the charged particle is needed as well. This violation of locality is of course due to the interaction of the accelerating charged particle with the electromagnetic field. Moreover, eqn (2.6) contains an intrinsic time scale of order $\bar{q}^2/(mc^3)$, which is the time that light would take to cross the classical radius of the charged particle. The presence of such an intrinsic time scale in the equation of motion is naturally expected to be inconsistent with the hypothesis of locality.

2.4 Acceleration Scales

Huygens' principle suggests that in general wave properties cannot be measured at one event. Various thought experiments involving measurement of electromagnetic wave phenomena by accelerated observers indicate that for wave phenomena the hypothesis of locality is valid only in the ray limit, where $\lambda/\mathcal{L}$ is negligible. Here λ is the characteristic wavelength of the phenomenon under observation and $\mathcal{L}$ is the relevant acceleration length. We therefore expect that in this case deviations from locality would be proportional to $\lambda/\mathcal{L}$ (Mashhoon 2008). If this ratio is sufficiently small compared to unity, $\lambda/\mathcal{L} \ll 1$, then the hypothesis of locality in relativistic physics can be a very good approximation. Moreover, if the physical processes of interest all involve *pointlike coincidences* of point particles and rays of radiation such that in effect $\lambda = 0$, then $\lambda/\mathcal{L} = 0$ and the accelerated observer may be considered *pointwise* inertial. That is, at each instant the observer's measurements depend only upon its position and velocity, but not upon its acceleration, which is indeed the locality postulate of the standard theory of relativity.

In general, an observer in special relativity has a translational acceleration length c^2/g_0 and a rotational acceleration length c/Ω. The acceleration lengths (c^2/g_0 and c/Ω) are familiar concepts in standard relativistic physics, since they indicate the spatial limitations associated with accelerated coordinate systems. However, these lengths emerge from the basic heuristic considerations regarding locality in relativity theory with a different fundamental significance: Lorentz invariance can be directly extended to accelerated observers only when $\mathcal{L} \gg \lambda$. The importance of this conceptual analysis lies in the establishment of the notion that the standard theory of relativity is generally valid so long as intrinsic wave phenomena are considered in the eikonal (JWKB) approximation.

2.4.1   Accelerated systems

Consider the coordinate-independent acceleration tensor $\Phi_{\hat{\alpha}\hat{\beta}}$ defined in eqn (1.41). If $\Phi_{\hat{\alpha}\hat{\beta}}(\tau) = 0$, the observer is *inertial*; otherwise, the observer is *accelerated*. Thus accelerated motion is in this sense absolute. To study accelerated systems in Minkowski spacetime, let us first consider the case of an arbitrary accelerated observer following a world line $x^\mu(\tau)$ in a global inertial frame with Cartesian coordinates (t, x, y, z), where we have set $c = 1$ for the sake of simplicity. Henceforth, we use units such that $c = 1$, unless specified otherwise. The observer carries its local orthonormal tetrad frame $\lambda^\mu{}_{\hat{\alpha}}(\tau)$ in accordance with eqn (1.41). The 4-velocity vector of the observer is given in (t, x, y, z) coordinates by

$$u^\mu = \lambda^\mu{}_{\hat{0}}(\tau) = \frac{dx^\mu}{d\tau} = (\gamma, \gamma\mathbf{v}). \tag{2.7}$$

It follows from $u \cdot u = -1$ that the observer's 4-acceleration vector $\mathfrak{a}^\mu$ is such that $u \cdot \mathfrak{a} = 0$. Hence $\mathfrak{a}^\mu(\tau)$ is spacelike and $\mathfrak{a} \cdot \mathfrak{a} = \tilde{g}^2 \geq 0$. Thus,

$$\mathfrak{a}^\mu = \frac{du^\mu}{d\tau} = (\gamma\frac{d\gamma}{dt}, \gamma\frac{d\gamma}{dt}\mathbf{v} + \gamma^2\frac{d\mathbf{v}}{dt}) \tag{2.8}$$

and

$$\frac{d\gamma}{dt} = \gamma^3\mathbf{v}\cdot\frac{d\mathbf{v}}{dt}. \tag{2.9}$$

It follows that the magnitude of the proper acceleration $\tilde{g}(\tau) > 0$ is given by

$$\tilde{g}^2 = \gamma^4\left|\frac{d\mathbf{v}}{dt}\right|^2 + \gamma^6\left(\mathbf{v}\cdot\frac{d\mathbf{v}}{dt}\right)^2. \tag{2.10}$$

Here, $d\mathbf{v}/dt$ is the observer's Newtonian acceleration as measured by inertial observers at rest in the global inertial frame. In terms of the local frame of the accelerated observer, we have

$$\mathfrak{a}^{\mu} = \mathfrak{a}^{\hat{\alpha}}(\tau)\,\lambda^{\mu}{}_{\hat{\alpha}}(\tau), \qquad \mathfrak{a}^{\hat{\alpha}}(\tau) = (0,\ \tilde{\mathbf{g}}), \tag{2.11}$$

where $\tilde{\mathbf{g}}(\tau)$ is the invariant translational acceleration of the observer and $\tilde{g} = |\tilde{\mathbf{g}}|$.

For accelerated motion that is linear, eqn (2.10) further simplifies to $\gamma^6\,(dv/dt)^2 = \tilde{g}^2$; that is,

$$\frac{dv}{d\tau} = \pm\tilde{g}(\tau)\,(1 - v^2)\,. \tag{2.12}$$

Let the motion be along the x direction, then we find that in (t, x, y, z) coordinates, the 4-velocity of the observer is

$$u^{\mu} = \lambda^{\mu}{}_{\hat{0}}(\tau) = (\cosh\Theta,\ \sinh\Theta,\ 0,\ 0). \tag{2.13}$$

Here,

$$\Theta(\tau) = \Theta_0 \pm \int_{\tau_0}^{\tau} \tilde{g}(\tau')d\tau', \tag{2.14}$$

where $\tanh\Theta_0 = v_0$ is the observer's initial velocity at τ_0. If the observer's natural spatial frame is initially aligned with the coordinates axes of the global inertial frame and is non-rotating; that is, it is Fermi–Walker transported along the world line, then

$$\lambda^{\mu}{}_{\hat{1}}(\tau) = (\sinh\Theta,\ \cosh\Theta,\ 0,\ 0), \tag{2.15}$$

$$\lambda^{\mu}{}_{\hat{2}} = (0,0,1,0), \qquad \lambda^{\mu}{}_{\hat{3}} = (0,0,0,1). \tag{2.16}$$

In this case, the acceleration tensor, given by eqn (1.41), simply reduces to translational acceleration along the x axis, $\Phi_{\hat{\alpha}\hat{\beta}} \mapsto (-\tilde{\mathbf{g}}, 0)$, where $\tilde{\mathbf{g}} = (\pm\tilde{g}, 0, 0)$, as expected.

Consider next an observer that revolves on a circle of radius r about the z axis with speed $v(t) = r\,\Omega(t)$. The path of the observer is given by $x^{\mu}(\tau) = (t, r\cos\varphi, r\sin\varphi, z_0)$, where z_0 is constant and φ is the azimuthal angle defined by

$$\varphi = \varphi_0 + \int_0^t \Omega(t')\,dt'. \tag{2.17}$$

The proper time of the rotating observer is

$$\tau = \int_0^t [1 - v^2(t')]^{1/2}\,dt'. \tag{2.18}$$

Let $\gamma = dt/d\tau$ be the corresponding Lorentz factor; then, the natural tetrad frame of the nonuniformly rotating observer, adapted to the rotating system, is given with respect to the inertial (t, x, y, z) coordinates by (Mashhoon 2004a)

$$\lambda^{\mu}{}_{\hat{0}} = \gamma(1,\ -v\sin\varphi,\ v\cos\varphi,\ 0), \tag{2.19}$$

$$\lambda^{\mu}{}_{\hat{1}} = (0,\ \cos\varphi,\ \sin\varphi,\ 0), \tag{2.20}$$

$$\lambda^{\mu}{}_{\hat{2}} = \gamma(v,\ -\sin\varphi,\ \cos\varphi,\ 0), \tag{2.21}$$

$$\lambda^{\mu}{}_{\hat{3}} = (0,\ 0,\ 0,\ 1). \tag{2.22}$$

Here, $d\varphi/d\tau = \gamma\,\Omega$ is the proper angular speed of the observer and it follows from eqn (1.41) that the invariant translational and rotational accelerations of a typical rotating observer are given by

$$\tilde{\mathbf{g}} = (-v\gamma^2\,\Omega,\ \gamma^2\,\frac{dv}{d\tau},\ 0), \tag{2.23}$$

$$\tilde{\mathbf{\Omega}} = (0,\ 0,\ \gamma^2\,\Omega), \tag{2.24}$$

where the components of these vectors are expressed here with respect to local spatial axes of the observer's frame $\lambda^\mu{}_{\hat{i}}, i = 1, 2, 3$, that indicate the radial, tangential and z directions, respectively. As expected, the translational acceleration vector has only centripetal $(\gamma^2\,v^2/r)$ and tangential $(\gamma^2\,dv/d\tau)$ components in this case.

Imagine now, for example, the special case of static noninertial observer on the z axis with $r = 0$. Such a rotating observer is at *rest* in the background inertial frame, but refers its measurements to *rotating* axes. The issue of whether the *pointlike* observer itself rotates is immaterial and, in any case, irrelevant to the physics at hand. Let us next imagine that the noninertial static observer in this example undergoes accelerated motion up or down the z axis. The acceleration tensor in this case has the simple form in which the translational and rotational acceleration vectors are *parallel*; that is, this situation illustrates the typical case where $\tilde{\mathbf{g}} \times \tilde{\mathbf{\Omega}} = 0$.

Relativistic kinematics of accelerated systems has been extensively studied for uniform rotation (Landau and Lifshitz 1971) and hyperbolic motion (Born 1909). The latter is the direct relativistic generalization of one-dimensional motion in Newtonian mechanics with constant acceleration. In these cases of uniformly accelerated motion, the acceleration tensor $\Phi_{\hat{\alpha}\hat{\beta}}$ is constant. It is possible to extend the definition of uniform acceleration to more general configurations (Friedman and Scarr 2013, 2015; Scarr and Friedman 2016).

2.4.2 Proper acceleration scales

The spacetime invariants $\tilde{\mathbf{g}}(\tau)$ and $\tilde{\mathbf{\Omega}}(\tau)$ of the acceleration tensor in general depend upon the instantaneous *speed* of the observer as well as the orientation of its spatial frame. Under a local Lorentz transformation of the observer's tetrad frame, $\Phi_{\hat{\alpha}\hat{\beta}}$ transforms as a tensor. It is therefore natural to define the observer's *proper acceleration scales* (of length and time) using the local Lorentz invariants $\tilde{I}$ and $\tilde{I}^*$ of the acceleration tensor (Mashhoon 1990b)

$$\tilde{I} = \frac{1}{4}\Phi_{\hat{\alpha}\hat{\beta}}\,\Phi^{\hat{\alpha}\hat{\beta}}, \qquad \tilde{I}^* = \frac{1}{4}\Phi^*_{\hat{\alpha}\hat{\beta}}\,\Phi^{\hat{\alpha}\hat{\beta}}. \tag{2.25}$$

Here $\Phi^*_{\hat{\alpha}\hat{\beta}}$ is the dual acceleration tensor given by

$$\Phi^*_{\hat{\alpha}\hat{\beta}} = \frac{1}{2}\epsilon_{\hat{\alpha}\hat{\beta}\hat{\gamma}\hat{\delta}}\,\Phi^{\hat{\gamma}\hat{\delta}}, \tag{2.26}$$

where $\epsilon_{0123} = 1$ in our convention. Thus we have

$$\tilde{I} = \frac{1}{2}\,(-\tilde{g}^2 + \tilde{\Omega}^2), \qquad \tilde{I}^* = -\tilde{\mathbf{g}} \cdot \tilde{\mathbf{\Omega}}. \tag{2.27}$$

For instance, for the (nonuniformly) rotating observer, $\tilde{I}^* = 0$. If the angular velocity of rotation is so chosen that $\tilde{I} = 0$ as well, then we have the case of null accelerated observers discussed in Mashhoon (2004a). Moreover, for the observer undergoing hyperbolic motion, $-2\,\tilde{I} = \tilde{g}^2$ and $\tilde{I}^* = 0$, so that the proper acceleration scale is $1/\tilde{g}$, while for the uniformly rotating observer, $2\,\tilde{I} = (\gamma\,\Omega)^2$ and $\tilde{I}^* = 0$, so that the proper acceleration scale is $1/(\gamma\,\Omega)$ in this case. Let us note that $\gamma\,\Omega$ is the observer's proper angular speed, so that due account has been taken of time dilation in this case.

It is interesting to note that (Mashhoon 2013a)

$$\Phi_{\hat{\alpha}\hat{\gamma}}\,\Phi_{\hat{\beta}}{}^{\hat{\gamma}} = \tilde{I}\,\eta_{\hat{\alpha}\hat{\beta}} + 4\pi\,\tilde{T}_{\hat{\alpha}\hat{\beta}}, \tag{2.28}$$

$$\Phi^*_{\hat{\alpha}\hat{\gamma}}\,\Phi_{\hat{\beta}}{}^{\hat{\gamma}} = \tilde{I}^*\,\eta_{\hat{\alpha}\hat{\beta}}, \tag{2.29}$$

where the symmetric and traceless tensor $\tilde{T}_{\hat{\alpha}\hat{\beta}}$ can be simply obtained from the electromagnetic energy–momentum tensor (Landau and Lifshitz 1971) by replacing $\mathbf{E}$ with $-\tilde{\mathbf{g}}$ and $\mathbf{B}$ with $\hat{\boldsymbol{\Omega}}$. The general significance of such relations for the electromagnetic field has been investigated in Hehl and Obukhov (2003).

It is possible to provide a detailed pointwise classification of the various standard forms of the acceleration tensor $\Phi_{\hat{\alpha}\hat{\beta}}$ in complete analogy with electrodynamics (Synge 1965; Landau and Lifshitz 1971). Let us denote an eigenvector and the corresponding eigenvalue of the acceleration tensor by $\Psi^{\hat{\alpha}}$ and $\tilde{\chi}$, respectively. Then,

$$\Phi_{\hat{\alpha}\hat{\beta}}\,\Psi^{\hat{\beta}} = \tilde{\chi}\,\Psi_{\hat{\alpha}}. \tag{2.30}$$

If $\tilde{\chi} \neq 0$, the associated eigenvector is null; moreover, it follows from eqn (2.30) that

$$\tilde{\chi}^2_{\pm} = -\tilde{I} \pm (\tilde{I}^2 + \tilde{I}^{*2})^{1/2}. \tag{2.31}$$

If the invariants $\tilde{I}$ and $\tilde{I}^*$ both vanish, then $\tilde{\chi} = 0$ and $\Phi_{\hat{\alpha}\hat{\beta}}$ represents the acceleration tensor of an observer with *null acceleration* (Mashhoon 2004a). If the observer's acceleration is not *null*, a local Lorentz boost can always render the translational and rotational acceleration vectors *parallel*.

It is natural to extend these general considerations regarding the acceleration tensor to the motion of an accelerated observer in a gravitational field via Einstein's principle of equivalence. In curved spacetime, or in curvilinear coordinates in Minkowski spacetime, differentiation of the tetrad in eqn (1.41) should be replaced by covariant differentiation. It is possible to construct *non-rotating* accelerated reference systems in Minkowski spacetime, but this is in general impossible in a gravitational field, where Fermi–Walker transport can be implemented only along a single observer's world line. The non-zero Riemannian curvature of spacetime in general prevents the extension of the criterion for non-rotation to a congruence of observers (Mashhoon 1987).

2.4.3 Limitations of accelerated systems

The domain of applicability of an extended system of coordinates that is constructed in the neighborhood of the world line of an accelerated observer is limited in general due to the existence of invariant acceleration scales. This will be explicitly demonstrated here for the quasi-inertial geodesic coordinates of Chapter 1. In these coordinates, the spacetime interval is $ds^2 = g_{\hat{\mu}\hat{\nu}} \, dX^{\hat{\mu}} \, dX^{\hat{\nu}}$, where from eqn (1.52) of Chapter 1, we have $g_{\hat{0}\hat{0}} = -\mathcal{P}^2 + \mathcal{Q}^2$, $g_{\hat{0}\hat{i}} = \mathcal{Q}_{\hat{i}}$ and $g_{\hat{i}\hat{j}} = \delta_{\hat{i}\hat{j}}$, with $\mathcal{P} = 1 + \tilde{\mathbf{g}} \cdot \mathbf{X}$ and $\mathcal{Q}_{\hat{i}} = (\tilde{\mathbf{\Omega}} \times \mathbf{X})_{\hat{i}}$. One can show that $\det(g_{\hat{\mu}\hat{\nu}}) := g = -\mathcal{P}^2$ and the inverse metric is given by

$$g^{\hat{0}\hat{0}} = \frac{1}{g}, \quad g^{\hat{0}\hat{i}} = -\frac{\mathcal{Q}^{\hat{i}}}{g}, \quad g^{\hat{i}\hat{j}} = \delta^{\hat{i}\hat{j}} + \frac{1}{g} \mathcal{Q}^{\hat{i}} \mathcal{Q}^{\hat{j}}. \tag{2.32}$$

It is useful to note that $\mathcal{Q}^2 = \tilde{\Omega}^2 \, X_{\hat{i}} \, X^{\hat{i}} - (\tilde{\Omega}_{\hat{i}} \, X^{\hat{i}})^2$. The geodesic coordinate system is admissible for $g_{\hat{0}\hat{0}} < 0$; therefore, the admissible region has a boundary surface given by $g_{\hat{0}\hat{0}} = -\mathcal{P}^2 + \mathcal{Q}^2 = 0$. This boundary surface can be written as a quadratic equation in the spatial coordinates, namely,

$$1 + 2\tilde{g}_{\hat{i}} \, (X^{\hat{0}}) \, X^{\hat{i}} + A_{\hat{i}\hat{j}} \, (X^{\hat{0}}) \, X^{\hat{i}} \, X^{\hat{j}} = 0. \tag{2.33}$$

It is possible to show in general that at any given time $X^{\hat{0}}$, this quadric surface is degenerate and has the form of a real quadric cone when matrix A is invertible (Mashhoon 2003a). In eqn (2.33), $(A_{\hat{i}\hat{j}})$ is a symmetric matrix with components

$$A_{\hat{i}\hat{j}} = \tilde{g}_{\hat{i}} \, \tilde{g}_{\hat{j}} + \tilde{\Omega}_{\hat{i}} \, \tilde{\Omega}_{\hat{j}} - \tilde{\Omega}^2 \, \delta_{\hat{i}\hat{j}}. \tag{2.34}$$

The eigenvalues of this matrix are ν_0 and $\nu_{\pm}$ given by

$$\nu_0 = -\tilde{\Omega}^2, \quad \nu_{\pm} = -\tilde{I} \pm (\tilde{I}^2 + \tilde{I}^{*2})^{1/2}. \tag{2.35}$$

We note that $\nu_+ = \tilde{\chi}_+^2 \geq 0$, $\nu_0 \leq 0$ and $\nu_- = \tilde{\chi}_-^2 \leq 0$. Thus,

$$\det(A_{\hat{i}\hat{j}}) = \nu_+ \, \nu_0 \, \nu_- = \tilde{\Omega}^2 \, (\tilde{\mathbf{g}} \cdot \tilde{\mathbf{\Omega}})^2. \tag{2.36}$$

By an orthogonal similarity transformation, matrix A can be rendered diagonal at any instant of time $X^{\hat{0}}$.

For the general case with $\det(A_{\hat{i}\hat{j}}) > 0$, one can show that the inverse of matrix A is given by

$$(A^{-1})_{\hat{i}\hat{j}} = -\frac{\tilde{I}^* \, (\tilde{g}_{\hat{i}} \, \tilde{\Omega}_{\hat{j}} + \tilde{g}_{\hat{j}} \, \tilde{\Omega}_{\hat{i}}) - 2\, \tilde{I} \, \tilde{\Omega}_{\hat{i}} \, \tilde{\Omega}_{\hat{j}} + \tilde{I}^{*2} \, \delta_{\hat{i}\hat{j}}}{\det(A_{\hat{i}\hat{j}})}. \tag{2.37}$$

It follows from this relation that

$$(A^{-1})_{\hat{i}\hat{j}} \, \tilde{g}^{\hat{i}} \, \tilde{g}^{\hat{j}} = 1. \tag{2.38}$$

Let $\tilde{M}$ be the orthogonal matrix such that the similarity transformation $\tilde{M}^{-1} A \, \tilde{M} = \tilde{A}$ leads to $\tilde{A} = \mathrm{diag}(\nu_+, \nu_0, \nu_-)$. Then, with $\tilde{X} := \tilde{M}^{-1} X$ and $\tilde{h} := \tilde{M}^{-1} \tilde{g}$, consider the translation to new spatial coordinates $(\tilde{\xi}_{\hat{1}}, \tilde{\xi}_{\hat{2}}, \tilde{\xi}_{\hat{3}})$ given by

$$\tilde{\xi}_{\hat{1}} = \tilde{X}_{\hat{1}} + \frac{\tilde{h}_{\hat{1}}}{\nu_+}, \quad \tilde{\xi}_{\hat{2}} = \tilde{X}_{\hat{2}} + \frac{\tilde{h}_{\hat{2}}}{\nu_0}, \quad \tilde{\xi}_{\hat{3}} = \tilde{X}_{\hat{3}} + \frac{\tilde{h}_{\hat{3}}}{\nu_-}. \tag{2.39}$$

Next, from $A^{-1} = \tilde{M}\,\tilde{A}^{-1}\,\tilde{M}^{-1}$ and eqn (2.38), we find $(\tilde{A}^{-1})_{\hat{i}\hat{j}}\,\tilde{h}^{\hat{i}}\,\tilde{h}^{\hat{j}} = 1$ or

$$\frac{\tilde{h}_{\hat{1}}^2}{\nu_+} + \frac{\tilde{h}_{\hat{2}}^2}{\nu_0} + \frac{\tilde{h}_{\hat{3}}^2}{\nu_-} = 1. \tag{2.40}$$

It now follows from eqns (2.39) and (2.40) that the quadric surface (2.33) is of the form

$$|\nu_+|\,\tilde{\xi}_{\hat{1}}^2 - |\nu_0|\,\tilde{\xi}_{\hat{2}}^2 - |\nu_-|\,\tilde{\xi}_{\hat{3}}^2 = 0, \tag{2.41}$$

which represents an *elliptic cone* (O'Neill 1966). Therefore, the spatial extent of validity of the geodesic coordinates is determined by the acceleration lengths that characterize the eigenvalues given in eqn (2.35).

For a singular matrix A, either $\tilde{\boldsymbol{\Omega}} = 0$, in which case the quadric boundary in eqn (2.33) degenerates to coincident planes orthogonal to $\tilde{\mathbf{g}}$, or $\tilde{\boldsymbol{\Omega}} \neq 0$ and $\tilde{\mathbf{g}} \cdot \tilde{\boldsymbol{\Omega}} = 0$. In the latter case, imagine a rotation of spatial axes in eqn (2.33) such that the new axes are in $\tilde{\mathbf{g}}$, $\tilde{\boldsymbol{\Omega}}$ and $\tilde{\mathbf{g}} \times \tilde{\boldsymbol{\Omega}}$ directions. Then, eqn (2.33) implies that for $\tilde{\Omega}^2 < \tilde{g}^2$, the boundary surface is a hyperbolic cylinder. It is a parabolic cylinder for $\tilde{\Omega}^2 = \tilde{g}^2$ and an elliptic cylinder for $\tilde{\Omega}^2 > \tilde{g}^2$. For $\tilde{\mathbf{g}} = 0$, the boundary surface is a circular cylinder of radius $\tilde{\Omega}^{-1}$ with its axis along $\tilde{\boldsymbol{\Omega}}$.

Let us observe that Earth-bound observers, though generally accelerated, experience no difficulty in receiving or sending signals to distant parts of the observable universe. The boundary hypersurface $\mathcal{P}^2 = \mathcal{Q}^2$ could in general be timelike, spacelike or null. Let $N_{\hat{\mu}} = \partial(-g_{\hat{0}\hat{0}})/\partial X^{\hat{\mu}}$ be the normal to the boundary hypersurface and consider $\tilde{N} = \frac{1}{4}g^{\hat{\alpha}\hat{\beta}}N_{\hat{\alpha}}N_{\hat{\beta}}$. Then, using eqns (2.32) and (2.33), we find that

$$\tilde{N} = -\tilde{W}^2 + 2\,\tilde{W}(\tilde{\mathbf{g}} \times \tilde{\boldsymbol{\Omega}}) \cdot \mathbf{X} + [\tilde{\boldsymbol{\Omega}} + (\tilde{\mathbf{g}} \cdot \tilde{\boldsymbol{\Omega}})\,\mathbf{X}]^2, \tag{2.42}$$

where

$$\tilde{W} = \dot{\tilde{\mathbf{g}}} \cdot \mathbf{X} - \frac{(\dot{\tilde{\boldsymbol{\Omega}}} \times \mathbf{X}) \cdot (\tilde{\boldsymbol{\Omega}} \times \mathbf{X})}{1 + \tilde{\mathbf{g}} \cdot \mathbf{X}} \tag{2.43}$$

and an overdot indicates differentiation with respect to $X^{\hat{0}}$.

For an arbitrary accelerated observer, $\tilde{N}$ could be negative, zero or positive at an event $X^{\hat{\alpha}}$ on the boundary hypersurface, indicating that the hypersurface is spacelike, null, or timelike at that event. If rotation is absent ($\tilde{\boldsymbol{\Omega}} = 0, \dot{\tilde{\mathbf{g}}} \neq 0$), then the hypersurface is in general spacelike, except for $\dot{\tilde{\mathbf{g}}} = 0$, in which case it becomes a null hypersurface. But this last case is unphysical, since an infinite amount of energy must be supplied by an external source to maintain complete hyperbolic motion (i.e. uniform translational acceleration of the observer for all time). On the other hand, if $\tilde{\mathbf{g}} = 0$, then $\tilde{N} = \tilde{\Omega}^2$ for uniform rotation ($\dot{\tilde{\boldsymbol{\Omega}}} = 0$), in which case the boundary hypersurface is timelike.

2.5 Applications of Locality

In the practical calculation of measurable quantities, the application of the locality postulate of the special relativity theory can be implemented in one of two equivalent ways. The first method, as already described in Section 2.1, involves making repeated Lorentz transformations to the instantaneous inertial rest frame of the observer. In the second method, one employs the local tetrad frame of the accelerated observer. That is, instead of a continuous infinity of different inertial frames, a tetrad field is deduced from the basis vectors of the corresponding inertial frames. In a fixed background inertial frame, the physical quantities measured by the accelerated observer—such as the electromagnetic field—are then the projections of various spacetime tensors on its tetrad frame.

Consider first the determination of the propagation vector of a monochromatic plane wave by an accelerated observer. As described in Section 2.1, the first method results in $k'_\alpha = k_\mu \Lambda^\mu{}_\alpha$, where k^μ is the propagation vector of the wave according to the static inertial observers in the background global inertial frame and $(\Lambda^\mu{}_\alpha)$ is the Lorentz matrix corresponding to the Poincaré transformation (1.1) that connects the background inertial frame to the instantaneous inertial frame of the accelerated observer. The second way involves $\lambda^\mu{}_{\hat{\alpha}}(\tau)$, which is the orthonormal tetrad frame along the world line of the accelerated observer. Here τ is the proper time, $\lambda^\mu{}_{\hat{0}} = dx^\mu/d\tau$ is the unit timelike vector tangent to the observer's path and $\lambda^\mu{}_{\hat{i}}$, $i = 1, 2, 3$, constitute the local spatial frame of the observer. According to the second method, the propagation vector measured by the accelerated observer is the projection of the propagation vector k^μ on the tetrad frame of the accelerated observer, namely, $k_{\hat{\alpha}} = k_\mu \lambda^\mu{}_{\hat{\alpha}}$.

In the instantaneous inertial rest frame of the accelerated observer, the tetrad frame of the corresponding fundamental observer is given by $h'^\mu{}_{\hat{\alpha}} = \delta^\mu_\alpha$, where the tetrad index of the Kronecker delta is not distinguished from the coordinate index here as the local frame of the fundamental observer agrees with the instantaneous global frame. Hence, it follows from the hypothesis of locality that in the background inertial frame, the tetrad frame of the accelerated observer at each instant τ is

$$\lambda^\mu{}_{\hat{\alpha}} = \Lambda^\mu{}_\nu \, h'^\nu{}_{\hat{\alpha}} = \Lambda^\mu{}_\alpha. \tag{2.44}$$

Therefore,

$$k_{\hat{\alpha}} = k_\mu \lambda^\mu{}_{\hat{\alpha}} = k_\mu \Lambda^\mu{}_\alpha = k'_\alpha, \tag{2.45}$$

which illustrates the equivalence of the two methods in this case. Henceforward, the second method will be employed throughout.

It is a direct consequence of the locality postulate that the projection of various tensorial quantities on the orthonormal tetrad frame of the accelerated observer can be physically interpreted as the measurement of these quantities by the observer. For instance, given an electromagnetic field $F_{\mu\nu}(x)$ in the background global inertial frame,

$$F_{\hat{\alpha}\hat{\beta}}(\tau) = F_{\mu\nu} \lambda^\mu{}_{\hat{\alpha}} \lambda^\nu{}_{\hat{\beta}} \tag{2.46}$$

is the field measured by the hypothetical momentarily comoving inertial observer at τ, which is also what the accelerated observer with orthonormal tetrad frame $\lambda^\mu{}_{\hat{\alpha}}$ would measure according to the standard local theory.

Extending eqn (2.46) to the world lines of a congruence of accelerated observers, one can attempt to find the frequency and wave vector content of an electromagnetic radiation field via Fourier analysis (Mashhoon 1987; Hauck and Mashhoon 2003). The Fourier content of the radiation field would then depend upon the extended spacetime domain; therefore, this approach is in part *nonlocal*. In general, the result of the field method will be different from the propagation vector given in eqn (2.45). We will explore the implications of the local field method in Chapter 3.

The practical establishment of the local tetrad frame $\lambda^\mu{}_{\hat{a}}$ is ultimately based on the standard clock and measuring rods that the accelerated observer may use for local spacetime determinations. To treat the fundamental problem of field measurement, we may tentatively assume the existence of ideal standard devices and return to a deeper examination of this issue once the nonlocal theory has been properly formulated.

2.6 Bohr–Rosenfeld Principle

The thought experiment of Section 2.2 in connection with the measurement of the frequency of an electromagnetic wave involves a process that is not instantaneous, so that the accelerated observer along its world line needs to measure the radiation field for some time before a determination of its frequency becomes possible. Therefore, as a matter of principle, a certain integration of data over the past world line of the observer is necessarily required in any measurement process involving electromagnetic waves.

A further significant step in the analysis of the measurement process involving an accelerated observer is the recognition that the electromagnetic field itself cannot be measured instantaneously. This general assertion, which applies to any electromagnetic field, is the content of the *Bohr–Rosenfeld principle* (Bohr and Rosenfeld 1933, 1950).

Consider Maxwell's electrodynamics in an inertial frame of reference in Minkowski spacetime. The electric and magnetic fields, $\mathbf{E}(t, \mathbf{x})$ and $\mathbf{B}(t, \mathbf{x})$, respectively, that satisfy Maxwell's equations are assumed to be fields measured instantaneously by the fundamental inertial observers at rest in the background global inertial frame. In 1933, Bohr and Rosenfeld pointed out that in fact only *spacetime averages* of these fields have immediate physical significance. That is, $\mathbf{E}(t, \mathbf{x})$ and $\mathbf{B}(t, \mathbf{x})$ occur in Maxwell's equations as idealizations (Bohr and Rosenfeld 1933). To illustrate this point, Bohr and Rosenfeld (1933) considered a simple situation involving the measurement of the electric field using a macrophysical object of volume V and typical spatial dimension L, $V \sim L^3$, with *uniform* volume electric charge density $\bar{\rho}_e$ and total linear momentum $\mathbf{P}$. When placed in the external electric field $\mathbf{E}(t, \mathbf{x})$, the object moves with respect to the fundamental inertial observers according to the Lorentz force law, namely,

$$\frac{d\mathbf{P}}{dt} = \bar{\rho}_e \int_V \mathbf{E}(t, \mathbf{x}) \, d^3x. \tag{2.47}$$

Suppose that the motion of the object is monitored over an interval of time $T_0 = t'' - t'$ and the momentum $\mathbf{P}$ is measured to be $\mathbf{P}'$ and $\mathbf{P}''$ at the initial and final instants, t' and t'', respectively, of the experiment. Then,

$$\mathbf{P}'' - \mathbf{P}' = \bar{\rho}_e \int_{t'}^{t''} \int_V \mathbf{E}(t, \mathbf{x}) \, dt \, d^3x = \bar{\rho}_e \, \langle \mathbf{E} \rangle \, T_0 \, V, \tag{2.48}$$

where the measured quantity is the spacetime average of the electric field, namely,

$$\langle \mathbf{E} \rangle = \frac{1}{\Delta} \int_{\Delta} \mathbf{E}(x) \, d^4x \tag{2.49}$$

with $\Delta = T_0 V$ and $x^\mu = (t, \mathbf{x})$. It is assumed here that the time intervals needed by the fundamental inertial observers for momentum measurements are $\ll T_0$ and the corresponding displacements caused by these measurements are $\ll L$.

  The gist of the Bohr–Rosenfeld argument is that physical fields cannot be measured instantaneously. While this argument appears to be relatively innocuous for classical field measurements via ideal *inertial* observers, it acquires the significance of a physical principle when extended to *accelerated* observers as a direct consequence of the existence of invariant acceleration scales. In the case of ideal inertial observers, their acceleration scales are all infinite and though their field measurements would involve an averaging process in accordance with the considerations of Bohr and Rosenfeld (1933, 1950), such averaging is essentially harmless in *classical field theory.* Acceleration-induced nonlocality is, however, essential due to the existence of the intrinsic acceleration scales. Approximating the world tube of the accelerated observer by a world line, we conclude in accordance with causality that it is necessary to take the past world line of the accelerated observer into account in any field determination.

2.7   Nonlocal Ansatz

It follows from the work of Bohr and Rosenfeld (1933, 1950) that the measurement of the electromagnetic field by ideal inertial observers involves an averaging process over a past spacetime domain. We expect that this is true as well for accelerated observers; however, a direct extension of the Bohr–Rosenfeld treatment to ideal accelerated observers appears to be a rather daunting task. Instead, we approach this basic problem indirectly as follows. Let us first note that for field measurements by actual (i.e. accelerated) observers, their finite acceleration scales must be taken into account as well. When extended to a pointlike noninertial observer, the Bohr–Rosenfeld principle simply implies that the memory of the field along the past world line of the observer cannot be ignored. In other words, the averaging process reduces to an integration over the past world line of the accelerated observer. The observer has no spatial extension by assumption and is thus represented only by its world line and the adapted frame. Thus in searching for a physical link between a noninertial observer and the class of hypothetical momentarily comoving inertial observers along its past world line, we must go beyond the pointwise condition (2.46) and consider a nonlocal relationship involving a certain average over the past world line of the observer. The averaging process in the Bohr–Rosenfeld principle is linear in the field; hence, we look for a general linear connection between the field as determined by the accelerated observer at proper time τ, $\mathcal{F}_{\hat{\alpha}\hat{\beta}}(\tau)$, and the field determinations $F_{\hat{\alpha}\hat{\beta}}$ of the infinite sequence of momentarily comoving inertial observers that the accelerated observer has passed through along its world line. *The most general linear relation consistent with causality is of the form* (Mashhoon 1993a)

$$\mathcal{F}_{\hat{\alpha}\hat{\beta}}(\tau) = F_{\hat{\alpha}\hat{\beta}}(\tau) + u(\tau - \tau_0) \int_{\tau_0}^{\tau} \tilde{K}_{\hat{\alpha}\hat{\beta}}{}^{\hat{\gamma}\hat{\delta}}(\tau, \tau') \, F_{\hat{\gamma}\hat{\delta}}(\tau') \, d\tau', \qquad (2.50)$$

where kernel $\tilde{K}$, *which is antisymmetric in its first and second pairs of indices, is expected to be directly related to the acceleration tensor* $\Phi_{\hat{\mu}\hat{\nu}}$. Here, τ_0 is the instant of proper time at which the observer's acceleration is initially turned on and $u(t)$ is the unit step function such that $u(t) = 0$ for $t < 0$ and $u(t) = 1$ for $t > 0$.

The nonlocal ansatz (2.50) has the form of a Volterra integral equation of the second kind. According to Volterra's theorem, the relationship between $\mathcal{F}_{\hat{\alpha}\hat{\beta}}(\tau)$ and $F_{\hat{\alpha}\hat{\beta}}(\tau)$ is unique in the space of continuous functions (Volterra 1959). This physically significant uniqueness result has been extended to the Hilbert space of square-integrable functions by Tricomi (1957). Indeed, it is possible to show that for physical fields of interest

$$F_{\hat{\alpha}\hat{\beta}}(\tau) = \mathcal{F}_{\hat{\alpha}\hat{\beta}}(\tau) + u(\tau - \tau_0) \int_{\tau_0}^{\tau} \tilde{R}_{\hat{\alpha}\hat{\beta}}{}^{\hat{\gamma}\hat{\delta}}(\tau, \tau') \, \mathcal{F}_{\hat{\gamma}\hat{\delta}}(\tau') \, d\tau', \qquad (2.51)$$

where $\tilde{R}$ is the resolvent kernel (Tricomi 1957; Davis 1930; Lovitt 1950). A derivation of the resolvent kernel via the method of successive substitutions is contained in Section 2.9.

To avoid unphysical situations—for instance, the expenditure of an infinite amount of energy by an external source in the case of complete hyperbolic motion—we always assume that the acceleration of the observer is turned on at some time τ_0 and then after a finite duration turned off at τ_f. For $\tau < \tau_0$, the motion is free of acceleration and $\mathcal{F}_{\hat{\alpha}\hat{\beta}}(\tau) = F_{\hat{\alpha}\hat{\beta}}(\tau)$ in conformity with standard practice in classical electrodynamics; that is, we formally ignore Bohr–Rosenfeld averaging in the absence of acceleration. Moreover, for $\tau > \tau_f$, when the motion is again free of acceleration and the observer's tetrad frame no longer varies with time, we expect that the nonlocal term

$$\int_{\tau_0}^{\tau} \tilde{K}_{\hat{\alpha}\hat{\beta}}{}^{\hat{\gamma}\hat{\delta}}(\tau, \tau') F_{\hat{\gamma}\hat{\delta}}(\tau') d\tau', \qquad (2.52)$$

which contains the memory of the past acceleration of the observer, would now correspond to a *constant* electromagnetic field. We recall that Maxwell's field eqns (1.68) are partial differential equations and are unchanged by the addition of a constant field. Thus for $\tau > \tau_f$, when the motion is free of acceleration, the electromagnetic field given by eqn (2.50) would again satisfy Maxwell's field eqns (1.68). Moreover, in a measuring device, the constant memory of past acceleration can be simply canceled when the device is reset. To ensure that these physical requirements are indeed satisfied and the quantity in display (2.52) is in fact constant in time for $\tau > \tau_f$, we assume that *acceleration kernel* $\tilde{K}(\tau, \tau')$ is independent of τ; that is, for $\tau > \tau_0$

$$\tilde{K}_{\hat{\alpha}\hat{\beta}}{}^{\hat{\gamma}\hat{\delta}}(\tau, \tau') = \tilde{k}_{\hat{\alpha}\hat{\beta}}{}^{\hat{\gamma}\hat{\delta}}(\tau'), \qquad (2.53)$$

where kernel $\tilde{k}$ vanishes whenever $(\Phi_{\hat{\mu}\hat{\nu}}) = 0$. In this way, we adopt a passive (or kinetic) memory of past acceleration, as opposed to an active (or dynamic) memory

(Chicone and Mashhoon 2002a). In the absence of compelling experimental evidence in support of the latter alternative, it is prudent to choose the kinetic memory option in the development of acceleration-induced nonlocality.

The nonlocal ansatz (2.50) expresses the sum of two terms that contribute to the measured field: the local part required by the locality postulate together with an "average" over the past world line of the accelerated observer. The nonlocal averaging involves a weight function given by kernel $\tilde{k}$ that vanishes in the absence of acceleration. For a radiation field $F_{\mu\nu}$ in the eikonal approximation, we expect that the nonlocal term in eqn (2.50) would be proportional to $\lambda/\mathcal{L}$, so that locality is recovered in the eikonal limit.

It is possible to think of eqn (2.50) as an expansion of the measured field in powers of $F_{\mu\nu}$, where the terms beyond the linear order have been simply neglected. The provisional character of our ansatz should thus be emphasized. Perhaps future observational data will provide the necessary motivation to go beyond the present linear theory.

2.8 Nonlocal Relativity

The approach to relativity theory in Minkowski spacetime outlined here leads to nonlocal special relativity, which is treated in the next chapter, where we assume that measuring devices are all standard in order that the explicit effects of acceleration would *only* appear in the nonlocal kernels (Mashhoon 2011a). These originate from the circumstance that the determination of physical fields is *not* instantaneous and requires measurements along the past world line of the observer. Thus in a measured field, the acceleration of the world line is explicitly represented by a universal kernel that acts as the weight function for the memory of past acceleration. The existence of the acceleration-induced nonlocal kernel for field measurements in Minkowski spacetime must be regarded as a purely *vacuum* effect of accelerated motion, since all measuring devices have been assumed to be *standard.* The basic extension of the locality postulate in eqn (2.50) can be naturally expressed for any field. The main task of nonlocal special relativity is then the determination of acceleration kernels for various fields. We emphasize that ansatz (2.50) is envisioned to be in accordance with measurements performed with *standard* devices; therefore, the nonlocal term in this ansatz is expected to be independent of any measuring device employed and is thus purely induced by the acceleration of the observer in Minkowski spacetime. The situation here is reminiscent of the correspondence between wave mechanics and classical mechanics. The linear memory of past acceleration is then a vacuum effect. One may contemplate further extension of these ideas; for instance, eqn (2.50) may have to be further generalized in the presence of a medium.

Ansatz (2.50) introduces a new basic element into relativity theory, namely, the acceleration-dependent kernel that weighs the significance of past events for the present time. The general physical content of eqn (2.50) for a basic field may be expressed as follows: for classical point particles and rays of radiation, the measurements of actual (accelerated) observers are pointlike and the locality hypothesis is valid. However, for field measurements, observers have *memory* (Mashhoon 1993a) and are in general nonlocal. As the infinite sequence of momentarily comoving inertial observers along the path are fictitious, one can interpret eqn (2.50) as a formal expression for the field as measured by the accelerated observer at τ given in terms of the projections of the field $F_{\mu\nu}(x)$ on the accelerated observer's tetrad frame along its past world line. The upshot of this interpretation is again that the locality postulate must be amended by the memory of past acceleration. Imagine now a *congruence* of accelerated observers and consider

$$\mathcal{F}_{\hat{\alpha}\hat{\beta}}(\tau) = \mathcal{F}_{\mu\nu}\,\lambda^{\mu}{}_{\hat{\alpha}}\,\lambda^{\nu}{}_{\hat{\beta}}, \tag{2.54}$$

where $\mathcal{F}_{\mu\nu}(x)$ is a *local* field that the accelerated observer measures at time τ by projecting this field on its local tetrad frame. Now eqn (2.51) and the resolvent kernel can be used for the congruence to express $F_{\mu\nu}(x)$ in terms of $\mathcal{F}_{\mu\nu}(x)$ via an integral equation. The electromagnetic field $F_{\mu\nu}(x)$ satisfies Maxwell's field equations; therefore, substituting this nonlocal relation into Maxwell's equations would result in partial integro-differential field equations for $\mathcal{F}_{\mu\nu}(x)$. This special character of the nonlocal ansatz should be emphasized: nonlocal special relativity actually involves only local fields that satisfy certain integro-differential field equations carrying the memory of past acceleration. Such nonlocal field equations are Lorentz invariant, as they originate from the manifestly Lorentz-invariant nonlocal ansatz.

The nonlocal ansatz (2.50) is reminiscent of the nonlocal characterization of certain constitutive properties of continuous media that exhibit history-dependent phenomena. The nonlocal treatment of the electrodynamics of media can be traced back to the investigations of Poisson (1823), Liouville (1837) and Hopkinson (1877). We emphasize that acceleration-induced nonlocality is associated with the vacuum state as perceived by accelerated observers; nevertheless, nonlocal Maxwell's equations for $\mathcal{F}_{\mu\nu}(x)$ are reminiscent of the partial integro-differential field equations of the nonlocal electrodynamics of media.

Is it possible to determine the acceleration kernel $\tilde{k}$ by means of available observational data? In the design of electric machines, the electromagnetic field experienced by a moving part is traditionally estimated by assuming that it is instantaneously motionless (Van Bladel 1976, 1984). Thus in customary engineering applications, the instantaneous velocity, acceleration, etc., at any given time t are all neglected, while in the locality hypothesis of the standard theory of relativity only the instantaneous velocity is taken into account. To obtain observational data that would pertain to an examination of the foundations of electrodynamics of accelerated systems, one must have access to experiments of rather high sensitivity. Unfortunately, however, further progress is currently hampered by a serious lack of reliable observational data: see Hehl and Obukhov (2003), Mashhoon (2008), Van Bladel (1984), Zhang (1997), Canovan and Tucker (2010) and the references cited therein.

In the absence of significant experimental results to guide the development of nonlocal special relativity, we must rely on certain general theoretical ideas that are described in Chapter 3.

2.9 Appendix: Resolvent Kernel

Consider the Volterra integral equation of the second kind given by

$$\phi(x) + \lambda \int_a^x \tilde{K}(x,y)\,\phi(y)\,dy = f(x), \tag{2.55}$$

where λ and a are constants (Davis 1930; Lovitt 1950; Tricomi 1957). We work in the space of continuous functions with $a \leq x \leq b$, where b is a constant, and assume that kernel $\tilde{K}(x,y)$ vanishes identically for $y > x$. To find $\phi(x)$, we use the method of *successive substitutions* due originally to Liouville, Neumann and Volterra. In other words, we first take the integral term to the right-hand side of eqn (2.55)

$$\phi(x) = f(x) - \lambda \int_a^x \tilde{K}(x,y)\,\phi(y)\,dy, \tag{2.56}$$

and then replace ϕ in the integrand of eqn (2.56) by its value given by this same equation, namely,

$$\phi(x) = f(x) - \lambda \int_a^x \tilde{K}(x,y)\,f(y)\,dy$$
$$+\lambda^2 \int_a^x \tilde{K}(x,y) \left[\int_a^y \tilde{K}(y,z)\,\phi(z)\,dz \right] dy. \tag{2.57}$$

Repeating this process results in an infinite series. If this series is uniformly convergent then we have a unique solution of the Volterra equation that can be written in the form

$$f(x) + \lambda \int_a^x \tilde{R}(x,y)\,f(y)\,dy = \phi(x), \tag{2.58}$$

where $\tilde{R}(x,y)$ is the resolvent kernel. The resolvent kernel here depends upon x, y and λ; however, we express this kernel as $\tilde{R}(x,y)$ for the sake of simplicity. To determine $\tilde{R}(x,y)$, let us first note that in the double integral in eqn (2.57), the integration is over the triangular region depicted in Fig. 2.1 and the order of integration can be interchanged. Therefore, instead of the order of double integration in eqn (2.57) that corresponds to summing over vertical strips in Fig. 2.1a, we can just as well sum over horizontal strips to get

$$\int_a^x \tilde{K}(x,y) \left[\int_a^y \tilde{K}(y,z)\,\phi(z)\,dz \right] dy = \int_a^x \left[\int_z^x \tilde{K}(x,y)\,\tilde{K}(y,z)\,dy \right] \phi(z)\,dz. \tag{2.59}$$

Thus eqn (2.57) can be written as

$$\phi(x) = f(x) + \lambda \int_a^x \tilde{K}_1(x,y)\,f(y)\,dy + \lambda^2 \int_a^x \tilde{K}_2(x,y)\,\phi(y)\,dy, \tag{2.60}$$

where we have employed iterated kernels $\tilde{K}_1$ and $\tilde{K}_2$. In general, we define iterated kernels $\tilde{K}_n$, for $n = 1, 2, 3, \ldots$, as

$$\tilde{K}_1(x,y) = -\tilde{K}(x,y), \qquad \tilde{K}_{n+1}(x,y) = \int_y^x \tilde{K}_n(x,z)\,\tilde{K}_1(z,y)\,dz. \tag{2.61}$$

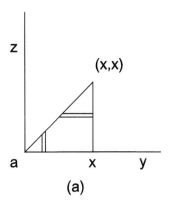

(a)

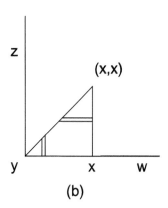

(b)

Fig. 2.1 The triangular regions in (a) and (b) are the domains of double integrations in eqns (2.59) and (2.68), respectively.

It is now possible to see that by continuing the substitution method in eqn (2.60) and using the general idea contained in eqn (2.59), the infinite series for $\phi(x)$ will take the form

$$\phi(x) = f(x) + \int_a^x \left[\sum_{n=1}^{\infty} \lambda^n \, \tilde{K}_n(x, y) \right] f(y) \, dy. \tag{2.62}$$

Therefore, $\tilde{R}$ is given in terms of iterated kernels by

$$\tilde{R}(x, y) = \sum_{n=1}^{\infty} \lambda^{n-1} \, \tilde{K}_n(x, y). \tag{2.63}$$

The basic reciprocity between eqns (2.55) and (2.58) should be noted. It implies that kernels $\tilde{K}$ and $\tilde{R}$ are reciprocal to each other. If, for instance, we interchange f with ϕ in eqn (2.55), we obtain eqn (2.58), provided kernel $\tilde{K}$ is replaced by its reciprocal, namely, kernel $\tilde{R}$. On the other hand, the substitution of eqn (2.55) into eqn (2.58) and vice versa would lead to the basic *reciprocity* integral equations

$$\tilde{K}(x, y) + \tilde{R}(x, y) = -\lambda \int_y^x \tilde{K}(x, z) \, \tilde{R}(z, y) \, dz = -\lambda \int_y^x \tilde{R}(x, z) \, \tilde{K}(z, y) \, dz. \tag{2.64}$$

It is possible to provide a simple and direct proof of these reciprocity equations by means of the formula (2.63) for $\tilde{R}$ and the general relation

$$\tilde{K}_{n+m}(x, y) = \int_y^x \tilde{K}_n(x, z) \, \tilde{K}_m(z, y) \, dz \tag{2.65}$$

that holds for $m = 1, 2, 3, \ldots$. To prove this last result we proceed by induction; indeed, eqn (2.65) holds for $m = 1$ by the definition of iterated kernels in eqn (2.61). Assuming the validity of eqn (2.65), we must show that

$$\tilde{K}_{n+m+1}(x, y) = \int_y^x \tilde{K}_n(x, z)\, \tilde{K}_{m+1}(z, y)\, dz. \tag{2.66}$$

To this end, we note that by definition

$$\tilde{K}_{n+m+1}(x, y) = \int_y^x \tilde{K}_{n+m}(x, z)\, \tilde{K}_1(z, y)\, dz. \tag{2.67}$$

Next, we can write the integrand using eqn (2.65) as

$$\tilde{K}_{n+m+1}(x, y) = \int_y^x \left[\int_z^x \tilde{K}_n(x, w)\, \tilde{K}_m(w, z)\, dw \right] \tilde{K}_1(z, y)\, dz. \tag{2.68}$$

Interchanging the order of integration in the double integral as before, we find

$$\tilde{K}_{n+m+1}(x, y) = \int_y^x \tilde{K}_n(x, w) \left[\int_y^w \tilde{K}_m(w, z)\, \tilde{K}_1(z, y)\, dz \right] dw. \tag{2.69}$$

The integration domain is depicted in Fig. 2.1b, where eqn (2.68) corresponds to summation of horizontal strips, while eqn (2.69) corresponds to the summation of vertical strips in Fig. 2.1b. By the definition of iterated kernels

$$\tilde{K}_{m+1}(w, y) = \int_y^w \tilde{K}_m(w, z)\, \tilde{K}_1(z, y)\, dz, \tag{2.70}$$

which, when substituted in eqn (2.69), leads to eqn (2.66) and this completes the proof.

If $\tilde{K}(x, y) = \tilde{k}(y)$, the iterated kernels $\tilde{K}_n$ for $n > 1$ and consequently the resolvent kernel are in general functions of both x and y. On the other hand, if $\tilde{K}(x, y) = \bar{k}(x-y)$, so that the kernel is of the convolution (Faltung) type, the resolvent kernel turns out to be of the convolution type as well. In fact, inspection of eqn (2.61) reveals that with $\tilde{K}_n(x, y) = \bar{k}_n(x - y)$ and $\bar{k}_1 = -\bar{k}$, we have

$$\bar{k}_{n+1}(t) = \int_0^t \bar{k}_n(t - u)\, \bar{k}_1(u)\, du, \tag{2.71}$$

where $x - y = t$ and $z - y = u$. Therefore, all of the iterated kernels are of the convolution type and can be obtained by successive convolutions of $\bar{k}$ with itself.

3
Acceleration Kernel

To develop the tools necessary for the determination of the acceleration kernel we first need to examine more closely the implications of the locality postulate in connection with the measurement of the electromagnetic field. Consider the reception of electromagnetic waves by an accelerated observer in a background global inertial frame. To measure the frequency of the incident wave the observer needs to register several oscillations before an adequate determination of the frequency becomes even possible. Thus an extended period of proper time τ is necessary for this purpose. On the other hand, as described in Chapter 2, for an incident wave with propagation vector $k^\alpha = (\omega, \mathbf{k})$, the hypothesis of locality implies that at each instant of proper time τ, the observer measures $\omega_D(\tau) = -k_\alpha \lambda^\alpha{}_{\hat{0}}$ via the Doppler effect. Thus, from a physical standpoint, such an instantaneous Doppler formula can be strictly valid only in the eikonal limit of rays of radiation.

There is an alternative way to apply the locality hypothesis to this situation. One can project the Faraday tensor of the wave on the local tetrad frame of the accelerated observer at τ to determine, via the locality postulate, the electromagnetic field that is presumably measured directly by the accelerated observer at τ. The Fourier analysis of this measured field in terms of proper time would then result in the frequency ω' measured by the accelerated observer. This approach, which we adopt in the following section, is in fact nonlocal in time insofar as it relies on Fourier analysis, but the field determination is still based on the locality assumption. However, we recall from Chapter 2 that, as pointed out by Bohr and Rosenfeld (1933, 1950), it is not physically possible to measure an electromagnetic field at one event; instead, a certain averaging process is required. Indeed, ignoring this fact leads to a basic difficulty; that is, we will find in Section 3.2 that by a mere rotation an observer can stand completely still with respect to an incident radiation field. This is an aspect of the hypothesis of locality that we need to abolish in the nonlocal theory. In this way, a physical principle can be formulated in order to determine the acceleration kernel. The end result is a nonlocal theory of special relativity.

To simplify matters, we will mainly work with the electromagnetic vector potential A_μ instead of the Faraday tensor $F_{\mu\nu}$. It turns out that for the determination of the frequency content of the field, both approaches give the same results. According to the hypothesis of locality, the field as determined by an accelerated observer is given by the projection of the incident field upon the tetrad frame of the observer, namely,

$$A_{\hat{\alpha}} = A_\mu \lambda^\mu{}_{\hat{\alpha}}. \tag{3.1}$$

Nonlocal Gravity. Bahram Mashhoon. © Bahram Mashhoon 2017. Published 2017 by Oxford University Press.

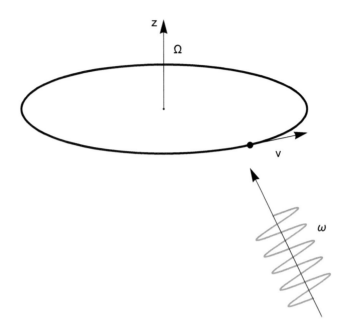

Fig. 3.1 Reception of incident electromagnetic radiation by rotating observer $\mathcal{O}$.

The initial step in our approach involves the determination of the frequency content of the incident radiation using eqn (3.1).

3.1 Local Field Determination

Let us first imagine the *local* determination of the electromagnetic vector potential of an incident radiation field by an observer $\mathcal{O}$ that for $t \geq 0$ is rotating uniformly in the positive sense with constant frequency Ω about the z axis of a background global inertial frame in Minkowski spacetime; see Fig. 3.1. The tetrad frame of observer $\mathcal{O}$ has been derived in Section 1.1. The accelerated observer carries the tetrad frame $\lambda^{\mu}{}_{\hat{\alpha}}$ given by eqns (1.33)–(1.36) and follows the world line $x^{\mu}(\tau) = (t, r\cos\varphi, r\sin\varphi, z_0)$. Here, $t = \gamma\tau$, r and z_0 are constants and $\varphi = \gamma\Omega\tau$ is the azimuthal angle of the observer. The observer's path for $\tau \geq 0$ is a circle of radius r in a plane that is orthogonal to the z axis at z_0. Let A and $\hat{A}$ represent the column vectors (A_{μ}) and $(A_{\hat{\mu}})$ along the world line of the rotating observer, respectively. Then, eqn (3.1) can be written as

$$\hat{A} = \begin{bmatrix} \gamma & -v\gamma\sin\varphi & v\gamma\cos\varphi & 0 \\ 0 & \cos\varphi & \sin\varphi & 0 \\ v\gamma & -\gamma\sin\varphi & \gamma\cos\varphi & 0 \\ 0 & 0 & 0 & 1 \end{bmatrix} A, \tag{3.2}$$

where $v = r\Omega$ and γ is the corresponding Lorentz factor.

The incident radiation field in the background inertial frame can be characterized by its gauge-dependent vector potential. We work in the *radiation* gauge, where $A^\mu = (0, \mathbf{A})$ and $\nabla \cdot \mathbf{A} = 0$ (Gottfried 1966). It then follows from eqn (3.2) that $A_{\hat{0}} = v\,A_{\hat{2}}$,

$$A_{\hat{1}} = \cos\varphi\, A_1 + \sin\varphi\, A_2, \tag{3.3}$$

$$A_{\hat{2}} = \gamma\,(-\sin\varphi\, A_1 + \cos\varphi\, A_2) \tag{3.4}$$

and $A_{\hat{3}} = A_3$.

3.1.1 Normal incidence

Consider, for the sake of simplicity, a plane circularly polarized monochromatic electromagnetic wave propagating along the z axis given by

$$\mathbf{A} = \mathrm{Re}\left[\mathfrak{A}(\hat{\mathbf{x}} \pm i\hat{\mathbf{y}})e^{-i\,\omega\,(t-z)}\right], \tag{3.5}$$

where the upper (lower) sign represents positive (negative) helicity radiation and $\mathfrak{A}$ is a constant complex amplitude. It is then straightforward to find from eqn (3.2) that along the world line of the observer,

$$A_{\hat{1}} = \mathrm{Re}\,(\mathfrak{A}\,e^{-i\,\omega'\,\tau + i\,\omega\,z_0}), \qquad A_{\hat{2}} = \gamma\,\mathrm{Re}\,(\pm i\,\mathfrak{A}\,e^{-i\,\omega'\,\tau + i\,\omega\,z_0}), \tag{3.6}$$

$A_{\hat{0}} = v\,A_{\hat{2}}$ and $A_{\hat{3}} = 0$. Here, ω' is the frequency of the wave as measured by the rotating observer and is given by

$$\omega' = \gamma\,(\omega \mp \Omega). \tag{3.7}$$

It is important to note that this is the result of the Fourier analysis of the measured field in terms of the proper time of the observer and is completely independent of r and z_0, so that eqn (3.7) is valid for the entire class of observers rotating uniformly in the positive sense about the z axis with frequency Ω. In eqn (3.7), the upper (lower) sign refers to positive (negative) helicity radiation. With $E = \hbar\omega$, our classical result (3.7) illustrates the phenomenon of spin–rotation coupling for spin-1 photons. It is important to note here that for incident positive-helicity radiation with $\omega = \Omega$, $\omega' = 0$ and the locally determined radiation field $A_{\hat{\alpha}}$ is constant in time but oscillatory in space as a consequence of the hypothesis of locality.

It proves instructive to explore briefly the physical difference between ω' and the Doppler frequency ω_D for the specific case of the uniformly rotating observer. Here, $\omega_D = \gamma\omega$ by the transverse Doppler effect, which follows in general from the invariance of the phase of the radiation under Lorentz transformation. In this case, the Lorentz factor takes due account of time dilation. On the other hand, the Fourier analysis of the field, pointwise "measured" by the observer via the hypothesis of locality, reveals that

$$\omega' = \omega_D(1 \mp \Omega/\omega). \tag{3.8}$$

The new result reflects time dilation as well as the coupling of photon helicity with the rotation of the observer. The nonlocal aspect of the helicity–rotation coupling is evident in eqn (3.8), where Ω/ω is the ratio of the reduced wavelength of the wave

$(1/\omega)$ to the acceleration length $\mathcal{L}$ of the observer $(1/\Omega)$. The deviation from locality is thus proportional to $\lambda/\mathcal{L}$, as expected. The transverse Doppler effect is recovered as $\Omega/\omega \to 0$ in the geometric optics (i.e. ray) limit. However, locality is expected to fail for positive-helicity radiation with $\Omega/\omega = 1$ in eqn (3.8), where $\omega' = 0$.

Aside from the presence of the Lorentz factor that is due to time dilation, it is simple to interpret the coupling of helicity with rotation in eqn (3.7) in terms of an "angular Doppler shift". In a positive (negative) helicity wave, the electromagnetic field rotates in the positive (negative) sense with frequency ω about the direction of propagation of the wave. The rotating observer therefore perceives incident positive (negative) helicity radiation with the electromagnetic field rotating with relative frequency $\omega - \Omega$ ($\omega + \Omega$) about the direction of wave propagation.

The helicity-dependent contribution to the transverse Doppler effect in eqn (3.7) has been verified for $\omega \gg \Omega$ via the GPS, where it is responsible for the phenomenon of *phase wrap-up* (Ashby 2003). For applications of eqn (3.7) in connection with magnetic resonance and related phenomena see Tejada *et al.* (2010) and Lendinez, Chudnovsky and Tejada (2010).

3.1.2 Oblique incidence

The calculation of the frequency of the incident radiation as locally determined by observer $\mathcal{O}$ becomes considerably more complicated for the case of oblique incidence. To simplify matters, we consider instead the class of static rotating observers $\tilde{\mathcal{O}}$ that are all at rest at fixed spatial positions in the background inertial frame, but refer their measurements to axes that rotate uniformly with angular speed Ω about the z axis; see Section 1.1. Specifically, we consider the *static rotating observer $\tilde{\mathcal{O}}$ that is at rest at the origin of spatial coordinates*.

The vector potential for an incident wave packet in the background frame is given by

$$\mathbf{A}(x) = \mathrm{Re} \sum_{\mathbf{k}} \mathbf{a_k}\, e^{-i\,(\omega\,t - \mathbf{k}\cdot\mathbf{x})}, \tag{3.9}$$

where $\mathbf{a_k}$ is a complex amplitude, $\mathbf{k}$ is the wave vector and $\omega = k = |\mathbf{k}|$. Moreover, $\mathbf{k} \cdot \mathbf{a_k} = 0$ by the transversality condition, which follows from $\nabla \cdot \mathbf{A} = 0$. The vector potential as determined by observer $\tilde{\mathcal{O}}$ can be obtained from eqn (3.2) with $v = 0$ and $\gamma = 1$, and we find

$$\tilde{A}_{\hat{\alpha}} = A_\mu\, \tilde{\lambda}^\mu{}_{\hat{\alpha}} = (0, \tilde{\mathbf{A}}), \tag{3.10}$$

where $\tilde{\mathbf{A}}(x')$ must be evaluated at the location of observer $\tilde{\mathcal{O}}$, which is the origin of spatial coordinates. However, let us first note that $\tilde{\mathbf{A}}(x')$ is obtained from $\mathbf{A}(x)$ via the standard passive transformation, $x \mapsto x'$, of the background inertial coordinate system to the rotating system of coordinates, namely, $t' = t$,

$$x' = x \cos\Omega t + y \sin\Omega t, \qquad y' = -x \sin\Omega t + y \cos\Omega t \tag{3.11}$$

and $z' = z$. This transformation, in terms of spherical polar coordinates (r, ϑ, φ) takes the form $(r, \vartheta, \varphi) \mapsto (r', \vartheta', \varphi')$, where $r = r'$, $\vartheta = \vartheta'$ and $\varphi = \varphi' + \Omega t$. Therefore, it proves useful to express plane waves in eqn (3.9) in terms of spherical waves,

$$e^{i\,\mathbf{k}\cdot\mathbf{x}} = 4\pi \sum_{lm} i^l \, j_l(kr)\, Y_{lm}^*(\hat{\mathbf{k}})\, Y_{lm}(\hat{\mathbf{x}}), \tag{3.12}$$

where j_l, $l = 0, 1, 2, \ldots$, are spherical Bessel functions; moreover, for each l, $m = l, l-1, \ldots, -l$ (Gottfried 1966; Jackson 1999). Here, the unit position vector $\hat{\mathbf{x}}$, for instance, represents the angular coordinates (ϑ, φ), etc. The incident vector potential can then be expressed in terms of vector spherical harmonics as

$$\mathbf{A}(x) = 4\pi \, \mathrm{Re} \sum_{\mathbf{k}JlM} i^l \, [\mathbf{Y}_{JlM}^*(\hat{\mathbf{k}}) \cdot \mathbf{a_k}] \, j_l(kr)\, \mathbf{Y}_{JlM}(\hat{\mathbf{x}})\, e^{-i\omega t}, \tag{3.13}$$

where J and M are the total angular momentum parameters of the field. Indeed, for a photon state of definite total angular momentum $\hbar J$, with $J = 1, 2, 3 \ldots$, the orbital angular momentum $\hbar l$ is such that $l = J+1$, $l = J$, or $l = J-1$, and the eigenvalues of the z component of the total angular momentum vector $\hbar M$ are such that $M = J, J-1, \ldots, -J$. Under the rotation $x \mapsto x'$, we have

$$\mathbf{Y}_{JlM}(\vartheta, \varphi) \mapsto e^{iM\Omega t} \, \mathbf{Y}_{JlM}(\vartheta', \varphi'). \tag{3.14}$$

Hence, the vector potential as locally determined by the static rotating observer $\tilde{O}$ is given by $\tilde{\mathbf{A}}(x')$, where

$$\tilde{\mathbf{A}}(x') = 4\pi \, \mathrm{Re} \sum_{\mathbf{k}JlM} i^l \, [\mathbf{Y}_{JlM}^*(\hat{\mathbf{k}}) \cdot \mathbf{a_k}] \, j_l(kr)\, \mathbf{Y}_{JlM}(\hat{\mathbf{x}}')\, e^{-i\,(\omega - M\,\Omega)\,t}, \tag{3.15}$$

which contains the spectrum of frequencies $\omega - M\,\Omega$, for $M = 0, \pm1, \pm2, \ldots$. We must now compute $\tilde{\mathbf{A}}(x')$ at $r = 0$, which is the location of observer $\tilde{O}$. For $\rho \to 0$,

$$j_l(\rho) \to \frac{\rho^l}{(2l+1)!!}, \tag{3.16}$$

where $(2l+1)!! = (2l+1)(2l-1)\cdots 3 \cdot 1$. Therefore, $\tilde{\mathbf{A}}(x')$ at $r = 0$ contains only $l = 0$, so that $J = 1$ and $M = 0, \pm1$ due to photon spin; as expected, the orbital angular momentum of the radiation field vanishes for the rotating observer at rest at the origin of spatial coordinates.

It follows from a more extensive calculation for the rotating observer O that moves on a circular orbit with proper time $\tau = t/\gamma$ that the orbital angular momentum of the radiation field is non-zero in general and time dilation should be taken into account as well, so that the frequencies measured by the uniformly rotating observer are given by the general formula

$$\omega' = \gamma\,(\omega - M\,\Omega), \tag{3.17}$$

where $\hbar M$, $M = 0, \pm1, \pm2, \ldots$, is the component of the total angular momentum of the electromagnetic radiation field along the axis of rotation of the observer (Mashhoon 2009). Equation (3.17) reduces to eqn (3.7) for radiation propagating along the rotation axis of the observer.

In the geometric optics (i.e. eikonal) approximation, eqn (3.17) may be written as $\omega' = \gamma(\omega - \mathbf{J} \cdot \mathbf{\Omega})$ with $\mathbf{J} = \mathbf{x} \times \mathbf{k} + s\,\hat{\mathbf{h}}$, where $\hbar\,\mathbf{J}$ is the total angular momentum vector, s is the spin, and $\hat{\mathbf{h}} = \pm\hat{\mathbf{k}}$ is the helicity vector of the radiation. Thus eqn (3.17) in the ray approximation reduces to the Doppler effect with $\mathbf{v} = \mathbf{\Omega} \times \mathbf{x}$ together with the term $-\gamma s\,\hat{\mathbf{h}} \cdot \mathbf{\Omega}$, which indicates helicity–rotation coupling; see eqn (3.19). The helicity–rotation phenomenon has been investigated in detail due to its basic significance (Mashhoon 1986, 1989; Mashhoon *et al.* 1998; Hauck and Mashhoon 2003). It has extensive observational support; see Mashhoon (2009) and the references cited therein. Moreover, its existence implies that the phase of the radiation is not in general a Lorentz-invariant quantity.

Finally, it is interesting to note two unusual features of the general formula $\omega' = \gamma\,(\omega - M\,\Omega)$: The measured frequency can be zero or negative. Indeed, $\omega' = 0$ for $\omega = M\,\Omega$, a situation that is discussed in the next section, while ω' can be negative for $\omega < M\,\Omega$. The latter circumstance does not pose any basic difficulty once it is recognized that the notion of relativity of motion does not extend to accelerated observers. Accelerated motion is absolute. For the rotating observer, the Hamiltonian is not bounded from below due to the absolute character of rotational motion.

3.1.3   Modification of Doppler effect and aberration due to helicity–rotation coupling

Consider first the reception of electromagnetic radiation with wave vector $k^\mu = (\omega, \mathbf{k})$ by the class of static observers $\tilde{\mathcal{O}}$ that refer their measurements to axes that rotate uniformly with frequency $\mathbf{\Omega}$. We work in the high-frequency approximation ($\omega \gg \Omega$) and assume that to measure these quantities in the rotating frame the net duration of measurement, which naturally extends over many periods of the incident radiation, must be much shorter than Ω^{-1}. In this case, observers $\tilde{\mathcal{O}}$ perceive $\tilde{k}^\mu = (\tilde{\omega}, \tilde{\mathbf{k}})$ given by (Mashhoon 1989; Hauck and Mashhoon 2003)

$$\tilde{\omega} = \omega - \hat{\mathbf{h}} \cdot \mathbf{\Omega}, \qquad \tilde{\mathbf{k}} = \mathbf{k}. \tag{3.18}$$

It can be shown that $\tilde{k}^\mu$ is the weighted average of the measured propagation vector by analogy with the quasi-classical approximation in wave mechanics (Hauck and Mashhoon 2003).

Next, we recall from Section 1.1 that the frame of the rotating observer $\mathcal{O}$ is locally related to the corresponding static observer $\tilde{\mathcal{O}}$ by a pure boost. It follows that the propagation vector of the incident radiation as perceived by the rotating observer $\mathcal{O}$, namely, $k'^\mu = (\omega', \mathbf{k}')$ is related to $\tilde{k}^\mu = (\tilde{\omega}, \tilde{\mathbf{k}})$ by the same Lorentz boost in this eikonal approximation. In this way, we find the modified expressions for the Doppler effect and aberration,

$$\omega' = \gamma[(\omega - s\,\hat{\mathbf{h}} \cdot \mathbf{\Omega}) - \mathbf{v} \cdot \mathbf{k}], \tag{3.19}$$

$$\mathbf{k}' = \mathbf{k} + \frac{\gamma - 1}{v^2}(\mathbf{v} \cdot \mathbf{k})\mathbf{v} - \gamma(\omega - s\,\hat{\mathbf{h}} \cdot \mathbf{\Omega})\mathbf{v}, \tag{3.20}$$

where $s = 1$ for the photon (Mashhoon 1987) and $s = 2$ for the graviton (Ramos and Mashhoon 2006). The consequences of these results for interferometry with polarized radiation in rotating frames have been studied in Mashhoon (1989) and for Doppler tracking of spacecraft in Mashhoon (2002). These references should be consulted for observational evidence in favor of helicity–rotation coupling in the radio, microwave and optical domains.

To illustrate the new terms in eqns (3.19) and (3.20), let us first note that the standard Doppler and aberration formulas are recovered for $\mathbf{k} \cdot \mathbf{\Omega} = 0$. Hence, let us consider a simple situation involving normal incidence with $\mathbf{k} = \omega \, \hat{\mathbf{\Omega}}$, where $\hat{\mathbf{\Omega}}$ is a unit vector along the direction of rotation. Then,

$$\omega' = \gamma(\omega \mp s\,\Omega), \qquad \mathbf{k}' - \mathbf{k} = -\gamma(\omega \mp s\,\Omega)\mathbf{v}. \tag{3.21}$$

The expression for frequency in eqn (3.21) happens to be exact in this case; for the aberration part of eqn (3.21), we note that with respect to the direction of incidence of the wave, $\mathbf{k}'_\| = \mathbf{k}$ and $\mathbf{k}'_\perp = -\gamma(\omega \mp s\,\Omega)\mathbf{v}$; hence the aberration angle $\tan^{-1}(|\mathbf{k}'_\perp|/|\mathbf{k}'_\||)$ is given in this case by $\tan^{-1}[\gamma v(1 \mp s\,\Omega/\omega)]$. The deviations from the standard results in eqn (3.21) are again proportional to Ω/ω, as expected.

Finally, we note that for (GPS) radio waves with frequency of order 1 GHz, $\Omega_\oplus/\omega \sim 10^{-14}$ for the rotation of the Earth about its axis, while for a receiver rotating at ~ 10 cps we find $\Omega/\omega \sim 10^{-8}$.

3.1.4 Spin–rotation coupling

The helicity–rotation coupling that has been briefly described above is an instance of the general phenomenon of the coupling of intrinsic spin with rotation that is due to the *inertia of intrinsic spin*. Mass and spin characterize the irreducible unitary representations of the Poincaré group (Wigner 1939). The state of a particle in spacetime is thus described by its mass and spin, which determine its inertial properties. Therefore, the inertial characteristics of a particle are determined by its inertial mass (Werner, Staudenmann and Colella 1979; Moorhead and Opat 1996; Werner 2008; Rauch and Werner 2015) as well as intrinsic spin (Mashhoon 1988; Mashhoon and Kaiser 2006). To illustrate the latter, we note that in a macroscopic body rotating in the positive sense with uniform angular velocity $\mathbf{\Omega}$, the spins of the constituent particles do not naturally participate in the rotation; instead, they all tend to stay essentially fixed with respect to the local inertial frame. Thus relative to the rotating body, the spins precess with angular velocity $-\mathbf{\Omega}$. The Hamiltonian corresponding to this motion is $-\hat{\boldsymbol{\sigma}} \cdot \mathbf{\Omega}$, where $\hat{\boldsymbol{\sigma}}$ is the operator of intrinsic spin. For a proper relativistic treatment of this subject and further developments see Hehl and Ni (1990), Mashhoon (1992), Damião Soares and Tiomno (1996), Ryder (1998), Singh and Papini (2000), Papini (2002), Kiefer and Weber (2005), Silenko and Teryaev (2007), Bini and Lusanna (2008), Obukhov, Silenko and Teryaev (2009, 2011, 2013, 2014, 2016), Randono (2010) and Jentschura and Noble (2014). A review of spin–rotation coupling and a more complete list of references is contained in Mashhoon (2006).

In general, the energy of an incident particle as measured by the rotating observer is given by

$$E' = \gamma\,(E - \hbar\,M\Omega), \tag{3.22}$$

where E is the energy of the incident particle in the inertial frame and M is the total (orbital plus spin) "magnetic" quantum number along the axis of rotation. In fact, $M = 0, \pm 1, \pm 2, \ldots$, for a scalar or a vector particle, while $M \mp \frac{1}{2} = 0, \pm 1, \pm 2, \ldots$, for a Dirac particle. Moreover, in the JWKB approximation, $E' = \gamma (E - \boldsymbol{\Omega} \cdot \mathbf{J})$, where $\mathbf{J} = \mathbf{x} \times \mathbf{P} + \boldsymbol{\sigma}$ is the total angular momentum of the particle and $\mathbf{P}$ is its momentum; hence, $E' = \gamma (E - \mathbf{v} \cdot \mathbf{P}) - \gamma \boldsymbol{\sigma} \cdot \boldsymbol{\Omega}$, where $\mathbf{v} = \boldsymbol{\Omega} \times \mathbf{x}$ is the velocity of the uniformly rotating observer with respect to the background inertial frame and γ is the Lorentz factor of the observer. The energy corresponding to spin–rotation coupling is naturally augmented by time dilation.

It is important to remark here that the spin–rotation coupling is completely independent of the inertial mass of the particle. Moreover, the associated spin–gravity coupling is an interaction of the intrinsic spin with the gravitomagnetic field of the rotating source that is also independent of the mass of the test particle (Mashhoon 1999, 2000; Konno and Takahashi 2012). For instance, free neutral Dirac particles with their spins up and down (i.e. parallel and antiparallel to the vertical direction, respectively) in general fall differently in the gravitational field of the rotating Earth (Mashhoon 1995). Similar phenomena are expected for massless particles as a consequence of helicity–gravity coupling (Mashhoon 1974, 1993b; Ramos and Mashhoon 2006; Frolov and Shoom 2011, 2012).

A general consequence of spin–rotation coupling involves the *energy shift* that would be induced when polarized radiation passes through a rotating spin flipper. Imagine, for instance, positive-helicity electromagnetic radiation of frequency ω that is normally incident on a half-wave plate that is uniformly rotating with frequency $\Omega \ll \omega$. Within the half-wave plate, the measured frequency of the radiation is $\omega - \Omega$, where we neglect time dilation for the sake of simplicity. The spacetime in a uniformly rotating system is stationary; therefore, the frequency of the radiation throughout the half-wave plate remains constant and equal to $\omega - \Omega$. The radiation that emerges has negative helicity and hence its frequency is given by $\omega - 2\,\Omega$ in accordance with eqn (3.7). In passing through the half-wave plate, the photon energy is thus downshifted by $-2\,\hbar\,\Omega$. The frequency-shift phenomenon was first discovered experimentally in the microwave regime (Allen 1966). Further discussion of optical phenomena associated with spin–rotation coupling is contained in Garetz and Arnold (1979), Garetz (1981), Bliokh *et al.* (2008), Bliokh (2009) and Bliokh and Aiello (2013). Furthermore, it is important to remark that nonlocality brings about a corresponding *amplitude shift* as well (Mashhoon 2012).

The spin–rotation coupling has recently been measured for neutrons via neutron polarimetry (Demirel, Sponar and Hasegawa 2015). Moreover, this general coupling has now been incorporated into the condensed-matter physics of spin mechanics and spin currents (Shen and He 2003; Matsuo *et al.* 2011a, 2011b; Matsuo *et al.* 2013; Chowdhury and Basu 2013; Matsuo, Ieda and Maekawa 2013; Ieda, Matsuo and Maekawa 2014; Lima and Moraes 2015; Hamada, Yokoyama and Murakami 2015). The effect of rotation on spin current is also expected to play a role in the emerging field of spintronics (Papini 2013).

There is now ample experimental evidence in support of the spin–rotation coupling for $E \gg \hbar M\Omega$. On the other hand, it is theoretically possible to have $E' = 0$ or $E' < 0$, though these situations do not appear to be easily accessible observationally. The possibility that the measured energy in eqn (3.22) can be zero is a serious difficulty and its circumvention in the nonlocal theory is important for the determination of the acceleration kernel, as described in the following sections. The situation is different for $E' < 0$ due to the absolute character of accelerated motion. A thought experiment involving conservation of energy as well as total angular momentum along the rotation axis of the observer is useful here to show that a negative E' is consistent with the observer-independence of the temporal order of events. Imagine a quantum system that, according to inertial observers at rest in a global background inertial frame, makes a transition from a stationary state (E_1, J_1, M_1) to a lower energy state (E_2, J_2, M_2) by emitting a photon of energy $E_1 - E_2 = \hbar\omega$. The photon is detected by the rotating observer $\mathcal{O}$ and is found to have energy $E' = \hbar\omega'$ and frequency $\omega' = \gamma(\omega - M\Omega)$. From the viewpoint of the rotating observer, the states of the quantum system have energies $E'_i = \gamma(E_i - \hbar M_i \Omega)$ for $i = 1, 2$, and $E'_1 - E'_2 = E' = \hbar\omega'$ with $M_1 - M_2 = M$. If $E' < 0$, the rotating observer might claim that the photon was actually emitted by its detector and later absorbed by the state of energy E'_1 causing a transition to a state of higher energy E'_2; however, the causal sequence of events would then be opposite to that perceived by the inertial observers. To maintain the same causal order of events for all observers, the possibility of negative energy states for accelerated observers must be admitted.

3.2 Can Light Stand Completely Still?

The exact formula $\omega' = \gamma(\omega - \Omega)$ for normally incident positive-helicity electromagnetic radiation has a remarkable consequence that is not easily accessible to experimental physics: The incident wave stands completely still relative to observers that rotate uniformly with frequency $\Omega = \omega$ about the direction of propagation of the wave; however, the locality postulate is expected to fail for $\Omega/\omega \sim 1$. That is, the hypothesis of locality has the consequence that a rotating observer can in principle be comoving with an electromagnetic wave; in fact, the wave appears to be oscillatory in space but stands completely still with respect to the rotating observer. The fundamental difficulty under consideration here is quite general, as it occurs for oblique incidence as well; that is, $\omega' = \gamma(\omega - M\Omega)$ vanishes for $\omega = M\Omega$ with $M \neq 0$, so that by a uniform rotation of frequency $\Omega = \omega/M$, $M > 0$, the rotating observer can stand completely still with respect to the incident radiation.

An important consequence of the Lorentz invariance of Maxwell's equations is that the propagation of electromagnetic radiation is independent of the motion of inertial observers. It is natural to expect that this circumstance would extend to all observers, but one encounters a difficulty due to the hypothesis of locality of the standard framework of special relativity theory. By a mere rotation, an observer can in principle stay completely at rest with respect to an electromagnetic wave. This circumstance is rather analogous to the difficulty with the pre-relativistic Doppler formula, where an inertial observer moving with speed c along a light beam would see a wave that is oscillatory in space but is otherwise independent of time and hence completely at rest. This issue, as is well known, played a part in Einstein's path to relativity, as mentioned in his autobiographical notes; see Einstein (1949, p. 53). The difficulty in that case was eventually removed by Lorentz invariance; however, in the present case, the problem has to do with the local way in which Lorentz invariance is extended to accelerated observers in Minkowski spacetime. In the special theory of relativity, Lorentz invariance is extended to accelerated systems via the *hypothesis of locality*, namely, the assumption that an accelerated observer is pointwise inertial. It is therefore important to formulate the theory of accelerated observers in such a way that an accelerated observer cannot stay completely at rest with respect to an electromagnetic wave. This can be implemented within the context of the nonlocal theory of accelerated observers and it plays an important role in the determination of the acceleration kernel.

To go beyond the locality postulate of special relativity theory, the past history of the observer must be taken into account. Thus the locality postulate must be supplemented by a certain average over the past world line of the observer. In this way, the observer retains the memory of its past acceleration. This averaging procedure involves a kernel (or weight function) that must be determined. To this end, we introduce the fundamental assumption that *a basic radiation field can never stand completely still with respect to any observer*. On this basis a nonlocal theory of accelerated observers can be developed (Mashhoon 1993a). Indeed, this is the main postulate that is used in nonlocal special relativity for the determination of the acceleration kernel; that is, the kernel is so chosen as to correct a perceived defect in the standard *local* special relativity theory. As discussed in Section 3.5, the nonlocal approach turns out to be in better correspondence with quantum theory than the standard treatment based on the hypothesis of locality (Mashhoon 2005).

3.3   Determination of the Kernel

Consider a basic radiation field $\psi(x)$ in a background global inertial frame in Minkowski spacetime and an accelerated observer that measures this field. The events along the world line of the observer are characterized by its proper time τ. The observer passes through an infinite sequence of hypothetical momentarily comoving inertial observers. Let $\hat{\psi}(\tau)$ be the field measured by these inertial observers. This local field definition is the result of the projection of the background radiation field onto the local frame of the observer. It follows from the hypothesis of locality that $\hat{\psi}(\tau) = \Lambda(\tau)\psi(\tau)$ along the world line of the accelerated observer, where Λ belongs to a matrix representation of the Lorentz group. For instance, Λ is unity for a scalar field. Equation (3.2) provides a nontrivial example of such a matrix relationship in the case of the electromagnetic vector potential.

The fundamental laws of microphysics have been formulated with respect to inertial observers. On the other hand, physical measurements are performed by observers that are, in general, accelerated. To interpret observation via theory, a connection must be established between the field $\hat{\Psi}(\tau)$ that is actually measured by the accelerated observer and $\hat{\psi}(\tau)$. The standard theory of relativity postulates that the accelerated observer is pointwise inertial and hence $\hat{\Psi}(\tau) = \hat{\psi}(\tau)$. This is, of course, the simplest possibility and has been quite successful as a first approximation. It is consistent with the physical principles of superposition and causality. It is important, however, to go beyond this relation in view of the limitations of the hypothesis of locality. As already emphasized in Chapter 2, the most general relationship between $\hat{\Psi}(\tau)$ and $\hat{\psi}(\tau)$ that preserves linearity and causality for $\tau \geq \tau_0$ is

$$\hat{\Psi}(\tau) = \hat{\psi}(\tau) + \int_{\tau_0}^{\tau} \hat{K}(\tau, \tau')\, \hat{\psi}(\tau')\, d\tau', \tag{3.23}$$

where τ_0 is the instant at which the observer's acceleration is turned on and kernel $\hat{K}$ vanishes in the absence of acceleration. The matrix form of our nonlocal ansatz goes beyond the locality assumption by virtue of an integral over the past world line of the observer. Nonlinear generalizations of eqn (3.23) may be contemplated, of course, but these appear unnecessary at the present stage of development. Equation (3.23) is a Volterra integral equation of the second kind and the existence of a unique relationship between $\hat{\Psi}$ and $\hat{\psi}$ in cases of physical interest is ensured by the Volterra–Tricomi theorem. For the physical fields under consideration here, we assume that $\hat{\psi}$ is indeed uniquely determined by $\hat{\Psi}$; that is,

$$\hat{\psi}(\tau) = \hat{\Psi}(\tau) + \int_{\tau_0}^{\tau} \hat{R}(\tau, \tau')\, \hat{\Psi}(\tau')\, d\tau', \tag{3.24}$$

where $\hat{R}$ is the resolvent kernel; see Chapter 2.

Equation (3.23) is reminiscent of the nonlocal characterization of certain constitutive properties of continuous media that exhibit memory-dependent phenomena ("after-effects"). However, it is important to remark that the nonlocality considered here is in the absence of any medium; rather, it is associated with the vacuum state as perceived by accelerated observers (Mashhoon 1993a).

How should kernel $\hat{K}$ be determined? A detailed examination of field determination by a uniformly rotating observer via the *locality* postulate in Section 3.1 has led to the conclusion that by a mere rotation the observer can in principle stay completely at rest with respect to an incident electromagnetic wave. We, therefore, raise a consequence of Lorentz invariance, that an *inertial* observer cannot stay at rest with respect to a fundamental radiation field, to the level of a postulate that must hold for all observers. Thus we assume that *a basic radiation field cannot stand completely still with respect to an accelerated observer*. To implement this requirement in the nonlocal theory, let us first recall an aspect of the Doppler formula in electrodynamics for an inertial observer moving with uniform velocity $\mathbf{v}$; namely, the inertial observer measures $\omega' = \gamma(\omega - \mathbf{v} \cdot \mathbf{k})$, where $\omega = |\mathbf{k}|$. We note that $\omega' = 0$ here only when $\omega = 0$, since $v < 1$; that is, if the moving observer encounters a constant wave field, then the field must already be constant for the inertial observers at rest. The generalization of this circumstance to accelerated observers would imply that if $\hat{\Psi}$ in eqn (3.23) turns out to be constant, then ψ must have been a constant field in the first place. It would then follow from the Volterra–Tricomi uniqueness theorem that for a realistic variable field $\psi(x)$, the measured field $\hat{\Psi}(\tau)$ would never be a constant. In this way, a basic radiation field can never stand completely still with respect to any observer (Mashhoon 1993a).

Our physical postulate leads to an integral equation for kernel $\hat{K}$ by means of the nonlocal ansatz (3.23). Writing $\hat{\psi}(\tau) = \Lambda(\tau)\psi(\tau)$ in (3.23), we find for $\tau \geq \tau_0$ the matrix relation

$$\hat{\Psi}(\tau) = \Lambda(\tau)\psi(\tau) + \int_{\tau_0}^{\tau} \hat{K}(\tau, \tau')\, \Lambda(\tau')\psi(\tau')\, d\tau', \tag{3.25}$$

where at $\tau = \tau_0$, $\hat{\Psi}(\tau_0) = \Lambda(\tau_0)\psi(\tau_0)$. If for $\tau \geq \tau_0$, $\hat{\Psi}$ is a constant, i.e. $\hat{\Psi}(\tau) = \hat{\Psi}(\tau_0)$, then ψ must be a constant as well by our basic physical assumption, i.e. $\psi(\tau) = \psi(\tau_0)$; therefore, eqn (3.25) reduces to

$$\Lambda(\tau_0) = \Lambda(\tau) + \int_{\tau_0}^{\tau} \hat{K}(\tau, \tau') \, \Lambda(\tau') \, d\tau'. \tag{3.26}$$

Given $\Lambda(\tau)$, this integral relation is not sufficient to determine the kernel uniquely. To proceed, a simplifying assumption would be appropriate. Two possibilities appear natural: (i) $\hat{K}(\tau, \tau')$ is only a function of $\tau - \tau'$ or (ii) $\hat{K}(\tau, \tau')$ is only a function of τ'. These lead to the same constant kernel for uniform acceleration. The convolution kernel in case (i) was initially adopted by analogy with nonlocal theories of continuous media (Mashhoon 1993a), but was later found to lead to divergences in cases of nonuniform acceleration (Chicone and Mashhoon 2002a). A detailed investigation (Chicone and Mashhoon 2002b) reveals that case (ii) provides the only physically acceptable solution of eqn (3.26), so that

$$\hat{K}(\tau, \tau') = \hat{k}(\tau'). \tag{3.27}$$

In this case, differentiation of eqn (3.26) results in

$$\hat{k}(\tau) = -\frac{d\Lambda(\tau)}{d\tau} \Lambda^{-1}(\tau). \tag{3.28}$$

This kernel is directly proportional to the acceleration of the observer. Hence, it follows that once the acceleration is turned off at τ_f, then for $\tau > \tau_f$, though the motion of the observer is uniform, there is a constant nonlocal contribution to the measured field in eqn (3.23) that contains the memory of past acceleration. This constant memory is measurable in principle but it is simply canceled in a measuring device whenever the device is reset.

With the kernel as in eqn (3.27), the nonlocal part of our main ansatz (3.23) takes the form of a weighted average over the past world line of the accelerated observer such that the weighting function is directly proportional to the acceleration of the observer. This circumstance is consistent with the Bohr–Rosenfeld viewpoint regarding field determination. Moreover, this general approach to acceleration-induced nonlocality appears to be consistent with quantum theory (Buchholz, Mund and Summers 2002). Substitution of eqns (3.27) and (3.28) in eqn (3.23) results in

$$\hat{\Psi}(\tau) = \hat{\psi}(\tau_0) + \int_{\tau_0}^{\tau} \Lambda(\tau') \frac{d\psi(\tau')}{d\tau'} \, d\tau'. \tag{3.29}$$

It follows from this relation that if the field $\psi(x)$ evaluated along the world line of the accelerated observer turns out to be a constant over a certain interval (τ_0, τ_1), then the variable nonlocal part of eqn (3.29) vanishes in this interval and the measured field turns out to be a constant as well for $\tau_0 < \tau < \tau_1$. This result plays an important role in the development of nonlocal field theory of electrodynamics in Section 3.6. That is, we have thus far considered radiation fields; however, in Section 3.6 we need to deal with limiting situations such as electrostatics and magnetostatics as well.

The discussion of spin–rotation coupling in the previous section leads to the conclusion that for a basic *scalar* (or pseudoscalar) field of frequency w in the background inertial frame, an observer rotating uniformly with frequency Ω measures $w' = \gamma(w - M\Omega)$, where $M = 0, \pm 1, \pm 2, \dots$. Thus $w' = 0$ for $w = M\Omega$ with $M > 0$, so that the scalar (or pseudoscalar) radiation field is oscillatory in space but stands completely still with respect to the rotating observer. This possibility is ruled out by our fundamental assumption that no observer can stay completely at rest with respect to a basic radiation field. However, for a scalar field $\Lambda = 1$ and it follows from eqn (3.28) that $\hat{k} = 0$, so that a basic scalar radiation field is purely local. Our main physical postulate therefore implies that a pure scalar (or pseudoscalar) radiation field does not exist. Nevertheless, scalar or pseudoscalar fields can be composites formed from other basic fields. In 2012, a scalar Higgs boson with a mass of about 125 GeV/c^2 was discovered by means of the Large Hadron Collider (LHC) at CERN (ATLAS Collaboration 2012; CMS Collaboration 2012). The Higgs boson is classified as an elementary particle, unless it can be shown that it has internal structure. At the time of writing it is not known whether the Higgs boson has any internal structure. Except for the Higgs boson, there is no trace of any other fundamental scalar (or pseudoscalar) field in the present experimental data. Our basic assumptions leading to eqn (3.28) would be consistent with a composite Higgs boson. Hence, it remains to be seen if this important implication of nonlocal theory is consistent with observation.

3.4 Nonlocal Field Determination

To illustrate the nonlocal theory of accelerated observers that results from our choice of kernel in eqn (3.28), let us consider the nonlocal generalization of eqn (3.1) involving the determination of the electromagnetic vector potential by an accelerated observer. We now assume that what the accelerated observer actually determines is $\mathcal{A}_{\hat{\alpha}}(\tau)$, which is given for $\tau \geq \tau_0$, in accordance with our nonlocal ansatz, by

$$\mathcal{A}_{\hat{\alpha}}(\tau) = A_{\hat{\alpha}}(\tau) + \int_{\tau_0}^{\tau} \tilde{k}_{\hat{\alpha}}{}^{\hat{\beta}}(\tau')\, A_{\hat{\beta}}(\tau')\, d\tau', \tag{3.30}$$

or in matrix notation

$$\hat{\mathcal{A}}(\tau) = \hat{A}(\tau) + \int_{\tau_0}^{\tau} \hat{k}(\tau')\, \hat{A}(\tau')\, d\tau', \tag{3.31}$$

where $\hat{k}$ is given by eqn (3.28). Writing eqn (3.28) as

$$\frac{d\Lambda}{d\tau} = -\hat{k}(\tau)\, \Lambda, \tag{3.32}$$

we recall that $\hat{A} = \Lambda\, A$ and Λ is the matrix that follows from the application of the hypothesis of locality; for instance, in the case of a uniformly rotating observer, Λ is given in eqn (3.2). In component form, we have $A_{\hat{\alpha}} = A_\mu \lambda^\mu{}_{\hat{\alpha}}$, as in eqn (3.1), so that Λ here is a 4×4 matrix with a typical element $\Lambda_{\hat{\alpha}\mu}$ that therefore corresponds to the tetrad component $\lambda^\mu{}_{\hat{\alpha}}$. Thus writing eqn (3.32) in terms of its matrix elements and then changing over to tetrad components via $\Lambda_{\hat{\alpha}\mu} \to \lambda^\mu{}_{\hat{\alpha}}$, we find that eqn (3.32) can be expressed as

$$\frac{d\lambda^\mu{}_{\hat{\alpha}}}{d\tau} = -\tilde{k}_{\hat{\alpha}}{}^{\hat{\beta}}(\tau)\,\lambda^\mu{}_{\hat{\beta}}. \tag{3.33}$$

The comparison of eqn (3.33) with the definition of the acceleration tensor in eqn (1.41) results in the general relation

$$\tilde{k}_{\hat{\alpha}\hat{\beta}} = -\Phi_{\hat{\alpha}\hat{\beta}}. \tag{3.34}$$

The remainder of this section is devoted to the determination of $\mathcal{A}_{\hat{\alpha}}(\tau)$ for the rotating observer $\mathcal{O}$ of Section 3.1 in the case of *normal incidence* of circularly polarized radiation with vector potential (3.5). A general treatment including the case of oblique incidence is contained in Mashhoon (2009).

We assume that observer $\mathcal{O}$ moves uniformly for $-\infty < t < 0$ on a straight world line $x^\mu(\tau) = (t, r, v\,t, z_0)$ with $\tau = t/\gamma$, but for $t \geq 0$ it is forced to move on a circular path of radius r about the z axis in the $z = z_0$ plane. We find from eqns (1.33)–(1.36) and eqn (1.41) that the only non-zero components of the acceleration tensor are given by

$$\Phi_{\hat{0}\hat{1}} = -\Phi_{\hat{1}\hat{0}} = -v\,\gamma^2\,\Omega, \tag{3.35}$$

$$\Phi_{\hat{1}\hat{2}} = -\Phi_{\hat{2}\hat{1}} = \gamma^2\,\Omega. \tag{3.36}$$

Therefore, eqn (3.30) can now be written out in component form and, with the help of $\mathcal{A}_{\hat{0}} = v\,\mathcal{A}_{\hat{2}}$, it is straightforward to show that we have $\mathcal{A}_{\hat{0}} = v\,\mathcal{A}_{\hat{2}}$,

$$\mathcal{A}_{\hat{1}} = A_{\hat{1}} - \Omega \int_0^\tau \mathcal{A}_{\hat{2}}(\tau')\,d\tau', \tag{3.37}$$

$$\mathcal{A}_{\hat{2}} = A_{\hat{2}} + \gamma^2\,\Omega \int_0^\tau \mathcal{A}_{\hat{1}}(\tau')\,d\tau' \tag{3.38}$$

and $\mathcal{A}_{\hat{3}} = A_{\hat{3}}$. These relations must be combined with eqns (3.3) and (3.4) for $A_{\hat{\alpha}}$ that are based on the hypothesis of locality. Moreover, we find from

$$\int_0^\tau e^{-i\,\omega'\,\tau' + i\,\omega\,z_0}\,d\tau' = \frac{i}{\omega'}\left(e^{-i\,\omega'\,\tau + i\,\omega\,z_0} - e^{i\,\omega\,z_0}\right) \tag{3.39}$$

that along the world line of the observer,

$$\mathcal{A}_{\hat{1}} = \mathrm{Re}\left(\mathfrak{A}^{-1}\,{}^1\mathfrak{f}_\pm\,e^{-i\,\omega'\tau + i\,\omega\,z_0}\right), \qquad \mathcal{A}_{\hat{2}} = \gamma\,\mathrm{Re}\left(\pm i\,\mathfrak{A}^{-1}\,{}^1\mathfrak{f}_\pm\,e^{-i\,\omega'\tau + i\,\omega\,z_0}\right), \tag{3.40}$$

$\mathcal{A}_{\hat{0}} = v\,\mathcal{A}_{\hat{2}}$ and $\mathcal{A}_{\hat{3}} = 0$, where ${}^1\mathfrak{f}_\pm$ is the $s = 1$ instance of the general factor

$$^s\mathfrak{f}_\pm = \frac{\omega \mp s\,\Omega\,e^{i\,\omega'\tau}}{\omega \mp s\,\Omega}. \tag{3.41}$$

The nonlocal results presented in display (3.40) must be compared and contrasted with the local results presented in display (3.6).

The Fourier analysis of the nonlocally determined field implies that the measured frequency is still $\omega' = \gamma (\omega \mp \Omega)$, unless $\omega' = 0$. Indeed, as $\omega \to \Omega$,

$$^1f_+ \to 1 - i\gamma\Omega\tau, \tag{3.42}$$

while

$$^1f_- \to e^{i\gamma\Omega\tau}\cos(\gamma\Omega\tau). \tag{3.43}$$

Thus for $\omega = \Omega$, the field is *not* static in the positive-helicity case; instead, it varies linearly with time, just as would be expected in a resonance situation. In fact, in this case the incident electromagnetic field rotates about the direction of propagation in the same sense as the rotation of the observer and with the same frequency; moreover, the constant amplitude of the incident plane wave is maintained over time. Hence, the measured field grows indefinitely with proper time. Such a divergence would be absent for any incident realistic *wave packet*.

Another consequence of the nonlocal theory is that the average of the measured field $\mathcal{A}_{\hat{a}}$ over time is non-zero and proportional to Ω/ω' for $\omega' \neq 0$, while the incident A_μ and $A_{\hat{a}}$ both have vanishing temporal averages.

Finally, it follows from the inspection of formula (3.41) for $^1f_\pm$ that the measured amplitude of positive-helicity radiation with $\omega > \Omega$ is *enhanced* by a factor of

$$\left(1 - \frac{\Omega}{\omega}\right)^{-1}, \tag{3.44}$$

while the measured amplitude of negative-helicity radiation is *diminished* by a factor of

$$\left(1 + \frac{\Omega}{\omega}\right)^{-1}. \tag{3.45}$$

In other words, the nonlocal theory predicts that the field strength, as measured by the uniformly rotating observer, will be higher (lower) when the electromagnetic field rotates in the same (opposite) sense as the rotation of the observer. In practice, this effect is very small; for example, for GHz radio waves incident on a system rotating at, say, 10^3 rounds per second, Ω/ω would be about 10^{-6}. It is important to verify these purely nonlocal effects experimentally. The nonlocal effects are rather small in practice; moreover, the task here is further complicated by the fact that the behavior of rotating measuring devices must be known. We therefore turn to a different approach based on the correspondence principle in non-relativistic quantum mechanics. The study of the orbital motion of electrons within the framework of quantum theory could shed light on the question of the correct classical theory of accelerated systems.

3.5 Confrontation with Experiment

The consequences of acceleration-induced nonlocality for spin–rotation coupling in electrodynamics should be tested experimentally. To this end, the behavior of rotating measuring devices must be known beforehand; that is, disentangling the effect under consideration from the response of the measuring devices under rotation could be rather complicated. Such issues of principle have received attention in connection with the emission of radiation by a rotating atomic system (Bialynicki-Birula and Bialynicka-Birula 1997). Moreover, as mentioned earlier, the available experimental results in support of the electrodynamics of accelerated systems are rather meager.

To proceed, one can contemplate indirect confirmation of the nonlocal theory of accelerated systems via the "accelerated" motion of electrons in the correspondence limit of quantum mechanics. For instance, electrons undergoing "circular" motion in the limit of large quantum numbers are expected to behave much like rotating observers. Therefore, we adopt an approach based on Bohr's correspondence principle and consider electrons in the correspondence regime to be qualitatively the same as classical accelerated observers. In other words, instead of a direct confrontation of the nonlocal theory of rotating systems with observation, which is in any case not feasible at present due to the absence of relevant experimental data, we study the behavior of orbiting electrons in quantum theory to see which classical theory is closer to quantum mechanics in the correspondence limit (Mashhoon 2005).

Let us first consider the possibility that by a mere rotation of frequency ω, an observer could stand completely still with respect to an incident positive-helicity wave of frequency ω in accordance with the hypothesis of locality. However, the nonlocal theory of accelerated systems predicts that in this case the field diverges linearly with time as in the case of resonance. To test this prediction, we imagine an electron with electric charge $-\bar{e}$ and mass m_e in a circular cyclotron "orbit" about a uniform magnetic field B that is along the z direction. Classically, the angular speed of the circular orbit is the cyclotron frequency $\Omega_c = \bar{e}B/(m_e c)$. We are interested in the transition of the stationary state of the electron to a state of higher energy as a result of the resonant absorption of a photon of frequency $\omega = \Omega_c$ that propagates along the z direction and is normally incident on the initial orbital plane of the electron. In this cylindrical configuration, there is translational symmetry along the direction of the magnetic field. The incident photon carries momentum $P = \hbar \Omega_c/c$ along the z axis; after absorption, the electron has an additional kinetic energy $P^2/(2m_e)$, which vanishes in the non-relativistic limit. To simplify our discussion, we henceforth ignore the motion of the electron along the z direction. Moreover, we neglect electron spin and work in the non-relativistic approximation, where $\bar{e}\hbar B \ll m_e^2 c^3$. The electron has energy eigenvalues $\hbar\Omega_c\left(N + \frac{1}{2}\right)$, where $\hbar\Omega_c \ll m_e c^2$, and $N = 0, 1, 2, \ldots$ is the principal quantum number in this case. The classical cyclotron motion of the electron is recovered in the correspondence regime, namely, $N \sim M \gg 1$, where $\hbar M$ is the eigenvalue of the z-component of the electron's orbital angular momentum. We then study the transition of the electron from a given stationary state to the next one as a consequence of absorption of a photon of frequency Ω_c and definite helicity that is incident along the direction of the uniform magnetic field, as this situation mimics the classical problem that is of interest here. It turns out that electric dipole transitions are possible for $(N, M) \to (N + 1, M + 1)$ due to incident positive-helicity photons, while $(N, M) \to (N+1, M-1)$ due to incident negative-helicity photons are forbidden (Mashhoon 2005). Thus resonance occurs only for a photon of positive helicity and that in the correspondence regime $\mathfrak{P}_+ \propto t^2$, while $\mathfrak{P}_- = 0$, where $\mathfrak{P}_+$ ($\mathfrak{P}_-$) is the probability of transition for an incident positive (negative) helicity photon based on first-order time-dependent perturbation theory for the ideal case of resonant absorption

(Landau and Lifshitz 1977). This result is in qualitative agreement with the nonlocal theory; that is, at resonance the field amplitude $^1f_+$ diverges linearly with time t such that $|^1f_+|^2 = 1 + \Omega^2 t^2$, while $|^1f_-|^2 = \cos^2 \Omega t$ is always less than, or equal to, unity, cf. eqns (3.42) and (3.43).

Next, we consider the reception of electromagnetic waves of frequency $\omega > \Omega$ by the rotating observer. The nonlocal theory predicts that, as determined by the rotating observer, the amplitude of the positive-helicity incident radiation with an electromagnetic field that rotates along the direction of propagation in the same sense as the observer is enhanced, while the amplitude of the corresponding negative-helicity incident wave is diminished. To imitate this situation, we study the helicity dependence of the photoeffect in the simple case of the hydrogen atom. The relative strength of the field amplitudes measured by the rotating observer for $\omega > \Omega$, namely, the enhancement factor of $(1 - \Omega/\omega)^{-1}$ in the positive-helicity case versus the reduction factor of $(1 + \Omega/\omega)^{-1}$ in the negative-helicity case, can be mimicked by considering the photoionization of the hydrogen atom when the electron is in a *circular state* with respect to the incident radiation.

The circular states of atomic hydrogen are stationary states with $n > 1$, $l = n - 1$ and $m = \pm l$, corresponding to classical circular orbits in the (x, y) plane. Here, n is the principal quantum number and (l, m) denote the angular momentum quantum numbers. We assume for the sake of simplicity that the proton is in effect fixed at the origin of spatial coordinates and we neglect electron spin. Imagine an initial counterclockwise circular state of energy $E_n = -m_e \bar{e}^4/(2\hbar^2 n^2)$ and $l = m = n - 1$. A photon of energy $\hbar\omega$, $\hbar\omega > -E_n$, is normally incident on the circular state and we are interested in the total cross section for the ionization of the hydrogen atom. It is interesting to digress briefly at this point and mention the *impulse approximation*, which was originally introduced by Fermi (1936) for treating certain problems in quantum scattering theory (Newton 1982; Goldberger and Watson 1964). It turns out that the impulse approximation is the quantum analog of the hypothesis of locality in this case. In the hypothesis of locality, the accelerated observer is replaced at each instant by an otherwise identical force-free comoving inertial observer; similarly, the impulse approximation replaces the corresponding *bound* electron in this case by a *free* electron of definite momentum (Gottfried 1966). In the impulse approximation, the cross section for photoionization is independent of the helicity of the incident radiation, in complete correspondence with the standard classical theory of relativity based on the hypothesis of locality. Indeed, the impulse approximation is valid when the energy of the incident photon is much larger than the binding energy of the electron, which corresponds, in the case of the locality postulate, to a negligibly small Ω/ω in comparison to unity.

The helicity dependence of the photoeffect appears when the Coulomb interaction is properly taken into account in the final state. Let $\varpi_n := \Omega_n/\omega$, where $\Omega_n = -2E_n/(\hbar n)$ is the Bohr frequency of the electron in the circular state. In the dipole approximation, $\omega r_n \ll c$, where $r_n = -\bar{e}^2/(2E_n)$, so that $r_n = \hbar^2 n^2/(m_e \bar{e}^2)$ and r_1 is the Bohr radius; therefore, $\varpi_n \gg (137 n)^{-1}$. On the other hand, $\hbar\omega > -E_n$ implies that $\varpi_n < 2/n$. In the non-relativistic approximation with $\hbar\omega \ll m_e c^2$, a detailed investigation (Mashhoon 2005) using the dipole approximation reveals that

$$\frac{\sigma_+}{\sigma_-} = \frac{2n[2(n-1)^2 + n(2n-1)\,\varpi_n]}{3(n-1) + 2n\,\varpi_n}, \tag{3.46}$$

where σ_+ (σ_-) is the *total photoionization cross section* when the electron moves about the direction of the incident photon in the same (opposite) sense as the photon helicity. For $n = 1$, the ground state of the hydrogen atom is spherically symmetric and hence $\sigma_+ = \sigma_-$, in agreement with eqn (3.46). However, for $n \geq 2$, it follows from eqn (3.46) that $\sigma_+ > \sigma_-$. This agrees qualitatively with the prediction of the nonlocal theory of accelerated observers.

3.6 Nonlocal Electrodynamics

The acceleration kernel for the electromagnetic vector potential has been determined in Section 3.4; therefore, it remains to determine the corresponding kernel $k_{\hat{\alpha}\hat{\beta}}{}^{\hat{\gamma}\hat{\delta}}(\tau)$ for the electromagnetic field tensor, where the field measured by an accelerated observer for $\tau \geq \tau_0$ is given by

$$\mathcal{F}_{\hat{\alpha}\hat{\beta}}(\tau) = F_{\hat{\alpha}\hat{\beta}}(\tau) + \int_{\tau_0}^{\tau} \tilde{k}_{\hat{\alpha}\hat{\beta}}{}^{\hat{\gamma}\hat{\delta}}(\tau')\, F_{\hat{\gamma}\hat{\delta}}(\tau')\, d\tau'. \tag{3.47}$$

How should the kernel be determined in this case? Consider an electromagnetic field $F_{\mu\nu}(x)$ and the corresponding gauge potential $A_\mu(x)$, $F_{\mu\nu} = \partial_\mu A_\nu - \partial_\nu A_\mu$, in the background global inertial frame in Minkowski spacetime. For the vector potential, we have chosen the kernel in accordance with eqn (3.28), namely, $\tilde{k}_{\hat{\alpha}}{}^{\hat{\beta}}(\tau) = -\Phi_{\hat{\alpha}}{}^{\hat{\beta}}(\tau)$. The general property of such a kernel is that a constant $A_\mu(x)$ will be determined to be constant by all accelerated observers. This circumstance poses no difficulty as the electromagnetic field vanishes for all observers in this case. However, the situation is quite different if the kernel in eqn (3.47) is chosen in accordance with eqn (3.28); then, constant electromagnetic fields in the laboratory, such as in electrostatics and magnetostatics, will always be measured to be constant by any accelerated observer. This conclusion contradicts the results of Kennard's experiment; see Kennard (1917), Pegram (1917), Swann (1920) and the references therein. In this experiment, a coaxial cylindrical capacitor is inserted into a region of constant magnetic field B_0. The direction of the magnetic field is parallel to the axis of the capacitor. In the static situation, no potential difference is measured between the inner cylinder of radius ρ_a and the outer cylinder of radius ρ_b. However, when the capacitor is set into rotation and a maximum rotation rate of $\Omega_{\max}$ is achieved, the potential difference (emf) in the rotating frame between the plates is found to be non-zero and in qualitative agreement with $\Omega_{\max} B_0 (\rho_b{}^2 - \rho_a{}^2)/2$, which is based on the hypothesis of locality. This result is consistent with the fact that observers at rest in the rotating frame experience the presence of a radial electric field (in cylindrical coordinates). According to the hypothesis of locality, the radial electric field should have a magnitude of $\gamma v B_0$ between the cylinders, where $v = \Omega\rho$ and Ω starts from zero and reaches $\Omega_{\max}$. Here $v \ll 1$; hence, v^2 effects can be neglected. Thus an accelerated observer in a constant magnetic field can in principle measure a variable electric field. It follows that the field kernel in eqn (3.47) cannot be the one obtained from eqn (3.28), since, according to the results of Kennard's experiment, a nonuniformly rotating observer in a constant magnetic field should measure a variable electric field. Thus the next step in the determination of the kernel would involve the consideration of all possible local combinations of the acceleration tensor, the Minkowski metric tensor and the Levi-Civita tensor that could generate kernels of the form needed in eqn (3.47).

Let us first note that in this case the kernel obtained via the general formula (3.28) is given by

$$\kappa_{\hat{\alpha}\hat{\beta}}{}^{\hat{\gamma}\hat{\delta}} = -\frac{1}{2}(\Phi_{\hat{\alpha}}{}^{\hat{\gamma}}\,\delta_{\hat{\beta}}^{\hat{\delta}} + \Phi_{\hat{\beta}}{}^{\hat{\delta}}\,\delta_{\hat{\alpha}}^{\hat{\gamma}} - \Phi_{\hat{\beta}}{}^{\hat{\gamma}}\,\delta_{\hat{\alpha}}^{\hat{\delta}} - \Phi_{\hat{\alpha}}{}^{\hat{\delta}}\,\delta_{\hat{\beta}}^{\hat{\gamma}}). \tag{3.48}$$

To simplify matters, we assume that kernel $\tilde{k}_{\hat{\alpha}\hat{\beta}}{}^{\hat{\gamma}\hat{\delta}}$ in eqn (3.47) is *linearly* dependent upon the acceleration of the observer; then, a detailed investigation reveals that this kernel must be a linear combination of $\kappa_{\hat{\alpha}\hat{\beta}}{}^{\hat{\gamma}\hat{\delta}}$ given in eqn (3.48) and its dual given by

$$\kappa^*_{\hat{\alpha}\hat{\beta}}{}^{\hat{\gamma}\hat{\delta}} = -\frac{1}{2}(\Phi^*_{\hat{\alpha}}{}^{\hat{\gamma}}\,\delta_{\hat{\beta}}^{\hat{\delta}} + \Phi^*_{\hat{\beta}}{}^{\hat{\delta}}\,\delta_{\hat{\alpha}}^{\hat{\gamma}} - \Phi^*_{\hat{\beta}}{}^{\hat{\gamma}}\,\delta_{\hat{\alpha}}^{\hat{\delta}} - \Phi^*_{\hat{\alpha}}{}^{\hat{\delta}}\,\delta_{\hat{\beta}}^{\hat{\gamma}}), \tag{3.49}$$

where $\Phi^*_{\hat{\alpha}\hat{\beta}}$ is the dual acceleration tensor. We can therefore write

$$\tilde{k}_{\hat{\alpha}\hat{\beta}}{}^{\hat{\gamma}\hat{\delta}} = \mathfrak{p}\,\kappa_{\hat{\alpha}\hat{\beta}}{}^{\hat{\gamma}\hat{\delta}} + \mathfrak{q}\,\kappa^*_{\hat{\alpha}\hat{\beta}}{}^{\hat{\gamma}\hat{\delta}}. \tag{3.50}$$

Finally, to simplify matters even further, we assume that $\mathfrak{p}$ and $\mathfrak{q}$ are *constant* dimensionless coefficients. Various properties of such a kernel and the implications of the resulting nonlocal electrodynamics have been discussed in Mashhoon (2007c, 2008, 2012, 2013a); clearly, the combination $\mathfrak{p} = 1$ and $\mathfrak{q} = 0$ is excluded, but we expect $\mathfrak{p} \geq 0$ and $0 < |\mathfrak{q}| \ll 1$, since the presence of $\mathfrak{q}$ indicates possible violations of invariance under time reversal and parity in an accelerated system.

The aim of this section has been the construction of the simplest tenable theory of nonlocal electromagnetic field; however, there is a lack of definitive experimental results that could guide such a development. We must therefore bear in mind the possibility that future experimental data may require a revision of the nonlocal theory presented in this section.

3.7 Nonlocal Special Relativity

The special theory of relativity is based on Lorentz invariance as well as the hypothesis of locality. The locality postulate is a good approximation whenever the intrinsic scale of the phenomenon under consideration is negligible in comparison with the relevant acceleration scales ($\lambda/\mathcal{L} \ll 1$); otherwise, the local theory breaks down. The situation is rectified in nonlocal special relativity, where the locality postulate is replaced with the nonlocal ansatz and an appropriate acceleration kernel is chosen. Thus far we have dealt with the electromagnetic field as well as its gauge potential; however, our nonlocal treatment can be naturally extended to the Dirac field (Mashhoon 2007b). The acceleration kernel in this case is again based on eqn (3.28). The spin–rotation coupling for a Dirac field leads to $E' = \gamma\,(E \mp s\,\hbar\Omega)$ for $s = \frac{1}{2}$ when observer $\mathcal{O}$ rotates in the positive sense about the direction of incidence of a positive-energy plane wave solution of the free Dirac equation. As before, the incident wave does not stand still with respect to $\mathcal{O}$ for $E = \hbar\Omega/2$; instead the spinor diverges linearly with time. Moreover, the amplitude of a positive-helicity spinor of energy $E > s\hbar\Omega$ is enhanced by a factor of $1/(1 - s\hbar\Omega/E)$, while the amplitude of the corresponding negative-helicity spinor is diminished by a factor of $1/(1 + s\hbar\Omega/E)$.

The essential formal elements of *nonlocal special relativity* have been presented in Mashhoon (2008). For electrodynamics, the purely nonlocal predictions of the theory in the case of helicity–rotation coupling have been compared with the non-relativistic orbital motion of the electrons in the correspondence regime. According to Bohr's correspondence principle, quantum mechanics in the limit of large quantum numbers can teach us about the physics of classical accelerated systems. Following this approach, a detailed investigation has revealed that the predictions of the nonlocal theory have closely related counterparts in quantum mechanics; therefore, nonlocal special relativity is in better qualitative agreement with quantum theory than the standard theory of special relativity (Mashhoon 2005). These encouraging results notwithstanding, it is important to subject nonlocal special relativity to direct experimental tests.

Various predictions of the nonlocal theory have been worked out in detail; however, the observational aspects of the results are not encouraging. For instance, the non-local electrodynamics of linearly accelerated systems has been studied in connection with accelerated plasmas generated by an incident femtosecond laser pulse (Mashhoon 2004b). Nonlocal effects may become directly detectable with the help of laser pulses that can induce linear electron accelerations of order 10^{24} cm s^{-2} using the chirped pulse amplification technique; see the review by Umstadter (2001) and the references cited therein. Moreover, Sauerbrey (1996) has employed such high-intensity femtosecond lasers to impart linear accelerations of order 10^{21} cm s^{-2} to small grains. A grain with a macroscopic mass of $\sim 10^{-12}$ g approximates a classical accelerated observer in such experiments. Unfortunately, the estimated nonlocal contribution turns out to be negligibly small for the experiments reported in Sauerbrey (1996). A second example involving the nonlocal electrodynamics of linearly accelerated systems has to do with the recent calculation of possible nonlocal contributions to the measurements of an observer that is uniformly accelerated with respect to the rest frame of a homogeneous and isotropic black body radiation field. The nonlocal effects in this case have been found to decay rapidly and are thus essentially transient (Bremm and Falciano 2015).

Observational consequences of acceleration-induced nonlocality involve effects that are expected to be very small under normal laboratory conditions and hence rather difficult to detect. At present, one can only hope that future experiments will directly verify the main tenets of nonlocal special relativity.

3.8 Nonlocal Field Equation

Imagine a congruence of accelerated observers in Minkowski spacetime. The purpose of this section is to extend the treatment of Section 3.3 to all of the accelerated observers in the congruence. In this generalization of our nonlocal ansatz from a single world line to the whole congruence, it becomes possible to develop nonlocal field equations in Minkowski spacetime. To see in detail how this comes about, let us return to our general statement of the nonlocal ansatz in Section 3.3 and choose the acceleration kernel in the special form (3.27). Suppose that $\hat{\Psi}(\tau)$ is the field that is actually measured by the accelerated observer at τ. The hypothesis of locality postulates that $\hat{\Psi}(\tau) = \hat{\psi}(\tau)$, where $\hat{\psi}(\tau)$, $\hat{\psi}(\tau) = \Lambda(\tau)\,\psi(\tau)$, is the field measured by the instantaneously comoving inertial observer and Λ is a matrix representation of the Lorentz group. To construct a nonlocal theory, we note that the most general relation between $\hat{\Psi}(\tau)$ and $\hat{\psi}(\tau)$ that is consistent with causality and the superposition principle is given for $\tau > \tau_0$ by

$$\hat{\Psi}(\tau) = \hat{\psi}(\tau) + \int_{\tau_0}^{\tau} \hat{k}(\tau')\,\hat{\psi}(\tau')\,d\tau', \qquad (3.51)$$

where τ_0 denotes the instant of proper time at which the observer's acceleration is turned on.

To avoid possible unphysical situations, we generally assume that the observer is accelerated only for a finite interval of proper time. Equation (3.51) involves spacetime scalars; thus, it is manifestly invariant under Poincaré transformations of the background spacetime. The kernel $\hat{k}$ is obtained from the acceleration of the observer. It vanishes for an inertial observer and the nonlocal part of eqn (3.51) vanishes in the JWKB limit.

In view of the results of Volterra (1959) and Tricomi (1957), we expect that under mild mathematical assumptions the relation between $\hat{\Psi}$ and $\hat{\psi}$ is unique; moreover, this uniqueness property would seem to be demanded on physical grounds. Indeed, for $\tau > \tau_0$,

$$\hat{\psi}(\tau) = \hat{\Psi}(\tau) + \int_{\tau_0}^{\tau} \hat{r}(\tau, \tau')\,\hat{\Psi}(\tau')\,d\tau', \qquad (3.52)$$

where $\hat{r}(\tau, \tau')$ is the resolvent kernel; see Section 2.9.

We now introduce a new function into this scheme; that is, we assume that a field Ψ exists such that for the accelerated observer under consideration here

$$\hat{\Psi}(\tau) = \Lambda(\tau)\,\Psi(\tau), \qquad (3.53)$$

which means that what is actually measured by the accelerated observer is the projection of the new field Ψ on its local tetrad frame. Substituting this relation into eqn (3.52), we find

$$\psi(\tau) = \Psi(\tau) + \int_{\tau_0}^{\tau} \mathfrak{r}(\tau, \tau')\,\Psi(\tau')\,d\tau', \qquad (3.54)$$

where $\mathfrak{r}$ is related to the resolvent kernel $\hat{r}$ via

$$\mathfrak{r}(\tau, \tau') = \Lambda^{-1}(\tau)\,\hat{r}(\tau, \tau')\,\Lambda(\tau'). \qquad (3.55)$$

Let us next consider a general congruence of accelerated observers and assume that eqn (3.54) is extended to the whole congruence so that ψ is related to a field Ψ by a nonlocal relation involving a suitable kernel $\mathfrak{R}$ via the integral equation

$$\psi(x) = \Psi(x) + \int \mathfrak{R}(x, x')\,\Psi(x')\,d^4x'. \qquad (3.56)$$

Suppose that the local field ψ satisfies a field equation; then, substituting eqn (3.56) for ψ in this field equation we obtain the corresponding field equation for the local field Ψ.

Thus if ψ satisfies a partial differential equation, then the local field Ψ satisfies the corresponding partial integro-differential equation. The general nonlocal ansatz is manifestly invariant under Poincaré transformations of the background spacetime; therefore, the resulting nonlocal field equation for Ψ has the same symmetry. Explicit nonlocal field equations have been obtained in some simple cases; in fact, nonlocal Maxwell's equations have been discussed in Mashhoon (2003b) and the nonlocal Dirac equation has been treated in Mashhoon (2007b, 2008). Moreover, it is in general possible to transform the nonlocal field equations to any other coordinate system; for instance, for the electromagnetic vector potential and the Faraday tensor, we can use the coordinate invariance of the 1-form $\mathcal{A}_\mu \, dx^\mu$ and the 2-form $\mathcal{F}_{\mu\nu} \, dx^\mu \wedge dx^\nu$, respectively, to express the corresponding nonlocal field equations in arbitrary admissible coordinate systems in Minkowski spacetime. An important feature of the fields that satisfy these nonlocal field equations must be noted: nonlocality survives—in the form of the memory of past acceleration—even after the acceleration has been turned off.

Under physically reasonable conditions, kernel $\mathfrak{R}$, which is determined by the acceleration of our congruence of observers, is such that Ψ is uniquely determined by ψ. This postulate plays a fundamental role in the formulation of a variational principle for the nonlocal field Ψ. In fact, the nonlocal equation of motion for Ψ can be derived from a variational principle of stationary action involving a nonlocal Lagrangian that is simply obtained by composing the local inertial Lagrangian for ψ with the nonlocal transformation of the field to the accelerated system. The implications of this approach for the electromagnetic and Dirac fields have been briefly discussed in Chicone and Mashhoon (2007) and Mashhoon (2008).

3.8.1   Uniform acceleration

It is useful to illustrate eqn (3.56) for the simple case of uniformly accelerated observers, namely, those for which the acceleration tensor $\Phi_{\hat\alpha\hat\beta}$ in eqn (1.41) is constant. To this end, we first turn to the acceleration-induced kernel $\mathfrak{r}$ in eqn (3.55).

Let us start with the primary kernel $\hat{k}(\tau)$ introduced in (3.51); moreover, we assume that this kernel is given by eqn (3.28). As is well known, a simple result of Lorentz invariance for inertial observers is that a basic radiation field can never stand completely still with respect to an inertial observer; this physical postulate—generalized to arbitrary accelerated observers—has been used to determine the primary kernel in eqn (3.28). We note that for the electromagnetic potential and the Dirac field, we have assumed that $\hat{k}$ is given by (3.28), while for the Faraday tensor the situation is more complicated; see Section 3.6.

It follows from eqn (3.28) that

$$\frac{d\Lambda(\tau)}{d\tau} = -\hat{k}(\tau)\,\Lambda(\tau). \tag{3.57}$$

It turns out that for *uniformly* accelerated observers, $\hat{k}$ is a *constant* matrix and thus eqn (3.57) has the solution

$$\Lambda(\tau) = e^{-(\tau-\tau_0)\,\hat{k}}\,\Lambda(\tau_0). \tag{3.58}$$

The next step is the determination of the resolvent kernel $\hat{r}(\tau, \tau')$. We use the method of successive substitutions described in Section 2.9. The resolvent kernel is the sum of iterated kernels $\hat{k}_n(\tau, \tau')$, namely,

$$\hat{r}(\tau, \tau') = \sum_{n=1}^{\infty} \hat{k}_n(\tau, \tau'), \tag{3.59}$$

in accordance with eqn (2.63). The iterated kernels can be easily computed in this case using eqn (2.61) and we find

$$\hat{k}_n(\tau, \tau') = (-1)^n \frac{(\tau - \tau')^{n-1}}{(n-1)!} \hat{k}^n. \tag{3.60}$$

It follows from eqns (3.59) and (3.60) that for a constant $\hat{k}$, the resolvent kernel $\hat{r}$ is given in general by a convolution-type kernel

$$\hat{r}(\tau, \tau') = -\hat{k} e^{-(\tau - \tau') \hat{k}}. \tag{3.61}$$

Using eqns (3.58) and (3.61), it is now straightforward to calculate $\mathfrak{r}(\tau, \tau')$ given by eqn (3.55). We find that $\mathfrak{r}(\tau, \tau')$ is in fact a *constant* matrix and can be expressed as

$$\mathfrak{r} = -\Lambda^{-1}(\tau_0) \, \hat{k} \, \Lambda(\tau_0). \tag{3.62}$$

Translational acceleration. To provide simple examples of congruences of uniformly accelerated observers, imagine the class of static observers at rest in a finite spatial region of the background global inertial frame. Each observer occupies an event $x^\mu = (t, \mathbf{x})$ and carries an orthonormal frame whose axes coincide with those of the background inertial frame. At $t = 0$, the whole class is accelerated from rest along the z axis with constant translational acceleration g_0; in this connection, see the discussion in Section 1.1 regarding observer $\hat{O}$ that undergoes uniform translational acceleration. The acceleration is turned off at $t = t_f$ and for $t > t_f$, the congruence moves uniformly with constant speed v_f along the z direction. We are interested in the relation between the fields ψ and Ψ for $t \in (0, t_f)$. At any time t, $0 < t < t_f$, in the congruence, an event (t, x, y, z) is occupied by an accelerated observer in hyperbolic motion that at $t = 0$ occupied the event $(0, x, y, z_0)$ such that

$$z_0 = z + \frac{1}{g_0} - \bar{\zeta}(t), \qquad \bar{\zeta}(t) := (t^2 + \frac{1}{g_0^2})^{1/2}. \tag{3.63}$$

The congruence is accelerated from rest and the proper time along each world line in the congruence is τ given by $\sinh g_0 \tau = g_0 t$, so that $\tau_0 = 0$; therefore, $\Lambda(\tau_0)$ is the identity matrix, $\mathfrak{r} = -\hat{k}$ is proportional to g_0 and

$$\psi(t, x, y, z) = \Psi(t, x, y, z) - \hat{k} \int_0^\tau \Psi(t', x, y, z') \, d\tau'. \tag{3.64}$$

Moreover, $g_0 \, d\tau = dt/\bar{\zeta}(t)$ and eqn (4.13) implies that $z' = z - \bar{\zeta}(t) + \bar{\zeta}(t')$, so that we finally have

$$\psi(t, x, y, z) = \Psi(t, x, y, z) - \frac{1}{g_0} \hat{k} \int_0^t \Psi(t', x, y, z - \bar{\zeta} + \bar{\zeta}') \frac{dt'}{\bar{\zeta}'}, \tag{3.65}$$

where $\bar{\zeta}$ and $\bar{\zeta}'$ stand for $\bar{\zeta}(t)$ and $\bar{\zeta}(t')$, respectively. A discussion of the field equation for $t > t_f$, when the congruence moves uniformly with $v_f = t_f/\bar{\zeta}(t_f)$ is contained in Mashhoon (2003b).

Rotation. Next, consider observers $\tilde{\mathcal{O}}$ that occupy a finite region of space and are always at rest in the background inertial frame. Thus $\tau = t$ for these observers and for $-\infty < t < 0$, they refer their measurements to standard inertial axes of the background frame. However, for $0 \leq t < t_f$, these noninertial observers refer their measurements to axes that rotate uniformly with frequency Ω about the z axis. The observers in this congruence are identical except for the fact that each occupies a different fixed position in space. The tetrad frame of these observers for $t \in (0, t_f)$ is given by eqns (1.25)–(1.28) with $\varphi = \Omega t$, or equivalently by eqns (1.33)–(1.36) with $\beta = 0$ and $\gamma = 1$. Thus the acceleration tensor vanishes except for $t \in (0, t_f)$. For $t > t_f$, the observers refer their measurements to inertial axes that are rotated about the z axis by a fixed angle Ωt_f, as expected. We are interested in these observers for $t \in (0, t_f)$, where $\tau_0 = 0$, $\Lambda(\tau_0)$ is the identity matrix, $\mathbf{r} = -\hat{k}$ is proportional to Ω and

$$\psi(t, \mathbf{x}) = \Psi(t, \mathbf{x}) - \hat{k} \int_0^t \Psi(t', \mathbf{x}) \, dt'. \tag{3.66}$$

The consequences of acceleration-induced nonlocality for spin–rotation coupling that have been considered in this chapter appear to be closer to reality—as provisionally defined by quantum mechanics—than the standard theory of accelerated systems. This circumstance provides the incentive to extend the nonlocal theory of accelerated systems to linearized gravitational waves as measured by accelerated observers in Minkowski spacetime. We have thus far discussed the acceleration-induced nonlocality of the electromagnetic and Dirac fields. Can nonlocal special relativity theory be extended to a nonlocal general theory of relativity? As a first step in this direction, it would be interesting to extend the nonlocal ansatz and the ideas developed in this section to linearized gravitational waves on Minkowski spacetime and explore some of the consequences of the resulting theory. This is done in the next chapter.

4
Toward Nonlocal Gravitation

Is gravitation history-dependent? Einstein's development of the general theory of relativity has indeed revealed a profound connection between *inertia* and *gravitation*. Following Einstein's basic insight, we would expect that gravitation would be nonlocal in much the same way that accelerated systems in Minkowski spacetime are nonlocal in the sense described in the last three chapters. What would be a natural way to develop a nonlocal theory of general relativity? Einstein's *local* principle of equivalence is the cornerstone of general relativity. Is it possible to formulate a natural nonlocal extension of Einstein's principle of equivalence?

The purpose of this chapter is to describe a tentative attempt at a *direct* nonlocal generalization of general relativity. The considerations of Chapter 3 appear to be provisionally applicable to linearized gravitational radiation in Minkowski spacetime. In this chapter, we extend acceleration-induced nonlocality to linearized gravitational waves. According to general relativity (GR), gravitational radiation linearized about flat spacetime behaves as a massless spin-2 field on Minkowski spacetime and it is interesting to extend the nonlocal ansatz to this field as well; that is, assuming that gravitation involves a basic radiation field, as predicted by GR, the nonlocal theory of accelerated observers can be extended to include linearized gravitational waves (Mashhoon 2007a). This approach and its implications are briefly discussed in this chapter. In particular, the nonlocal modifications of helicity–rotation coupling for linearized gravitational radiation are pointed out and a nonlocal wave equation is presented for a special class of uniformly rotating observers.

The acceleration-induced nonlocality of gravitational waves raises the question of whether the gravitational field is intrinsically nonlocal, since an observer in a gravitational field is equivalent, by Einstein's heuristic principle of equivalence, to a certain accelerated observer in Minkowski spacetime. Einstein's principle of equivalence is incorporated into general relativity theory in a strictly *local* manner. Nevertheless, the intrinsic nonlocality of the gravitational interaction is a distinct possibility, since gravitation is a universal interaction that is qualitatively different from the other *local* interactions. As a first step toward a nonlocal theory of the gravitational field, we study in this chapter the acceleration-induced nonlocal wave equation for linearized gravitational radiation in Minkowski spacetime.

The existence of gravitational radiation is a significant prediction of GR. There is *indirect* evidence for the existence of gravitational waves via the orbital decay of binary pulsars (Blanchet 2014); for instance, the rate of orbital decay of the Hulse–Taylor binary pulsar is consistent with the prediction of GR regarding the emission of

Nonlocal Gravity. Bahram Mashhoon. © Bahram Mashhoon 2017. Published 2017 by Oxford University Press.

gravitational waves by the binary system. Recently, the first *direct* detection of gravitational waves by the Laser Interferometer Gravitational-Wave Observatory (LIGO)
has been reported (Abbott *et al.* 2016a): On 14 September 2015, the two detectors
of LIGO simultaneously observed a transient gravitational-wave signal containing a
range of frequencies from about 35 Hz to about 250 Hz and a peak amplitude of
about 10^{-21}. The signal, GW150914, was interpreted to originate from the coalescence
of two black holes. Furthermore, on 26 December 2015, the twin detectors of LIGO
observed a second coincident signal, GW151226, that contained frequencies that ranged
from about 35 Hz to about 450 Hz and a peak amplitude of about 3.4×10^{-22}. This
second signal has also been interpreted as originating from the merger of two black
holes (Abbott *et al.* 2016b).

4.1   Linearized Gravitational Radiation in GR

In the linear approximation, general relativity (GR) can be treated on a background
global Minkowski spacetime; that is, the corresponding Lorentzian spacetime may
be regarded as a slightly perturbed Minkowski spacetime. The spacetime metric can
therefore be expressed as $g_{\mu\nu} = \eta_{\mu\nu} + h_{\mu\nu}(x^{\alpha})$, where $(\eta_{\mu\nu}) = \mathrm{diag}(-1, 1, 1, 1)$ is the
Minkowski metric tensor and $(h_{\mu\nu})$ is a sufficiently small symmetric perturbation. In
other words, we assume that the absolute magnitudes of the non-zero components of
the gravitational potentials $h_{\mu\nu}$ are so small in comparison to unity that the linear
weak-field approximation is valid. In terms of the gravitational potentials, the Riemann
curvature tensor, which represents the gravitational field in GR is given by

$$^0R_{\mu\nu\rho\sigma} = \frac{1}{2}(h_{\mu\sigma,\nu\rho} + h_{\nu\rho,\mu\sigma} - h_{\nu\sigma,\mu\rho} - h_{\mu\rho,\nu\sigma}). \tag{4.1}$$

It proves useful to introduce the trace-reversed potentials $\bar{h}_{\mu\nu}$,

$$\bar{h}_{\mu\nu} = h_{\mu\nu} - \frac{1}{2}\,\eta_{\mu\nu}\,h, \tag{4.2}$$

where $h = \eta^{\mu\nu} h_{\mu\nu}$ and $\bar{h} = \eta^{\mu\nu} \bar{h}_{\mu\nu} = -h$. In terms of $\bar{h}_{\mu\nu}$ the Einstein tensor,
$^0G_{\mu\nu} := {}^0R_{\mu\nu} - \frac{1}{2} g_{\mu\nu}\,{}^0R$, is then given in the linear approximation by

$$^0G_{\mu\nu} = -\frac{1}{2}\Box\,\bar{h}_{\mu\nu} + \bar{h}^{\rho}{}_{(\mu,\nu)\rho} - \frac{1}{2}\eta_{\mu\nu}\,\bar{h}^{\rho\sigma}{}_{,\rho\sigma}, \tag{4.3}$$

where $\Box := \eta^{\alpha\beta}\partial_{\alpha}\partial_{\beta}$. The source-free Einstein's field equation is $^0G_{\mu\nu} = 0$; therefore,
the wave equation for free gravitational waves in the linear approximation is given by

$$\Box\,\bar{h}_{\mu\nu} - 2\,\bar{h}^{\rho}{}_{(\mu,\nu)\rho} + \eta_{\mu\nu}\,\bar{h}^{\rho\sigma}{}_{,\rho\sigma} = 0. \tag{4.4}$$

Under an infinitesimal transformation of inertial coordinates given by $x^{\mu} \mapsto x'^{\mu} =
x^{\mu} - \epsilon^{\mu}(x)$, the gravitational potentials are subject to the gauge transformation $h_{\mu\nu} \mapsto
h'_{\mu\nu}$, where

$$h'_{\mu\nu} = h_{\mu\nu} + \epsilon_{\mu,\nu} + \epsilon_{\nu,\mu}. \tag{4.5}$$

However, the curvature tensor and the gravitational field equation remain invariant
under this gauge transformation. The gauge freedom of the gravitational potentials

may be used to impose the transverse gauge condition $\bar{h}^{\mu\nu}{}_{,\nu} = 0$; then, the source-free gravitational field equation reduces to the wave equation $\Box \bar{h}_{\mu\nu} = 0$. The remaining gauge freedom is usually restricted by introducing the transverse–traceless (TT) gauge in which the conditions $h = 0$ and $h_{0\mu} = 0$ are further imposed on the gravitational potentials; see Section 9.2 for a detailed discussion of the TT gauge in GR.

It is well known that the treatment of gravitational waves outlined earlier, namely, the linear approximation of general relativity for source-free gravitational fields on a Minkowski spacetime background, admits of an alternative interpretation: It can be regarded as a Lorentz-invariant theory of a free linear massless spin-2 field in special relativity. This latter approach—to which the nonlocal theory of accelerated systems is directly applicable—is adopted in the rest of this chapter. Therefore, we will henceforth regard $h_{\mu\nu}(x)$ and ${}^0R_{\mu\nu\rho\sigma}(x)$ as fields defined in a global inertial frame in Minkowski spacetime. This circumstance is the spin-2 analog of the massless spin-1 field in special relativity that involves the gauge potential $A_\mu(x)$ and the corresponding electromagnetic field $F_{\mu\nu}(x)$.

It is important to recognize that the nonlocal ansatz can be applied either to the gravitational field (${}^0R_{\mu\nu\rho\sigma}$) or the gravitational wave potential ($h_{\mu\nu}$ or $\bar{h}_{\mu\nu}$) resulting in two distinct but closely related approaches. The situation here is completely analogous to the electromagnetic case discussed in Chapter 3. For the sake of simplicity, we choose $h_{\mu\nu}$ in what follows. According to the hypothesis of locality, the potential as measured by an arbitrary accelerated observer in Minkowski spacetime is given by the projection of the potential on the orthonormal tetrad frame of the observer, namely,

$$h_{\hat{\alpha}\hat{\beta}} = h_{\mu\nu} \lambda^\mu{}_{\hat{\alpha}} \lambda^\nu{}_{\hat{\beta}}. \tag{4.6}$$

Our nonlocal ansatz (3.51) for the gauge potential $h_{\mu\nu}$ then takes the Lorentz-invariant form

$$\mathbb{H}_{\hat{\alpha}\hat{\beta}}(\tau) = h_{\hat{\alpha}\hat{\beta}}(\tau) + \int_{\tau_0}^\tau \mathfrak{K}_{\hat{\alpha}\hat{\beta}}{}^{\hat{\gamma}\hat{\delta}}(\tau') \, h_{\hat{\gamma}\hat{\delta}}(\tau') \, d\tau', \tag{4.7}$$

where $\mathbb{H}_{\hat{\alpha}\hat{\beta}}$ is the symmetric gravitational wave amplitude as measured by the accelerated observer and the acceleration kernel $\mathfrak{K}_{\hat{\alpha}\hat{\beta}\hat{\gamma}\hat{\delta}}$ is a tensor that is symmetric in its first and second pairs of indices.

In general, the symmetric tensor $h_{\mu\nu}$ has ten independent components. We arrange these in a column vector ψ such that eqn (4.6) can be written as $\hat{\psi}(\tau) = \Lambda(\tau)\,\psi(\tau)$, where Λ is a 10×10 matrix. Specifically, $\hat{\psi}_A = \Lambda_A{}^B \psi_B$, where the indices A and B belong to the set $\{00, 01, 02, 03, 11, 12, 13, 22, 23, 33\}$.

It is worthwhile to work out explicitly the nonlocal theory of linearized gravitational waves for a congruence of accelerated observers. We focus in this chapter on uniformly rotating observers. The results are approximately applicable to Earth-based gravitational wave antennas that rotate with the Earth. Current efforts to detect gravitational waves in Earth-bound laboratories involve incident radiation of frequency $\gtrsim 1$ Hz. By comparison, the corresponding rotation frequency of the Earth is nearly uniform and about 10^{-5} Hz, so that $\omega \gg \Omega$ for the current laboratory experiments (Ramos and Mashhoon 2006).

4.2   Uniformly Rotating Observer

Consider observer $\mathcal{O}$ that for $t < 0$ moves uniformly along the y axis such that its position is given by $x = r$, $y = r\Omega t$ and $z = z_0$ in an inertial system of coordinates, where $r > 0$, $\Omega > 0$ and z_0 are constants. At $t = 0$, observer $\mathcal{O}$ is forced to move on a circle of radius r with constant angular speed Ω about the z axis such that its position for $t \geq 0$ is given by $x = r\cos\varphi$, $y = r\sin\varphi$ and $z = z_0$. Here $\varphi = \Omega t = \gamma\Omega\tau$, where $\gamma = t/\tau$ is the observer's Lorentz factor that corresponds to $\beta = r\Omega$. For $\tau \geq 0$, the observer's orthonormal tetrad frame is given by eqns (1.33)–(1.36); see Section 1.1. We are interested in the reception of gravitational radiation by the uniformly rotating observer for $t \geq 0$.

It follows from the postulate of locality that the field measured by the uniformly rotating observer is given by eqn (4.6), which may be written out in component form using eqns (1.33)–(1.36). We find

$$h_{\hat{0}\hat{0}} = \gamma^2[h_{00} - \beta(\sin\varphi\, h_{01} - \cos\varphi\, h_{02})]$$
$$+\gamma^2\beta^2(\sin^2\varphi\, h_{11} - \sin 2\varphi\, h_{12} + \cos^2\varphi\, h_{22}), \tag{4.8}$$

$$h_{\hat{0}\hat{1}} = \gamma(\cos\varphi\, h_{01} + \sin\varphi\, h_{02})$$
$$+\frac{1}{2}\gamma\beta(-\sin 2\varphi\, h_{11} + 2\cos 2\varphi\, h_{12} + \sin 2\varphi\, h_{22}), \tag{4.9}$$

$$h_{\hat{0}\hat{2}} = \gamma^2[\beta\, h_{00} + (1 + \beta^2)(-\sin\varphi\, h_{01} + \cos\varphi\, h_{02})]$$
$$+\gamma^2\beta(\sin^2\varphi\, h_{11} - \sin 2\varphi\, h_{12} + \cos^2\varphi\, h_{22}), \tag{4.10}$$

$$h_{\hat{0}\hat{3}} = \gamma[h_{03} + \beta(-\sin\varphi\, h_{13} + \cos\varphi\, h_{23})], \tag{4.11}$$

$$h_{\hat{1}\hat{1}} = \cos^2\varphi\, h_{11} + \sin 2\varphi\, h_{12} + \sin^2\varphi\, h_{22}, \tag{4.12}$$

$$h_{\hat{1}\hat{2}} = \gamma\beta(\cos\varphi\, h_{01} + \sin\varphi\, h_{02})$$
$$+\frac{1}{2}\gamma(-\sin 2\varphi\, h_{11} + 2\cos 2\varphi\, h_{12} + \sin 2\varphi\, h_{22}), \tag{4.13}$$

$$h_{\hat{1}\hat{3}} = \cos\varphi\, h_{13} + \sin\varphi\, h_{23}, \tag{4.14}$$

$$h_{\hat{2}\hat{2}} = \gamma^2[\beta^2 h_{00} - 2\beta(\sin\varphi\, h_{01} - \cos\varphi\, h_{02})]$$
$$+\gamma^2(\sin^2\varphi\, h_{11} - \sin 2\varphi\, h_{12} + \cos^2\varphi\, h_{22}), \tag{4.15}$$

$$h_{\hat{2}\hat{3}} = \gamma(\beta h_{03} - \sin\varphi\, h_{13} + \cos\varphi\, h_{23}), \tag{4.16}$$

$$h_{\hat{3}\hat{3}} = h_{33}. \tag{4.17}$$

These results can be used to construct the 10×10 matrix Λ, which is needed for the determination of kernel $\hat{k}$.

Imagine next the reception of a normally incident plane monochromatic gravitational wave by observer $\mathcal{O}$. The incident radiation of frequency ω and definite helicity propagates along the z axis. In the TT gauge, the incident wave amplitude $h_{\mu\nu}$ is such that $h_{0\mu} = 0$ and (h_{ij}) is a symmetric and traceless 3×3 matrix given by the real part of

$$\mathfrak{A}_{gw}\,(e_\oplus \pm i e_\otimes)e^{-i\omega(t-z)}. \tag{4.18}$$

Here $\mathfrak{A}_{gw}$ is a constant complex amplitude of the gravitational wave, the upper (lower) sign corresponds to positive (negative) helicity radiation and the two independent linear polarization states are given by

$$e_\oplus = \begin{bmatrix} 1 & 0 & 0 \\ 0 & -1 & 0 \\ 0 & 0 & 0 \end{bmatrix}, \quad e_\otimes = \begin{bmatrix} 0 & 1 & 0 \\ 1 & 0 & 0 \\ 0 & 0 & 0 \end{bmatrix}. \tag{4.19}$$

For the wave functions that we consider in this section, the complex representation will be employed throughout as all operations involving gravitational waves are linear. Thus only the real parts of the fields are of physical interest.

It follows from eqns (4.8)–(4.17) that the incident radiation field (4.18), as measured by the rotating observer, is given by

$$(h_{\hat\alpha\hat\beta}) = \mathfrak{A}_{gw} \begin{bmatrix} -\beta^2\gamma^2 & \pm i\beta\gamma & -\beta\gamma^2 & 0 \\ \pm i\beta\gamma & 1 & \pm i\gamma & 0 \\ -\beta\gamma^2 & \pm i\gamma & -\gamma^2 & 0 \\ 0 & 0 & 0 & 0 \end{bmatrix} e^{-i\,\omega'\tau + i\,\omega\,z_0}, \tag{4.20}$$

where ω' is given by

$$\omega' = \gamma\,(\omega \mp 2\,\Omega). \tag{4.21}$$

Equation (4.20) should be compared and contrasted with the measured components of the Riemann tensor given in this case by Ramos and Mashhoon (2006), who worked out the spin–rotation–gravity coupling in detail for linearized gravitational waves with $\omega \gg \Omega$. In particular, it can be demonstrated by means of the curvature tensor that a plane monochromatic gravitational wave of frequency ω propagating in Minkowski spacetime has indeed frequency ω' as measured by an observer rotating uniformly with frequency Ω about the direction of propagation of the incident radiation. More generally,

$$\omega' = \gamma\,(\omega - M\,\Omega), \tag{4.22}$$

where $M = 0, \pm1, \pm2, \ldots$, is the total (orbital plus spin) angular momentum parameter in the case of oblique incidence (Ramos and Mashhoon 2006).

Equations (4.21) and (4.22) are the spin-2 analogs of similar results for electromagnetic radiation that have been discussed in detail in Chapter 3. In eqn (4.22), ω' can be zero or negative. A negative ω' cannot be excluded due to the absolute character of the observer's rotation. However, in the case of $\omega' = 0$, there is no experimental evidence to suggest that a basic radiation field could ever stand completely still with respect to any observer. In the derivation of eqns (4.21) and (4.22), the standard theory of relativity based upon the hypothesis of locality has been used. A consequence of this assumption is that the gravitational wave could stand completely still for $\omega = M\Omega$, $M > 0$, in eqn (4.22). For instance, by a mere rotation of frequency $\omega/2$ in the positive sense about the direction of propagation of a normally incident positive-helicity gravitational wave, the field becomes completely static in accordance with eqn (4.21). According to linearized GR, however, a gravitational radiation field can never stand completely still with respect to any *inertial* observer. Generalizing this circumstance to all observers, the nonlocal theory of accelerated systems can be extended to linearized gravitational radiation and the acceleration kernel may then be tentatively chosen in accordance with eqn (3.28). We are therefore able to proceed to the calculation of $\mathbb{H}_{\hat\alpha\hat\beta}$ for uniformly rotating observers.

4.2.1   Static observer with rotating frame

To simplify matters, the general case will not be treated here; instead, we concentrate on observer $\tilde{\mathcal{O}}$ that is at rest on the z axis at $z = z_0$ and for $t \geq 0$ refers its measurements to uniformly rotating axes, that is, $r = 0$ and $\varphi = \Omega t$, so that $\beta = 0$ and $\gamma = 1$ in eqns (1.33) and (1.35) of the corresponding tetrad frame. To work out the measured field $\mathbb{H}_{\hat{\alpha}\hat{\beta}}$ according to observer $\tilde{\mathcal{O}}$ using eqn (3.51), we first need to find the components of the acceleration kernel $\mathfrak{K}_{\hat{\alpha}\hat{\beta}}{}^{\hat{\gamma}\hat{\delta}}$ defined in eqn (4.7). We represent this kernel in matrix form by the 10×10 matrix $\hat{k}$ and assume that, just as in the case of the electromagnetic gauge potential, $\hat{k}$ is given via eqn (3.28). In this case Λ has a block diagonal form, $\Lambda = \text{diag}(1, R, 1, S, 1)$, where R is the 2×2 rotation matrix

$$R(\varphi) = \begin{bmatrix} \cos\varphi & \sin\varphi \\ -\sin\varphi & \cos\varphi \end{bmatrix} \tag{4.23}$$

and S is the 5×5 matrix

$$S(\varphi) = \begin{bmatrix} \cos^2\varphi & \sin 2\varphi & 0 & \sin^2\varphi & 0 \\ -\frac{1}{2}\sin 2\varphi & \cos 2\varphi & 0 & \frac{1}{2}\sin 2\varphi & 0 \\ 0 & 0 & \cos\varphi & 0 & \sin\varphi \\ \sin^2\varphi & -\sin 2\varphi & 0 & \cos^2\varphi & 0 \\ 0 & 0 & -\sin\varphi & 0 & \cos\varphi \end{bmatrix}. \tag{4.24}$$

We note that $\det R = \det S = 1$ and

$$R^{-1}(\varphi) = R(-\varphi), \qquad S^{-1}(\varphi) = S(-\varphi). \tag{4.25}$$

The acceleration kernel $\hat{k}$ can now be easily determined in this case via eqn (3.28), since $\Lambda^{-1} = \text{diag}(1, R(-\varphi), 1, S(-\varphi), 1)$. As expected, $\hat{k}$ turns out to be a constant matrix proportional to Ω. We represent the elements of the 10×10 matrix $\hat{k}$ by $\hat{k}_{AB}$, with $A, B = 1, \ldots, 10$. The non-zero elements of matrix $\hat{k}$ are given by

$$\hat{k}_{23} = \hat{k}_{68} = \hat{k}_{79} = -\Omega, \tag{4.26}$$

$$\hat{k}_{32} = \hat{k}_{65} = \hat{k}_{97} = \Omega, \tag{4.27}$$

$$-\hat{k}_{56} = \hat{k}_{86} = 2\,\Omega. \tag{4.28}$$

Using these results, the general nonlocal relationship, reflected in the ansatz (4.7), between the components $\mathbb{H}_{\hat{\alpha}\hat{\beta}}$ of the gravitational potential measured by $\tilde{\mathcal{O}}$ and the components $h_{\hat{\alpha}\hat{\beta}}$ that are obtained from the hypothesis of locality can be expressed as

$$\mathbb{H}_{\hat{0}\hat{0}} = h_{\hat{0}\hat{0}}, \quad \mathbb{H}_{\hat{0}\hat{3}} = h_{\hat{0}\hat{3}}, \quad \mathbb{H}_{\hat{3}\hat{3}} = h_{\hat{3}\hat{3}}, \tag{4.29}$$

$$\mathbb{H}_{\hat{0}\hat{1}} = h_{\hat{0}\hat{1}} - \Omega \int_0^t h_{\hat{0}\hat{2}}\, dt', \tag{4.30}$$

$$\mathbb{H}_{\hat{0}\hat{2}} = h_{\hat{0}\hat{2}} + \Omega \int_0^t h_{\hat{0}\hat{1}}\, dt', \tag{4.31}$$

$$\mathbb{H}_{\hat{1}\hat{1}} + \mathbb{H}_{\hat{2}\hat{2}} = h_{\hat{1}\hat{1}} + h_{\hat{2}\hat{2}}, \tag{4.32}$$

$$\mathbb{H}_{\hat{1}\hat{2}} = h_{\hat{1}\hat{2}} + \Omega \int_0^t (h_{\hat{1}\hat{1}} - h_{\hat{2}\hat{2}}) \, dt', \tag{4.33}$$

$$\mathbb{H}_{\hat{1}\hat{1}} - \mathbb{H}_{\hat{2}\hat{2}} = h_{\hat{1}\hat{1}} - h_{\hat{2}\hat{2}} - 4\Omega \int_0^t h_{\hat{1}\hat{2}} \, dt', \tag{4.34}$$

$$\mathbb{H}_{\hat{1}\hat{3}} = h_{\hat{1}\hat{3}} - \Omega \int_0^t h_{\hat{2}\hat{3}} \, dt', \tag{4.35}$$

$$\mathbb{H}_{\hat{2}\hat{3}} = h_{\hat{2}\hat{3}} + \Omega \int_0^t h_{\hat{1}\hat{3}} \, dt'. \tag{4.36}$$

These results can be employed to determine the nonlocal modifications of helicity–rotation coupling for gravitational waves incident on the special rotating observer $\tilde{O}$ that is at rest on the z axis.

4.2.2 Helicity–rotation coupling for gravitational waves

For the incident radiation field (4.18), $h_{\hat{\alpha}\hat{\beta}}$ for observer $\tilde{O}$ can be determined from eqn (4.20) with $\beta = 0$ and $\gamma = 1$. The result is

$$h_{\hat{\alpha}\hat{\beta}} = e^{\pm 2i\Omega t} h_{\alpha\beta}. \tag{4.37}$$

That is, $h_{\hat{\alpha}\hat{\beta}}$ can be formally obtained from $h_{\alpha\beta}$ at the position of observer $\tilde{O}$ by simply replacing ω with ω'. Moreover, eqns (4.29)–(4.36) imply that in this case

$$\mathbb{H}_{\hat{\alpha}\hat{\beta}} = {}^2\mathfrak{f}_\pm(t) \, h_{\hat{\alpha}\hat{\beta}}, \tag{4.38}$$

where

$$ {}^2\mathfrak{f}_\pm(t) = \frac{\omega \mp 2\Omega \, e^{i\omega't}}{\omega \mp 2\Omega}, \tag{4.39}$$

with $\omega' = \omega \mp 2\Omega$ and $t = \tau$, is simply the spin-2 instance of the function ${}^s\mathfrak{f}_\pm(\tau)$, defined in eqn (3.41), that has been discussed in detail in connection with nonlocal electrodynamics in Chapter 3. Specifically, for the case of resonance involving an incident positive-helicity wave of frequency $\omega \to 2\Omega$, we find that as $\omega' \to 0$, ${}^2\mathfrak{f}_+ \to 1 - 2i\Omega t$; as before, this linear divergence with time can be avoided with a finite incident wave packet. On the other hand, for an incident negative-helicity wave of $\omega = 2\Omega$, $\omega' = 4\Omega$ and ${}^2\mathfrak{f}_- = \exp(2i\Omega t)\cos(2\Omega t)$. Another direct consequence of nonlocality, evident in the factor ${}^2\mathfrak{f}_\pm$, is that the amplitude of an incident positive-helicity gravitational wave of frequency $\omega > 2\Omega$ as measured by the rotating observer is enhanced by a factor of $\omega/(\omega - 2\Omega)$, while that of a negative-helicity wave is diminished by a factor of $\omega/(\omega + 2\Omega)$.

The results that have been obtained thus far for the static rotating observer $\tilde{O}$ fixed on the z axis may be simply extended to a whole class of such static observers that are fixed in space and differ from each other only through their spatial positions. It is clearly simpler to deal with this class of uniformly rotating observers than the class of observers O whose tetrads are given by eqns (1.33)–(1.36) of Section 1.1. Therefore, in the following section, we present the nonlocal gravitational wave equation for the class of spatially fixed noninertial observers $\tilde{O}$.

4.3   Nonlocal Gravitational Wave Equation

Imagine observers $\tilde{\mathcal{O}}$ that are always at rest in a global inertial frame and refer their measurements to the standard inertial axes for $-\infty < t < 0$; however, for $t \geq 0$ they employ axes that rotate uniformly about the z axis with constant frequency Ω. Thus for $t \geq 0$, each such observer carries a tetrad frame given by eqns (1.33)–(1.36) with $\beta = 0$ and $\gamma = 1$. The purpose of this section is to develop the Lorentz-invariant nonlocal gravitational wave equation for this special congruence of noninertial observers.

It is a general consequence of eqn (4.7) that for $\tau > \tau_0$,

$$h_{\alpha\beta}(\tau) = \mathbb{H}_{\alpha\beta}(\tau) + \int_{\tau_0}^{\tau} \mathfrak{r}_{\alpha\beta}{}^{\gamma\delta}(\tau, \tau') \, \mathbb{H}_{\gamma\delta}(\tau') \, d\tau', \tag{4.40}$$

where $\mathfrak{r}$ is a variant of the resolvent kernel defined in eqn (3.55). It has been shown in Section 3.8 that if $\hat{k}$ is a constant kernel, then $\mathfrak{r}$ is constant as well and in matrix form is given by $\mathfrak{r} = -\Lambda^{-1}(\tau_0) \, \hat{k} \, \Lambda(\tau_0)$; see eqn (3.62).

For the special class of rotating observers $\tilde{\mathcal{O}}$ under consideration here, $\tau = t$, $\tau_0 = 0$, $\hat{k}$ is a constant matrix and its non-zero elements are given in eqns (4.26)–(4.28). Moreover, it is clear from eqns (4.23)–(4.24) that $\Lambda(0)$ is the identity matrix; hence, it follows that in this case $\mathfrak{r} = -\hat{k}$. Thus the explicit form of the ten independent equations contained in eqn (4.40) may be obtained from eqns (4.29)–(4.36) by making the formal replacement $(\mathbb{H}_{\hat{\alpha}\hat{\beta}}, h_{\hat{\alpha}\hat{\beta}}, \Omega) \mapsto (h_{\alpha\beta}, \mathbb{H}_{\alpha\beta}, -\Omega)$. It is interesting to note that the form of eqns (4.29)–(4.36) remains the same if the field indices are raised; the same is true for the explicit form of eqn (4.40) in the case under consideration here.

To express eqn (4.40) for the *special class of rotating observers* $\tilde{\mathcal{O}}$, it proves convenient to write

$$h^{\alpha\beta}(t, \mathbf{x}) = \mathbb{H}^{\alpha\beta}(t, \mathbf{x}) + \mathfrak{r}^{\alpha\beta}{}_{\gamma\delta} \int_0^t \mathbb{H}^{\gamma\delta}(t', \mathbf{x}) \, dt' \tag{4.41}$$

for $t > 0$, since the observers are fixed in space. Here the components of $\mathfrak{r} = -\hat{k}$ are all constants proportional to Ω, cf. eqns (4.26)–(4.28). The substitution of $h^{\alpha\beta}(t, \mathbf{x})$ in the equations that it satisfies would then result, via eqn (4.41), in the corresponding equations for the nonlocal wave amplitude $\mathbb{H}^{\alpha\beta}(t, \mathbf{x})$.

The wave function $h^{\alpha\beta}(t, \mathbf{x})$ is subject to the gauge condition

$$\left(h^{\alpha\beta} - \frac{1}{2} \eta^{\alpha\beta} h \right)_{,\beta} = 0 \tag{4.42}$$

and satisfies the wave equation

$$\Box \, h^{\alpha\beta} = 0, \tag{4.43}$$

which follows from $\Box \, \bar{h}_{\alpha\beta} = 0$. Thus for $t > 0$, $\mathbb{H}^{\alpha\beta}$ is subject to the gauge condition

$$\left(\mathbb{H}^{\alpha\beta} + \mathfrak{r}^{\alpha\beta}{}_{\gamma\delta} \int_0^t \mathbb{H}^{\gamma\delta}(t', \mathbf{x}) \, dt' \right)_{,\beta} = \frac{1}{2} \eta^{\alpha\beta} \mathbb{H}_{,\beta}, \tag{4.44}$$

where $\mathbb{H} = \eta_{\alpha\beta} \mathbb{H}^{\alpha\beta}$ and it turns out that $\mathbb{H} = h$ in this case. Moreover, for $t > 0$, $\mathbb{H}^{\alpha\beta}$

satisfies the nonlocal wave equation

$$\Box \, \mathbb{H}^{\alpha\beta} = \mathfrak{r}^{\alpha\beta}{}_{\gamma\delta} \left(\frac{\partial}{\partial t} \mathbb{H}^{\gamma\delta} - \nabla^2 \int_0^t \mathbb{H}^{\gamma\delta}(t', \mathbf{x}) \, dt' \right),$$ (4.45)

where ∇^2 is the Laplacian operator.

The approach developed here for the special congruence of rotating observers can be extended to arbitrary accelerated systems.

4.4 Acceleration-Induced Nonlocality and Gravitation

Following the approach presented in Chapter 3 for electrodynamics, it is in principle possible to develop nonlocal field equations for linear gravitational waves in Minkowski spacetime. This has been done in the present chapter for a rather simple class of uniformly rotating observers. Invoking Einstein's principle of equivalence, the results of this chapter may be considered to be a first step in the direction of a nonlocal classical theory of gravitation. In other words, Einstein's heuristic principle of equivalence may be employed to argue intuitively that acceleration-induced nonlocality should extend to purely gravitational situations as well. As incorporated into GR, however, Einstein's principle of equivalence is extremely *local* and its sphere of validity certainly does not extend to acceleration-induced nonlocality. Nevertheless, following Einstein's fundamental insight that the principle of equivalence of inertial and gravitational masses implies a deep connection between inertia and gravitation, acceleration-induced nonlocality provides the incentive to look for a nonlocal generalization of Einstein's theory of gravitation.

What would be the next step that could lead to a nonlocal extension of general relativity resulting in nonlocal as well as nonlinear gravitational field equations? A direct physically motivated approach might well involve a nonlocal generalization of Einstein's principle of equivalence. However, it is not clear how to implement this idea in a natural manner.

Frustrated with finding a direct innate path toward the nonlocal extension of GR, we turn to an indirect approach based on a certain subtle analogy with electrodynamics.

4.5 Nonlocal Gravity: Analogy with Electrodynamics

General relativity is a field theory of gravitation that has been modeled after Maxwell's field theory of electrodynamics. The latter, in the presence of a material medium, is the only nonlocal field theory that has firm observational support. In other words, Maxwell's equations in a medium in an inertial frame can be expressed in terms of the antisymmetric field tensors $F_{\mu\nu} \mapsto (\mathbf{E}, \mathbf{B})$ and $H_{\mu\nu} \mapsto (\mathbf{D}, \mathbf{H})$ as

$$\partial_{[\rho} F_{\mu\nu]} = 0, \qquad \partial_\nu H^{\mu\nu} = \frac{4\pi}{c} \bar{j}^\mu,$$ (4.46)

where $\bar{j}^\mu$ is the current 4-vector associated with *free* electric charges. To complete the theory, a constitutive relation between $F_{\mu\nu}$ and $H_{\mu\nu}$ is required. If we impose the local relation $H_{\mu\nu} = F_{\mu\nu}$, we recover Maxwell's equations in vacuum. However, in a medium, the constitutive relation is in general nonlocal (Landau and Lifshitz 1960; Jackson 1999); for instance, the relation between $\mathbf{D}$ and $\mathbf{E}$ (or $\mathbf{H}$ and $\mathbf{B}$) could be of the general form,

$$D_i(x) = \int \varepsilon_{ij}(x')\, E^j(x - x')\, d^4x', \tag{4.47}$$

thus leading to the nonlocal electrodynamics of media; see Jackson (1999, p. 14). An example involving a temporally nonlocal relationship is

$$\mathbf{D}(t) = \mathbf{E}(t) + \int_0^\infty f(t')\, \mathbf{E}(t - t')\, dt', \tag{4.48}$$

where the scalar permeability f is a function of time as well as the properties of the medium—see Landau and Lifshitz (1960, p. 249). This relation can be expressed, by a simple change of variable, as the Volterra integral equation

$$\mathbf{D}(t) = \mathbf{E}(t) + \int_{-\infty}^t f(t - \tau)\, \mathbf{E}(\tau)\, d\tau. \tag{4.49}$$

Is it possible to develop a nonlocal generalization of GR in analogy with the nonlocal electrodynamics of media? To do so, one would first need to represent the Einstein field equation in a form that more closely resembles Maxwell's original field equations. In this connection, it turns out that GR has an equivalent tetrad formulation, $\mathrm{GR}_{\parallel}$, within the framework of *teleparallelism*; that is, $\mathrm{GR}_{\parallel}$ is the teleparallel equivalent of GR (Blagojević and Hehl 2013; Aldrovandi and Pereira 2013; Maluf 2013). Teleparallelism has a long history; its application to gravitational physics has been considered by many authors starting with Einstein in 1928; see, for example, Hayashi and Shirafuji (1979) and the references cited therein. $\mathrm{GR}_{\parallel}$ is the gauge theory of the Abelian group of spacetime translations (Cho 1976; Hehl, Nitsch and Von der Heyde 1980); therefore, its field equations can be expressed in a form that closely resembles Maxwell's original field equations (4.46); see Hehl, Nitsch and Von der Heyde (1980). This circumstance led F.W. Hehl to suggest that one should attempt to formulate a nonlocal $\mathrm{GR}_{\parallel}$ theory in close analogy with nonlocal electrodynamics of media in order to arrive indirectly at a nonlocal generalization of GR. This fruitful suggestion has led to the approach to nonlocal gravity developed in this book. It is possible to present such a nonlocal gravity theory within the framework of gauge theories of gravitation (Hehl and Mashhoon 2009a, 2009b). There is, however, a complementary approach that emphasizes an extended GR framework. We find it convenient to work within the latter approach in this book. The conceptual basis for the extension of GR that we need for nonlocal gravity is presented in Chapter 5.

5
Extension of General Relativity

The aim of this chapter is to present an extension of general relativity that is based on the *fundamental observers*, namely, a preferred set of global observers whose orthonormal tetrads are parallel and carry the gravitational degrees of freedom. In general relativity (GR), the metric tensor $g_{\mu\nu}$ carries the *ten* degrees of freedom of the gravitational field. However, in the extended GR framework, the fundamental observers' orthonormal tetrad frames $e^\mu{}_{\hat\alpha}(x)$ carry the *sixteen* gravitational degrees of freedom. Given any smooth orthonormal tetrad field $\lambda_\mu{}^{\hat\alpha}(x)$ on spacetime, the spacetime metric is then defined by

$$g_{\mu\nu} = \lambda_\mu{}^{\hat\alpha}\lambda_\nu{}^{\hat\beta}\,\eta_{\hat\alpha\hat\beta}, \tag{5.1}$$

which is the orthonormality relation for the tetrad field. Of the sixteen degrees of freedom of the orthonormal tetrad field, ten are thus fixed by the metric and the other six specify the tetrad field with respect to a fiducial orthonormal frame field. In other words, at each event x, the different orthonormal tetrad fields at x are related to each other by an element of the *local* Lorentz group that can be characterized by three boost speeds and three rotation angles which would in general depend upon x.

To determine the ten components of the metric tensor in GR, we employ ten nonlinear partial differential equations that are contained in the gravitational field equation. Similarly, to determine the sixteen components of the fundamental tetrad field in nonlocal gravity, we have sixteen nonlinear partial integro-differential equations that are contained in the field equation of nonlocal gravity discussed in Chapter 6. Thus in nonlocal gravity, the gravitational field equation is expected to determine a congruence of fundamental observers in spacetime with a globally parallel frame field that is unique up to transformations under the global Lorentz group.

The universality of the gravitational interaction is the unique feature of this force and leads to its geometric description. In GR, the gravitational field is described by the Riemannian curvature of spacetime. An observer in GR follows a future-directed timelike world line and carries an orthonormal tetrad frame. Any physics experiment is subject to the gravitational influence of the total mass–energy content of the observable universe. In particular, in a gravitational experiment, all instruments employed by observers in the measurement process are subject to the influence of the gravitational field as well. As a consequence of this circumstance it is in general necessary to specify how an observable, which is a scalar invariant in GR, is actually measured in practice. It therefore appears reasonable to suppose that the orthonormal tetrad frame field of a hypothetical set of observers throughout spacetime could carry the gravitational degrees of freedom due to the ubiquity of gravitation.

Nonlocal Gravity. Bahram Mashhoon. © Bahram Mashhoon 2017. Published 2017 by Oxford University Press.

Einstein's theory of gravitation is fully capable of dealing with the measurement of tensorial and spinorial quantities and it therefore appears at first sight that the notion of economy of thought would lead one to reject our intended extension of general relativity based on observer dynamics. On the other hand, it was first pointed out by Møller that such an extension can solve the problem of gravitational energy in GR; see Møller (1961), Pellegrini and Plebanski (1963), Hayashi and Shirafuji (1979) and the references cited therein. As is well known, the notion of gravitational energy and its distribution, defined in GR via an algebraic function of $g_{\mu\nu}$ and its derivatives, is not physically satisfactory and a gravitational analog of Poynting's theorem does not in general exist. The main aim of the present chapter is to describe an extension of the general relativistic formalism in which a global tetrad field plays an essential dynamic role. The tetrad approach is then used in Chapter 6 to formulate a *nonlocal* generalization of GR.

In the standard geometric formulation of Einstein's theory of gravitation, the global inertial frames of Minkowski spacetime are replaced by the local inertial frames of curved spacetime manifold. In this way, the *local* distinction between inertial and accelerated systems is retained. The tensor fields on local tangent spaces are connected via the Levi-Civita connection. In the extended GR framework, we have a pseudo-Riemannian metric with *two* metric-compatible connections. In other words, it is possible to extend the pseudo-Riemannian (i.e. Lorentzian) structure of GR in a natural way by adding a non-symmetric connection due to Weitzenböck (1923). The Weitzenböck connection is related to congruent frames adapted to fundamental observers. The standard Levi-Civita connection $({}^0\Gamma^\mu_{\alpha\beta})$ is symmetric and hence torsion-free but gives rise to the Riemannian curvature of spacetime that characterizes the gravitational field in GR. On the other hand, the Weitzenböck connection $(\Gamma^\mu_{\alpha\beta})$ is curvature-free, but has torsion. Thus at each event in spacetime, the curvature and torsion tensors both characterize the gravitational field. In fact, the curvature of the Levi-Civita connection and the torsion of the Weitzenböck connection are complementary representations of the gravitational field in extended GR. To see how this comes about, we first consider the fundamental observers in Minkowski spacetime.

5.1   Fundamental Observers in Minkowski Spacetime

Imagine a global inertial frame of reference with standard inertial Cartesian coordinates $X^{\hat\alpha}$ and the corresponding *fundamental* observers, namely, the ideal inertial observers that are all at rest in this frame and each has a future-directed timelike world line and an associated orthonormal tetrad frame $\bar E^{\hat\mu}{}_{\hat\alpha} = \delta^{\hat\mu}_{\hat\alpha}$ with unit axes that coincide with the global Cartesian axes of the background frame. In the special case of the fundamental observers, their local frames can be identified with the corresponding global inertial rest frame; therefore, hatted indices have been employed here for the global frame as well. The fundamental laws of nongravitational physics have been formulated with respect to the hypothetical fundamental observers.

The ideal inertial observers are free to choose any admissible system of coordinates $x^\mu = (t, x^i)$; indeed, the implementation of this possibility requires the use of tensor calculus, but no new physical assumption is involved. In the new coordinate system, the spacetime metric tensor is $g_{\mu\nu}$ instead of $\eta_{\hat\mu\hat\nu}$; in fact,

$$ds^2 = \eta_{\hat\alpha\hat\beta}dX^{\hat\alpha}dX^{\hat\beta} = g_{\mu\nu}dx^\mu dx^\nu. \tag{5.2}$$

The law of inertia—namely, the equation of motion for a free test particle, $d^2X^{\hat\alpha}/ds^2 = 0$—takes the form of the *geodesic equation* in the new coordinate system

$$\frac{d^2x^\mu}{ds^2} + \Gamma^\mu_{\alpha\beta}\frac{dx^\alpha}{ds}\frac{dx^\beta}{ds} = 0, \tag{5.3}$$

where the Christoffel symbols are given by

$$\Gamma^\mu_{\alpha\beta} = \frac{\partial^2 X^{\hat\rho}}{\partial x^\alpha \partial x^\beta}\frac{\partial x^\mu}{\partial X^{\hat\rho}}. \tag{5.4}$$

In this global inertial frame, the motion of a free test particle is such that its 4-velocity is parallel transported along its path; hence, we recognize that in eqn (5.3), the Christoffel symbols constitute the symmetric (Levi-Civita) connection for parallel transport in arbitrary coordinates. Thus we define the covariant differential of a vector A^μ as $DA^\mu = (\nabla_\alpha A^\mu)\,dx^\alpha$, where

$$\nabla_\alpha A^\mu = \partial_\alpha A^\mu + \Gamma^\mu_{\alpha\rho}A^\rho. \tag{5.5}$$

Following the standard rules of tensor analysis, one can simply check that the Levi-Civita connection is compatible with the metric, namely, $\nabla_\alpha g_{\mu\nu} = 0$, so that one finds

$$\Gamma^\mu_{\alpha\beta} = \frac{1}{2}g^{\mu\nu}(g_{\nu\alpha,\beta} + g_{\nu\beta,\alpha} - g_{\alpha\beta,\nu}), \tag{5.6}$$

where a comma denotes partial differentiation.

Under an arbitrary change of local spacetime coordinates from X to x, the tetrad frames $\bar E^{\hat\mu}{}_{\hat\alpha}$ of the fundamental observers transform to $e^\mu{}_{\hat\alpha}$, namely,

$$e^\mu{}_{\hat\alpha} = \frac{\partial x^\mu}{\partial X^{\hat\nu}}\bar E^{\hat\nu}{}_{\hat\alpha} = \frac{\partial x^\mu}{\partial X^{\hat\alpha}} \tag{5.7}$$

and similarly

$$e_\mu{}^{\hat\alpha} = \frac{\partial X^{\hat\nu}}{\partial x^\mu}\bar E_{\hat\nu}{}^{\hat\alpha} = \frac{\partial X^{\hat\alpha}}{\partial x^\mu}. \tag{5.8}$$

These maintain a global network of four Cartesian axes such that each global axis is specified by a *parallel* sequence of the corresponding unit axes; that is, one can easily check using eqn (5.5) that

$$\nabla_\nu e^\mu{}_{\hat\alpha} = 0. \tag{5.9}$$

In fact, the symmetric Christoffel symbols given by eqn (5.4) can be expressed in terms of the fundamental tetrads as

$$\Gamma^\mu_{\alpha\beta} = e^\mu{}_{\hat\rho}\partial_\alpha e_\beta{}^{\hat\rho}. \tag{5.10}$$

It is important to note that in arbitrary admissible coordinates in Minkowski spacetime, the connection has two equivalent expressions: eqn (5.6) in terms of the metric and eqn (5.10) in terms of the tetrads of the fundamental observers. The fundamental frame field is orthonormal,

$$g_{\mu\nu} e^\mu{}_{\hat\alpha} e^\nu{}_{\hat\beta} = \eta_{\hat\alpha\hat\beta}, \tag{5.11}$$

which by eqn (5.7) is equivalent to the invariance of the spacetime interval given by eqn (5.2).

An arbitrary accelerated observer can employ at any event in Minkowski spacetime an arbitrary orthonormal tetrad frame $\lambda^\mu{}_{\hat\alpha}$ subject to $g_{\mu\nu} \lambda^\mu{}_{\hat\alpha} \lambda^\nu{}_{\hat\beta} = \eta_{\hat\alpha\hat\beta}$. The essential coordinate-independent physical property of Minkowski spacetime under scrutiny here is the global existence of an ensemble of *parallel* orthonormal tetrad frame fields $e^\mu{}_{\hat\alpha}$ associated with fundamental observers; moreover, one can go from one such global ensemble of *parallel* tetrad fields to another via a constant six-parameter element of the global Lorentz group. Under such a transformation of the fundamental frames, for instance, the Christoffel symbols (5.10) remain invariant.

The basic laws of microphysics have been formulated for the fundamental observers and the corresponding standard (quantum) measurement theory determines what the fundamental observers measure. For instance, observables are generally obtained from the projection of a physical quantity on the tetrad frame of the observer. Consider, for example, the measurement of the energy–momentum tensor by the fundamental observers; in this case, we have,

$$T^{\hat\alpha\hat\beta} = T^{\mu\nu} e_\mu{}^{\hat\alpha} e_\nu{}^{\hat\beta}, \tag{5.12}$$

which is equivalent, via eqn (5.8), to the rule for transforming tensors under a change of coordinates.

The fundamental frame field $e^\mu{}_{\hat\alpha}$ provides the Minkowski spacetime manifold with a useful scaffolding. To illustrate the physical significance of this scaffolding, let us imagine a distribution of mass–energy characterized in Minkowski spacetime by the total energy–momentum tensor $T^{\hat\alpha\hat\beta}(X)$, as expressed in standard inertial Cartesian coordinates $X^{\hat\mu}$. Gravity is turned off. The net amount of energy–momentum contained in a closed spatial volume V at time t can be written as

$$P^{\hat\alpha} = \int_V T^{\hat\alpha\hat0} d^3 X. \tag{5.13}$$

Now imagine expressing such a nonlocal quantity, which transforms as a vector under the global Lorentz group, in arbitrary local spacetime coordinates x^μ. In arbitrary admissible coordinates,

$$T^{\hat\alpha\hat\beta} = \frac{\partial X^{\hat\alpha}}{\partial x^\mu} \frac{\partial X^{\hat\beta}}{\partial x^\nu} T^{\mu\nu} \tag{5.14}$$

and the energy–momentum conservation law takes the form $\nabla_\nu T^{\mu\nu} = 0$. Employing standard notation for hypersurface elements, we note that the quantity integrated in eqn (5.13) can be represented in tensorial form as

$$T^{\hat\alpha\hat\beta} d^3\Sigma_{\hat\beta} = \left(\frac{\partial X^{\hat\alpha}}{\partial x^\mu} \frac{\partial X^{\hat\beta}}{\partial x^\nu} T^{\mu\nu} \right) \left(\frac{\partial x^\rho}{\partial X^{\hat\beta}} d^3\sigma_\rho \right). \tag{5.15}$$

The right-hand side of eqn (5.15) can be written as a spacetime scalar, namely,

$$e_\mu{}^{\hat\alpha}\, T^{\mu\nu}\, d^3\sigma_\nu, \tag{5.16}$$

due to the existence of the fundamental frame field. Therefore, the appropriate generalization of eqn (5.13) in arbitrary coordinates is

$$P^{\hat\alpha} = \int_{\mathcal{D}} e_\mu{}^{\hat\alpha}\, T^{\mu\nu}\, d^3\sigma_\nu, \tag{5.17}$$

where $\mathcal{D}$ is the hypersurface domain corresponding to the original volume V at time t. Equation (5.17) reduces to eqn (5.13) when standard inertial Cartesian coordinates are employed. This circumstance illustrates the crucial significance of our fundamental observers and their tetrads: the global existence of four *parallel* vector fields throughout domain $\mathcal{D}$ has made it possible to find a proper geometric expression for our *nonlocal* eqn (5.13).

In general, the differential form of the energy–momentum conservation law, $\partial\, T^{\hat\alpha\hat\beta}/\partial X^{\hat\beta} = 0$, becomes $\nabla_\nu\, T^{\mu\nu} = 0$ in arbitrary coordinates. Furthermore, it follows from this relation and eqn (5.9) that

$$\nabla_\nu\left(e_\mu{}^{\hat\alpha}\, T^{\mu\nu}\right) = 0. \tag{5.18}$$

This is a current conservation law and can be written as

$$\frac{\partial}{\partial x^\nu}\left(\sqrt{-g}\, e_\mu{}^{\hat\alpha}\, T^{\mu\nu}\right) = 0. \tag{5.19}$$

In this way, it is straightforward to obtain the integral conservation law for energy–momentum using Gauss's theorem, namely,

$$\oint_{\partial V} e_\mu{}^{\hat\alpha}\, T^{\mu\nu}\, d^3\sigma_\nu = 0, \tag{5.20}$$

where ∂V is the closed boundary of V, which can be any compact and oriented region of the spacetime manifold.

The approach to the physics of flat spacetime, briefly illustrated in this section, can be extended to the curved spacetime of GR.

5.2   Fundamental Observers in Curved Spacetime

The symmetric Christoffel symbols of arbitrary coordinate systems in flat spacetime could be expressed either in terms of the metric or the tetrad frame of the fundamental observers. These equivalent expressions in flat spacetime become inequivalent in curved spacetime; that is, the degeneracy is removed by the Riemannian curvature of spacetime. We then have in *extended general relativity* one pseudo-Riemannian metric with two metric-compatible connections corresponding to the Levi-Civita and Weitzenböck connections.

In the curved spacetime of general relativity (GR), spacetime is a smooth four-dimensional manifold with a Lorentzian metric such that the invariant spacetime interval ds is given by

$$ds^2 = g_{\mu\nu}\, dx^\mu\, dx^\nu. \tag{5.21}$$

The path of a test particle of constant inertial mass m is obtained as usual from

$$\delta \int -m\, ds = 0, \tag{5.22}$$

which results in the geodesic equation of motion for the test particle

$$\frac{d^2 x^\mu}{ds^2} + {}^0\Gamma^\mu_{\alpha\beta}\, \frac{dx^\alpha}{ds}\, \frac{dx^\beta}{ds} = 0. \tag{5.23}$$

Similarly, rays of radiation follow null geodesics of the spacetime manifold. The 4-velocity vector of a test particle, $u^\mu = dx^\mu/ds$, is parallel transported along a geodesic via the Levi-Civita connection that is given by the Christoffel symbols

$$ {}^0\Gamma^\mu_{\alpha\beta} = \frac{1}{2}\, g^{\mu\nu}(g_{\nu\alpha,\beta} + g_{\nu\beta,\alpha} - g_{\alpha\beta,\nu}). \tag{5.24}$$

In curved spacetime, we use a left superscript "0" to denote geometric quantities related to the Levi-Civita connection. This symmetric connection is torsion-free but has Riemannian curvature

$$ {}^0R^\alpha{}_{\mu\beta\nu} = \partial_\beta\, {}^0\Gamma^\alpha_{\nu\mu} - \partial_\nu\, {}^0\Gamma^\alpha_{\beta\mu} + {}^0\Gamma^\alpha_{\beta\sigma}\, {}^0\Gamma^\sigma_{\nu\mu} - {}^0\Gamma^\alpha_{\nu\sigma}\, {}^0\Gamma^\sigma_{\beta\mu}. \tag{5.25}$$

The gravitational field equation in GR is given by ${}^0G_{\mu\nu} = \kappa\, T_{\mu\nu}$, where $\kappa := 8\pi G/c^4$ is a constant, $T_{\mu\nu}$ is the *symmetric* energy–momentum tensor of matter and

$$ {}^0G_{\mu\nu} = {}^0R_{\mu\nu} - \frac{1}{2}\, g_{\mu\nu}\, {}^0R \tag{5.26}$$

is the Einstein tensor.

Each observer in spacetime carries an orthonormal tetrad frame $\lambda^\mu{}_{\hat{a}}(x)$, where $\lambda^\mu{}_{\hat{0}}$ is the observer's unit temporal direction and $\lambda^\mu{}_{\hat{i}}$, $i = 1, 2, 3$, constitute its local spatial frame. The projection of tensor fields on an observer's tetrad frame indicates the local measurement of the corresponding physical quantities by the observer. Spacetime indices are raised and lowered via the metric tensor $g_{\mu\nu}$, while the hatted tetrad indices—that is, the local Lorentz indices—are raised and lowered via the Minkowski metric tensor $\eta_{\hat{\mu}\hat{\nu}}$ given by $\mathrm{diag}(-1, 1, 1, 1)$ in our convention. The orthonormality condition for the tetrad frame $\lambda^\mu{}_{\hat{a}}(x)$ can be expressed as

$$ g_{\mu\nu}(x) = \eta_{\hat{\alpha}\hat{\beta}}\, \lambda_\mu{}^{\hat{\alpha}}(x)\, \lambda_\nu{}^{\hat{\beta}}(x), \tag{5.27}$$

so that we can write eqn (5.21) as

$$ ds^2 = \eta_{\hat{\alpha}\hat{\beta}}\, \varkappa^{\hat{\alpha}}\, \varkappa^{\hat{\beta}}, \tag{5.28}$$

where $\varkappa^{\hat{\alpha}} = \lambda_\mu{}^{\hat{\alpha}}\, dx^\mu$. Thus the tetrad provides the local connection between spacetime quantities and local Lorentz quantities for the observer.

For each $\hat{\alpha} = \hat{0}, \hat{1}, \hat{2}, \hat{3}$, the differential form $\varkappa^{\hat{\alpha}} = \lambda_\mu{}^{\hat{\alpha}} dx^\mu$ is in general not exact, since an exact form would imply that it is integrable, namely, that there exists a function $X^{\hat{\alpha}}(x)$ such that $\varkappa^{\hat{\alpha}} = dX^{\hat{\alpha}}$ and hence $\lambda_\mu{}^{\hat{\alpha}} = \partial X^{\hat{\alpha}}/\partial x^\mu$, cf. eqn (5.8) of the previous section. If there are four such exact 1-forms, then it would follow from eqn (5.27) that we are in *flat* Minkowski spacetime with ${}^0 R_{\mu\nu\rho\sigma} = 0$ and the four functions represent the inertial Cartesian spacetime coordinates of a global inertial frame. Indeed, the family of observers with such a frame field would be the corresponding *fundamental* observers, namely, static inertial observers with tetrad frames that are all parallel and point along the Cartesian coordinate axes of the global inertial frame with coordinates $X^{\hat{\alpha}}$. If ${}^0 R_{\mu\nu\rho\sigma} \neq 0$, then the 1-forms $\varkappa^{\hat{\alpha}}$ for $\hat{\alpha} = \hat{0}, \hat{1}, \hat{2}, \hat{3}$ are not all integrable and, at each event, such 1-forms will constitute a noncoordinate or anholonomic Lorentz basis. Therefore, to change a holonomic spacetime index of a tensor into an anholonomic local Lorentz index or vice versa, one can simply project the tensor onto the corresponding local tetrad frame.

In GR, the gravitational field is identified with spacetime curvature; moreover, one traditionally works with admissible coordinate systems (Bini, Chicone and Mashhoon 2012). Coordinate bases are *holonomic*, while noncoordinate bases are *anholonomic*. In differential geometry, one can work with either holonomic or anholonomic bases. We find it convenient to work primarily with holonomic bases in this book.

In flat spacetime, the fundamental observers carry globally parallel frames. As explained in the previous section, flat spacetime contains an equivalence class of such *parallel* frame fields that are related to each other by constant elements of the six-parameter global Lorentz group. This useful parallelism disappears in curved spacetime of GR. In other words, given any smooth orthonormal tetrad field $\lambda^\mu{}_{\hat{\alpha}}(x)$ adapted to an observer family in curved spacetime, it is not possible to render the frame field parallel in any spacetime domain due to the presence of the Riemannian curvature of the Levi-Civita connection. In order to have access to a global system of parallel axes in the presence of gravitation, GR can be extended by the introduction of a second (Weitzenböck) connection, which is so defined as to render a smooth orthonormal frame field parallel in extended GR. Therefore, of all possible smooth frame fields on the pseudo-Riemannian spacetime, one system can be chosen in order to define a global system of fundamental parallel axes that are, however, specified up to global Lorentz transformations. In extended GR, the chosen parallel frame field is adapted to a *fundamental* family of observers. Henceforth, a fundamental observer family in extended GR is one for which the frame field is globally parallel via the Weitzenböck connection. The fundamental observer family in curved spacetime is then unique up to elements of the global Lorentz group. This circumstance is reminiscent of the fundamental observers in flat spacetime. More precisely, in a patchwork of admissible coordinate charts in curved spacetime, consider a smooth orthonormal tetrad frame field $e^\mu{}_{\hat{\alpha}}(x)$ corresponding to a *fundamental* set of observers. We use this fundamental tetrad system to define a new linear *Weitzenböck connection* (Weitzenböck 1923)

$$\Gamma^\mu{}_{\alpha\beta} := e^\mu{}_{\hat{\rho}}\, \partial_\alpha\, e_\beta{}^{\hat{\rho}}. \tag{5.29}$$

It is interesting to compare and contrast this definition with eqn (5.10) of the previous section.

In the absence of Riemannian curvature, the Weitzenböck connection coincides with the symmetric Levi-Civita connection; however, when $^0R_{\mu\nu\rho\sigma} \neq 0$, the Weitzenböck connection is not symmetric. Let us first briefly digress here and explain our conventions regarding the use of a non-symmetric connection.

5.2.1 Non-symmetric connections

In our convention, the covariant derivative associated with a general non-symmetric connection $\Gamma^\mu_{\alpha\beta}$ is defined for vector fields A^μ and B_μ as

$$\nabla_\alpha A^\mu = \partial_\alpha A^\mu + \Gamma^\mu_{\alpha\beta} A^\beta, \qquad \nabla_\alpha B_\mu = \partial_\alpha B_\mu - \Gamma^\beta_{\alpha\mu} B_\beta. \tag{5.30}$$

Under an arbitrary transformation of coordinates $x^\mu \mapsto x'^\mu$, a general linear connection transforms as

$$\Gamma'^\mu_{\alpha\beta} = \frac{\partial x'^\mu}{\partial x^\nu} \frac{\partial x^\gamma}{\partial x'^\alpha} \frac{\partial x^\delta}{\partial x'^\beta} \Gamma^\nu_{\gamma\delta} + \frac{\partial x'^\mu}{\partial x^\nu} \frac{\partial^2 x^\nu}{\partial x'^\alpha \partial x'^\beta}. \tag{5.31}$$

Therefore, the difference between two linear connections on the same spacetime manifold is a tensor. In this way, we have the *torsion* tensor

$$C_{\alpha\beta}{}^\mu = \Gamma^\mu_{\alpha\beta} - \Gamma^\mu_{\beta\alpha} \tag{5.32}$$

and the *contorsion* tensor

$$K_{\alpha\beta}{}^\mu = {}^0\Gamma^\mu_{\alpha\beta} - \Gamma^\mu_{\alpha\beta}. \tag{5.33}$$

Thus for any covariant vector field A_μ we have in general

$$\nabla_\mu A_\nu - \nabla_\nu A_\mu = \partial_\mu A_\nu - \partial_\nu A_\mu - C_{\mu\nu}{}^\alpha A_\alpha. \tag{5.34}$$

In particular, for a scalar field S, $\nabla_\alpha S = \partial_\alpha S$ and we find

$$(\nabla_\alpha \nabla_\beta - \nabla_\beta \nabla_\alpha) S = C_{\alpha\beta}{}^\mu \partial_\mu S. \tag{5.35}$$

Moreover, the *Ricci identity* takes the form

$$(\nabla_\alpha \nabla_\beta - \nabla_\beta \nabla_\alpha) A_\mu = R^\gamma{}_{\mu\alpha\beta} A_\gamma + C_{\alpha\beta}{}^\nu \nabla_\nu A_\mu. \tag{5.36}$$

Here,

$$R^\gamma{}_{\mu\alpha\beta} = -R^\gamma{}_{\mu\beta\alpha} \tag{5.37}$$

is the general curvature tensor given by

$$R^\alpha{}_{\mu\beta\nu} = \partial_\beta \Gamma^\alpha_{\nu\mu} - \partial_\nu \Gamma^\alpha_{\beta\mu} + \Gamma^\alpha_{\beta\sigma} \Gamma^\sigma_{\nu\mu} - \Gamma^\alpha_{\nu\sigma} \Gamma^\sigma_{\beta\mu}. \tag{5.38}$$

As is well known, in the curved spacetime of GR, at an event with coordinates $\bar{x}^\mu$ one can introduce locally geodesic coordinates in the neighborhood of $\bar{x}^\mu$ such that in the new coordinates the Christoffel symbols all vanish at $\bar{x}^\mu$ and geodesic world lines that pass through $\bar{x}^\mu$ are rendered locally straight. This circumstance is due to the

Fig. 5.1 Schematic representation of an infinitesimal parallelogram. Here $(BD)^\mu = A^\mu - \Gamma^\mu_{\alpha\beta}(P)B^\alpha A^\beta$, while $(AC)^\mu = B^\mu - \Gamma^\mu_{\alpha\beta}(P)A^\alpha B^\beta$. Hence $(CD)^\mu = (\Gamma^\mu_{\alpha\beta} - \Gamma^\mu_{\beta\alpha})A^\alpha B^\beta$. Reprinted with permission from Bini, D. and Mashhoon, B., 2015, *Phys. Rev. D* **91**, 084026. DOI: 10.1103/PhysRevD.91.084026

fact that the Levi-Civita connection is symmetric and hence torsion-free. In a similar way, consider the coordinate transformation $x^\mu \mapsto x'^\mu$ in extended GR,

$$x'^\mu = x^\mu - \bar{x}^\mu + \frac{1}{2}\left(\Gamma^\mu_{\alpha\beta}\right)_{\bar{x}}(x^\alpha - \bar{x}^\alpha)(x^\beta - \bar{x}^\beta) + \cdots, \tag{5.39}$$

which can clearly involve only the symmetric part of the connection. Thus in the new local coordinate system only the *symmetric* part of the general linear connection vanishes at $\bar{x}^\mu$ and $\Gamma^\mu_{[\alpha\beta]} = \frac{1}{2}C_{\alpha\beta}{}^\mu$ in general remains non-zero. In this case, the corresponding autoparallels passing through $\bar{x}^\mu$ are rendered locally straight. Thus at each event in our extended GR framework, the curvature and torsion tensors both characterize the gravitational field. In fact, the symbiotic relationship between the Riemann curvature and the Weitzenböck torsion of the spacetime manifold turns out to be crucial for the nonlocal generalization of GR.

Torsion, like curvature, is a basic tensor associated with a linear connection. In the presence of torsion, infinitesimal parallelograms do not close. To illustrate this point, consider two *infinitesimal* vectors A^μ and B^μ at an event P in spacetime. Suppose that A^μ is parallel transported along B^μ via a general connection Γ and B^μ is in turn parallel transported along A^μ as in Fig. 5.1. The resulting infinitesimal parallelogram in general suffers from a lack of closure if the connection is not symmetric; in fact, as illustrated in Fig. 5.1, $(CD)^\mu = C_{\alpha\beta}{}^\mu(P)A^\alpha B^\beta$.

It is possible to introduce a coordinate system in the neighborhood of event P such that the symmetric part of the connection vanishes; see eqn (5.39). That is, in the new system of coordinates $\Gamma^\mu_{(\alpha\beta)}(P) = 0$. However, the antisymmetric part of the connection corresponds to the torsion tensor. If the torsion tensor does not vanish at P, then this fact is independent of any coordinate system and infinitesimal parallelograms based at P do not close. The situation is different, however, for non-infinitesimal parallelograms, whose closure, or lack thereof, would crucially depend on the detailed circumstances at hand and the nature of the spacetime under consideration.

Let us now return to the specific case of Weitzenböck's connection, which is the new element in the extension of general relativistic framework.

5.2.2 Weitzenböck's torsion

It follows from the definition of Weitzenböck's connection in eqn (5.29) that

$$\nabla_\nu e_\mu{}^{\hat\alpha} = 0, \tag{5.40}$$

where ∇_ν here denotes covariant differentiation with respect to the Weitzenböck connection. In other words, the Weitzenböck connection is so constructed as to render the fundamental frame field *parallel*. Moreover, it can be checked directly by substituting eqn (5.29) in eqn (5.38) that Weitzenböck's non-symmetric connection is curvature-free. That is, $R_{\mu\nu\rho\sigma} = 0$; see Section 6.1. This circumstance leads to *teleparallelism*; that is, distant vectors can be considered parallel if they have the same components with respect to their local fundamental frames. The Levi-Civita and Weitzenböck connections are both compatible with the spacetime metric tensor; indeed, the latter is a consequence of $\nabla_\nu g_{\alpha\beta} = 0$, which follows from eqn (5.40) and the orthonormality relation $g_{\mu\nu} = e_\mu{}^{\hat\alpha} e_\nu{}^{\hat\beta} \eta_{\hat\alpha\hat\beta}$.

The Weitzenböck torsion tensor is given by

$$C_{\alpha\beta}{}^\mu = e^\mu{}_{\hat\rho}\left(\partial_\alpha e_\beta{}^{\hat\rho} - \partial_\beta e_\alpha{}^{\hat\rho}\right). \tag{5.41}$$

From the compatibility of the Weitzenböck connection with the metric, namely, $\nabla_\gamma g_{\alpha\beta} = 0$, we find

$$g_{\alpha\beta,\gamma} = \Gamma^\mu{}_{\gamma\alpha}\, g_{\mu\beta} + \Gamma^\mu{}_{\gamma\beta}\, g_{\mu\alpha}, \tag{5.42}$$

which can be substituted in the Christoffel symbols (5.24) to show that the contorsion tensor (5.33) is linearly related to the torsion tensor via

$$K_{\alpha\beta\gamma} = \frac{1}{2}(C_{\alpha\gamma\beta} + C_{\beta\gamma\alpha} - C_{\alpha\beta\gamma}). \tag{5.43}$$

The torsion tensor is antisymmetric in its first two indices, while the contorsion tensor is antisymmetric in its last two indices.

Let us now express the torsion tensor in the form

$$C_{\mu\nu}{}^{\hat\alpha} = e_\rho{}^{\hat\alpha} C_{\mu\nu}{}^\rho = \partial_\mu e_\nu{}^{\hat\alpha} - \partial_\nu e_\mu{}^{\hat\alpha} \tag{5.44}$$

and note that for each $\hat\alpha = \hat0, \hat1, \hat2, \hat3$ in eqn (5.44) we have a quantity much like the electromagnetic field tensor defined in terms of vector potential $e_\mu{}^{\hat\alpha}$. As in electrodynamics, the field variables all vanish if $e_\mu{}^{\hat\alpha}$ is only characterized by a pure gauge, namely, if there exist four functions $X^{\hat\alpha}$ such that $e_\mu{}^{\hat\alpha} = \partial_\mu X^{\hat\alpha}$, which by the orthonormality relation $g_{\mu\nu} = e_\mu{}^{\hat\alpha} e_\nu{}^{\hat\beta} \eta_{\hat\alpha\hat\beta}$ means that we are back in flat Minkowski spacetime. As is well known, Riemann showed that this circumstance is equivalent to ${}^0R_{\mu\nu\rho\sigma} = 0$. Therefore, so long as the Riemannian curvature tensor ${}^0R_{\mu\nu\rho\sigma}$ associated with ${}^0\Gamma^\mu{}_{\alpha\beta}$ and $g_{\mu\nu}$ is non-zero, the torsion tensor (5.44) does not vanish. Henceforth, we identify the gravitational potentials with the sixteen components of the parallel frame field $e^\mu{}_{\hat\alpha}$

and the gravitational field with the torsion tensor $C_{\mu\nu}{}^{\alpha}$, provided ${}^0R_{\mu\nu\rho\sigma}$ is non-zero. Thus in teleparallelism, gravitation is still identified with the Riemannian curvature of spacetime and, as discussed in the next section, the link between the gravitational *field* $C_{\mu\nu\rho}$ and the curvature ${}^0R_{\mu\nu\rho\sigma}$ can be exploited to find an operational way to measure, albeit indirectly, the torsion tensor. However, if ${}^0R_{\mu\nu\rho\sigma} = 0$, then gravitation is turned off and we revert back to the considerations of the previous section; in particular, we note that for the tetrad frame field of an arbitrary set of accelerated observers in Minkowski spacetime the corresponding torsion tensor is in general non-zero.

It is important to emphasize here the subtle correlation between the curvature of the Levi-Civita connection and the torsion of the Weitzenböck connection. To illustrate this point, let us first imagine that $C_{\alpha\beta\gamma} = 0$. This is mathematically equivalent, via eqn (5.41), to the requirement that $d\left(e_{\mu}{}^{\hat{\alpha}}\, dx^{\mu}\right) = 0$. On a smoothly contractible spacetime domain, every closed form is exact in accordance with the Poincaré lemma. In this case, there are thus four functions $X^{\hat{\alpha}}(x)$ such that $e_{\mu}{}^{\hat{\alpha}}\, dx^{\mu} = dX^{\hat{\alpha}}$ or $e_{\mu}{}^{\hat{\alpha}} = \partial X^{\hat{\alpha}}/\partial x^{\mu}$. As before, it follows from the orthonormality condition that we are back in Minkowski spacetime where our fundamental observers are the static inertial observers of a global inertial frame with coordinates $X^{\hat{\alpha}}$ such that the tetrad axes are all parallel with the corresponding Cartesian coordinate axes. Therefore, $C_{\alpha\beta\gamma} = 0$ implies that ${}^0R_{\mu\nu\rho\sigma} = 0$, so that there is no gravitational field. In the presence of gravitation, however, ${}^0R_{\mu\nu\rho\sigma} \neq 0$ and this implies that $C_{\alpha\beta\gamma} \neq 0$. It thus appears that in *curved* spacetime, one can characterize the gravitational field via the torsion tensor as well.

In extended GR, the parallel frame field defined by the Weitzenböck connection is the natural generalization of the fundamental frame field of special relativity to the curved spacetime of general relativity. Let us recall that in the standard GR framework, a parallel (or non-rotating) frame field may be defined via parallel (or Fermi–Walker) transport using the Levi-Civita connection along a timelike world line; however, it *cannot* in general be extended to a finite region, as this is obstructed by the Riemannian curvature of spacetime (Mashhoon 1987). The introduction of the Weitzenböck connection remedies this situation. The Levi-Civita connection has curvature but not torsion, while the Weitzenböck connection has torsion but not curvature; in the extended GR framework, the curvature of the Levi-Civita connection and the torsion of the Weitzenböck connection are complementary aspects of the gravitational field. For other approaches to extending GR via the Weitzenböck connection see, for example, Bel (2008, 2016) and the references cited therein.

The Riemann curvature tensor can be expressed in terms of the Christoffel symbols and their derivatives; therefore, eqn (5.33) can be used to write the Riemann curvature tensor in terms of the torsion tensor. After detailed but straightforward calculations, it is then possible, for instance, to write the Einstein field equation in terms of the torsion tensor. The nature of the gravitational field equation in the context of extended GR framework is the subject of Chapter 6. In the rest of this section, we present a number of important formulas regarding the Weitzenböck torsion and contorsion tensors.

5.2.3 Torsion and contorsion

We recall from the definition of the determinant that for $\mathfrak{M} = \det(\mathfrak{M}_{\alpha\beta})$, we have $\delta\mathfrak{M} = \mu^{\alpha\beta}\,\delta\mathfrak{M}_{\alpha\beta}$, where $\mu^{\alpha\beta}$ is the minor associated with $\mathfrak{M}_{\alpha\beta}$. Let $(\mathfrak{M}^{\alpha\beta})$ be the corresponding inverse matrix; then, $\mathfrak{M}^{\alpha\beta} = \mu^{\beta\alpha}/\mathfrak{M}$. Hence, in general, $\delta\mathfrak{M} = \mathfrak{M} \mathfrak{M}^{\beta\alpha} \delta \mathfrak{M}_{\alpha\beta}$. Using this result for the symmetric metric tensor $g_{\mu\nu}$ and the definition of the Christoffel symbols (5.24), we find

$$^0\Gamma^\alpha_{\ \beta\alpha} = \frac{1}{\sqrt{-g}} \frac{\partial}{\partial x^\beta}(\sqrt{-g}), \qquad g^{\mu\nu}\,^0\Gamma^\alpha_{\ \mu\nu} = -\frac{1}{\sqrt{-g}} \frac{\partial}{\partial x^\beta}\left(\sqrt{-g}g^{\alpha\beta}\right). \tag{5.45}$$

It then follows from the symmetries of Weitzenböck's torsion and contorsion tensors that

$$\Gamma^\alpha_{\ \alpha\beta} = \Gamma^\alpha_{\ \beta\alpha} + C_\beta, \qquad \Gamma^\alpha_{\ \beta\alpha} = {}^0\Gamma^\alpha_{\ \beta\alpha} = \frac{1}{\sqrt{-g}} \frac{\partial}{\partial x^\beta}(\sqrt{-g}), \tag{5.46}$$

where the *torsion vector* C_α is the trace of the torsion tensor, namely,

$$C_\alpha := C_{\beta\alpha}{}^\beta = -C_\alpha{}^\beta{}_\beta. \tag{5.47}$$

Moreover, it is possible to introduce a *torsion pseudovector* $\check{C}_\alpha$ via the totally anti-symmetric part of the torsion tensor $C_{[\alpha\beta\gamma]}$. Indeed, this axial vector is given by the dual of $C_{[\alpha\beta\gamma]}$, namely,

$$\check{C}_\alpha = -\frac{1}{6}E_{\alpha\beta\gamma\delta}\,C^{[\beta\gamma\delta]}, \qquad C_{[\alpha\beta\gamma]} = -E_{\alpha\beta\gamma\delta}\,\check{C}^\delta, \tag{5.48}$$

where $E_{\alpha\beta\gamma\delta} = \sqrt{-g}\,\epsilon_{\alpha\beta\gamma\delta}$ is the Levi-Civita tensor and $\epsilon_{\alpha\beta\gamma\delta}$ is the alternating symbol with $\epsilon_{0123} = 1$ in our convention.

The torsion tensor, defined in eqn (5.41) in terms of the fundamental frame field $e^\mu{}_{\hat{a}}(x)$ has twenty-four independent components. It is therefore possible to introduce a *reduced torsion tensor* $T_{\alpha\beta\gamma} = -T_{\beta\alpha\gamma}$ with sixteen independent components by subtracting out from $C_{\alpha\beta\gamma}$, in an appropriate fashion, its vector and pseudovector parts. In fact, the torsion tensor can be decomposed as

$$C_{\alpha\beta\gamma} = -\frac{1}{3}(C_\alpha\,g_{\beta\gamma} - C_\beta\,g_{\alpha\gamma}) + C_{[\alpha\beta\gamma]} + T_{\alpha\beta\gamma}. \tag{5.49}$$

It is straightforward to check from this *definition* of the reduced torsion tensor that $T_{\alpha\beta\gamma}$ is totally traceless and $T_{[\alpha\beta\gamma]} = 0$.

It proves useful to introduce an *auxiliary torsion tensor*

$$\mathfrak{C}_{\alpha\beta\gamma} := K_{\gamma\alpha\beta} + C_\alpha\,g_{\gamma\beta} - C_\beta\,g_{\gamma\alpha}. \tag{5.50}$$

Then, employing the definition of the Weitzenböck contorsion tensor (5.43), we find

$$K_{[\alpha\beta\gamma]} = \mathfrak{C}_{[\alpha\beta\gamma]} = -\frac{1}{2}C_{[\alpha\beta\gamma]}. \tag{5.51}$$

Furthermore, we define the auxiliary torsion vector via

$$g^{\mu\nu}\mathfrak{C}_{\sigma\mu\nu} := -\mathfrak{C}_\sigma \tag{5.52}$$

and note that

$$-\frac{1}{2}\,\mathfrak{C}^\sigma = g^{\mu\nu} K_{\mu\nu}{}^\sigma = C^\sigma. \tag{5.53}$$

Using the decomposition of the torsion tensor, we can find the corresponding decompositions for the contorsion tensor

$$K_{\alpha\beta\gamma} = -\frac{1}{3}(C_\beta\, g_{\alpha\gamma} - C_\gamma\, g_{\alpha\beta}) + K_{[\alpha\beta\gamma]} + \frac{1}{2}(T_{\alpha\gamma\beta} + T_{\beta\gamma\alpha} - T_{\alpha\beta\gamma}) \tag{5.54}$$

and the auxiliary torsion tensor

$$\mathfrak{C}_{\alpha\beta\gamma} = \frac{2}{3}(C_\alpha\, g_{\beta\gamma} - C_\beta\, g_{\alpha\gamma}) + \mathfrak{C}_{[\alpha\beta\gamma]} + \frac{1}{2}(T_{\alpha\beta\gamma} + T_{\alpha\gamma\beta} - T_{\beta\gamma\alpha}). \tag{5.55}$$

Let us recall from the definition of the Weitzenböck connection that

$$\partial_\alpha e_\beta{}^{\hat{\gamma}} = \Gamma^\mu_{\alpha\beta}\, e_\mu{}^{\hat{\gamma}}. \tag{5.56}$$

Multiplying both sides of this equation by $g^{\alpha\beta}$, we get

$$\partial_\alpha\left(g^{\alpha\beta}\, e_\beta{}^{\hat{\gamma}}\right) - \left(\partial_\alpha g^{\alpha\beta}\right) e_\beta{}^{\hat{\gamma}} = g^{\alpha\beta}\, \Gamma^\mu_{\alpha\beta}\, e_\mu{}^{\hat{\gamma}}. \tag{5.57}$$

Next, we find from the definition of the contorsion tensor that

$$g^{\mu\nu}\, \Gamma^\alpha_{\mu\nu} = -C^\alpha - \frac{1}{\sqrt{-g}}\frac{\partial}{\partial x^\beta}\left(\sqrt{-g}\, g^{\alpha\beta}\right). \tag{5.58}$$

Using this result in eqn (5.57), we have the interesting relation

$$\frac{1}{\sqrt{-g}}\frac{\partial}{\partial x^\mu}\left(\sqrt{-g}\, e^\mu{}_{\hat{\alpha}}\right) = -C_{\hat{\alpha}}. \tag{5.59}$$

Finally, it s straightforward to show using $2\,K_{\mu\nu\beta} + C_{\mu\nu\beta} = C_{\mu\beta\nu} + C_{\nu\beta\mu}$ that

$$K_\alpha{}^{\mu\nu} K_{\mu\nu\beta} = -\frac{1}{2}K_\alpha{}^{\mu\nu}C_{\mu\nu\beta} = -K_\alpha{}^{\mu\nu}\,\Gamma^\gamma_{\mu\nu}\, g_{\gamma\beta}. \tag{5.60}$$

Weitzenböck invariants. Out of the torsion tensor, one can form three independent algebraic Weitzenböck invariants

$$I_1 = C_{\alpha\beta\gamma}C^{\alpha\beta\gamma}, \quad I_2 = C_{\alpha\beta\gamma}C^{\gamma\beta\alpha}, \quad I_3 = C_\alpha C^\alpha, \tag{5.61}$$

such that

$$\mathfrak{C}_{\alpha\beta\gamma}C^{\alpha\beta\gamma} = \frac{1}{2}I_1 + I_2 - 2I_3 \tag{5.62}$$

and

$$-K_{\alpha\beta\gamma}C^{\gamma\beta\alpha} = 2K_{\alpha\beta\gamma}K^{\beta\alpha\gamma} = \frac{1}{2}I_1 + I_2. \tag{5.63}$$

5.3 Measurement of Weitzenböck's Torsion

Torsion and curvature are the two fundamental differential geometric notions associated with a linear connection, or the corresponding covariant differentiation, on a manifold. Torsion has to do with the lack of symmetry of the connection and curvature is related to the lack of commutativity of covariant differentiation (Beem, Ehrlich and Easley 1996). The Gaussian curvature of a surface is its most significant property and is easy to visualize (O'Neill 1966; Harrison 2000). In general, spacetime curvature can be *operationally* defined, for instance, via geodesic deviation (the Jacobi equation and its generalizations) or via parallel vector fields (holonomy). Torsion, on the other hand, can be visualized as the failure of an infinitesimal parallelogram to close, a concept related to the presence of dislocations in continuous media (Hehl and Obukhov 2003; Maluf, Ulhoa and Faria 2009). However, in contrast to the case of curvature, there is no general operational definition for the torsion of spacetime. In particular, the torsion tensor apparently has no relation with the torsion of a curve in space (Hicks 1965).

It appears that the measurement of spacetime torsion depends upon the physical theory in which torsion plays a significant role. In the Poincaré gauge theory of gravitation, for instance, Cartan's torsion couples to intrinsic spin and its measurement has been discussed in that context; see Hehl (1971), Lämmerzahl (1997), Hehl, Obukhov and Puetzfeld (2013) and the references cited therein. Moreover, another approach involves the motion of extended bodies in the context of nonminimal theories, where torsion couplings can be important (Puetzfeld and Obukhov 2014). On the other hand, physical aspects of Weitzenböck's torsion have been previously studied in the context of teleparallelism; see Maluf, Ulhoa and Faria (2009) and the references cited therein.

Within the framework of teleparallelism, the metric is connected to the fundamental parallel frame field ("scaffolding") via orthonormality; furthermore, the Weitzenböck torsion is naturally related to the Riemannian curvature of spacetime. The elements necessary for the establishment of metric geometry, namely, infinitesimal rods, clocks, light signals, etc., may then be employed to provide an *indirect* operational definition of Weitzenböck's torsion. This is illustrated in this section via a specific example involving a frame field in an arbitrary geodesic (Fermi) coordinate system (Mashhoon 2015; Bini and Mashhoon 2015).

5.3.1 Structure functions

Let us consider the frame components of the Weitzenböck torsion with respect to the fundamental orthonormal frame $e_{\hat{\alpha}} = e^{\mu}{}_{\hat{\alpha}} \, \partial_{\mu}$ with dual $\omega^{\hat{\alpha}}$ such that $\omega^{\hat{\alpha}}(e_{\hat{\beta}}) = \delta^{\hat{\alpha}}_{\hat{\beta}}$; that is,

$$C_{\hat{\alpha}\hat{\beta}}{}^{\hat{\gamma}} = e^{\mu}{}_{\hat{\alpha}} \, e^{\nu}{}_{\hat{\beta}} \left(\partial_{\mu} \, e_{\nu}{}^{\hat{\gamma}} - \partial_{\nu} \, e_{\mu}{}^{\hat{\gamma}} \right). \tag{5.64}$$

These are measurable in principle and are essentially the structure functions of the fundamental frame $e_{\hat{\alpha}} = e^{\mu}{}_{\hat{\alpha}} \, \partial_{\mu}$; that is,

$$[e_{\hat{\alpha}}, e_{\hat{\beta}}] = -C_{\hat{\alpha}\hat{\beta}}{}^{\hat{\gamma}} \, e_{\hat{\gamma}}. \tag{5.65}$$

Equivalently, these components can be obtained by evaluating the exterior derivative of the frame 1-forms $\omega^{\hat{\gamma}} = e_{\mu}{}^{\hat{\gamma}} \, dx^{\mu}$ according to the relation

$$dω^{\hatγ} = \frac{1}{2} C_{\hatα\hatβ}{}^{\hatγ} ω^{\hatα} \wedge ω^{\hatβ},$$ (5.66)

or the Lie derivative of the frame vectors along each other

$$\mathcal{L}_{e_{\hatα}} e_{\hatβ} = [e_{\hatα}, e_{\hatβ}] = -C_{\hatα\hatβ}{}^{\hatγ} e_{\hatγ}$$ (5.67)

and its "dual" relation

$$\mathcal{L}_{e_{\hatα}} ω^{\hatβ} = C_{\hatα\hatγ}{}^{\hatβ} ω^{\hatγ}.$$ (5.68)

At any event in spacetime, two orthonormal frames are related to each other by an element of the local Lorentz group; therefore, $C_{\hatα\hatβ}{}^{\hatγ}$ transforms as a third-rank tensor under local Lorentz transformations. Moreover, these structure functions satisfy the Jacobi identity,

$$[e_{\hatα}, [e_{\hatβ}, e_{\hatγ}]] + [e_{\hatβ}, [e_{\hatγ}, e_{\hatα}]] + [e_{\hatγ}, [e_{\hatα}, e_{\hatβ}]] = 0,$$ (5.69)

which is equivalent to $d^2ω^{\hatα} = 0$. It follows from the Jacobi identity that

$$\partial_{[\hatα} C_{\hatβ\hatγ]}{}^{\hatμ} + C_{\hatσ[\hatα}{}^{\hatμ} C_{\hatβ\hatγ]}{}^{\hatσ} = 0,$$ (5.70)

where $\partial_{\hatα} := e_{\hatα}$ is the Pfaffian derivative associated with $e_{\hatα}$.

The main purpose of this section is to calculate the structure functions $C_{\hatα\hatβ}{}^{\hatγ}$ in a general and physically transparent setting and study their physical properties. Therefore, we consider next the structure functions in the physically meaningful Fermi coordinates in a general gravitational field.

5.3.2   Weitzenböck's torsion in Fermi coordinates

To gain physical insight into the structure of Weitzenböck's torsion, we consider an arbitrary gravitational field in extended GR and establish a geodesic (Fermi) coordinate system in a cylindrical spacetime region along the world line of an arbitrary accelerated observer. This coordinate system reduces, in the absence of gravitation, to the local geodesic coordinate system for accelerated observers in Minkowski spacetime discussed in Section 1.2. Fermi coordinates are invariantly defined and constitute the natural general-relativistic generalization of inertial Cartesian coordinates. We then define the frame field of *static* observers in the Fermi coordinate system and calculate explicitly their measured torsion tensor $C_{\hatα\hatβ}{}^{\hatγ}$.

Imagine an accelerated observer following the reference world line $\bar{x}^μ(τ)$, where $x^μ = (t, x^i)$ is an admissible system of spacetime coordinates and $τ$ is the proper time along the reference observer's trajectory. The observer carries an orthonormal tetrad frame $λ^μ{}_{\hatα}(τ)$ along its path in accordance with

$$\frac{{}^0Dλ^μ{}_{\hatα}}{dτ} = {}^0Φ_{\hatα}{}^{\hatβ}(τ) λ^μ{}_{\hatβ}.$$ (5.71)

Here, ${}^0Φ_{\hatα\hatβ} = -{}^0Φ_{\hatβ\hatα}$ is the acceleration tensor of the fiducial observer. In apt analogy with the Faraday tensor, we can decompose the acceleration tensor into its "electric" and "magnetic" components, namely, ${}^0Φ_{\hatα\hatβ} \mapsto (-\mathbf{a}, \mathbf{Ω})$; that is, the translational acceleration vector **a** is given by the frame components of the 4-acceleration vector associated with the 4-velocity vector $\lambda^\mu{}_{\hat{0}} = d\bar{x}^\mu/d\tau$ of the observer and $\boldsymbol{\Omega}$ is the angular velocity of the rotation of the observer's local spatial triad $\lambda^\mu{}_{\hat{i}}$, $i = 1, 2, 3$, with respect to the locally non-rotating (i.e. Fermi–Walker transported) triad.

Let us next establish an extended Fermi normal coordinate system in a world tube along $\bar{x}^\mu(\tau)$. The Fermi coordinates are scalar invariants by construction and are indispensable for the interpretation of measurements in GR; see Synge (1971), Mashhoon (1977), Ni and Zimmermann (1978), Chicone and Mashhoon (2002c, 2006) and the references cited therein. Consider the class of spacelike geodesics that are orthogonal to the world line of the fiducial accelerated observer at each event τ along $\bar{x}^\mu(\tau)$. These form a local hypersurface. For an event with coordinates x^μ on this hypersurface, let there be a *unique* spacelike geodesic of proper length σ that connects x^μ to $\bar{x}^\mu(\tau)$. Then, x^μ is assigned Fermi coordinates $X^{\hat{\mu}} = (T, X^{\hat{i}})$, where

$$T = \tau, \qquad X^{\hat{i}} = \sigma\,\xi^\mu\,\lambda_\mu{}^{\hat{i}}(\tau). \tag{5.72}$$

Here, ξ^μ is the *unit* vector at $\bar{x}^\mu(\tau)$ that is tangent to the spacelike geodesic segment from $\bar{x}^\mu(\tau)$ to x^μ. Thus the reference observer is always at the spatial origin of the Fermi coordinate system.

The construction of such coordinates in Minkowski spacetime is depicted in Fig. 1.3, where at each instant of proper time τ of the fiducial observer, $\bar{x}^\mu(\tau)$ and x^μ can always be connected by a straight line in the simultaneity hyperplane. In curved spacetime, however, x^μ and $\bar{x}^\mu(\tau)$ must be sufficiently close to each other such that the geodesic connecting $\bar{x}^\mu(\tau)$ to x^μ, which lies on the local hypersurface, is always unique.

In curved spacetime, the coordinate transformation $x^\mu \mapsto (X^{\hat{0}} = T, X^{\hat{a}})$ can only be specified implicitly in general; hence, it is useful to express the spacetime metric in Fermi coordinates as a Taylor expansion in powers of the spatial distance σ away from the reference world line. For our present purposes, we can write the metric in Fermi coordinates as

$$
\begin{aligned}
g_{\hat{0}\hat{0}} &= -\mathcal{P}^2 + \mathcal{Q}^2 - R_{\hat{0}\hat{i}\hat{0}\hat{j}}X^{\hat{i}}X^{\hat{j}} + O(|\mathbf{X}|^3), \\
&= -\mathcal{P}^2 + \mathcal{Q}^2 - 2\,{}^F\Phi + O(|\mathbf{X}|^3), \\
g_{\hat{0}\hat{i}} &= \mathcal{Q}_{\hat{i}} - \frac{2}{3}R_{\hat{0}\hat{j}\hat{i}\hat{k}}X^{\hat{j}}X^{\hat{k}} + O(|\mathbf{X}|^3), \\
&= \mathcal{Q}_{\hat{i}} - 2\,{}^F\!A_{\hat{i}} + O(|\mathbf{X}|^3), \\
g_{\hat{i}\hat{j}} &= \delta_{\hat{i}\hat{j}} - \frac{1}{3}R_{\hat{i}\hat{k}\hat{j}\hat{l}}X^{\hat{k}}X^{\hat{l}} + O(|\mathbf{X}|^3), \\
&= \delta_{\hat{i}\hat{j}} - 2\,\Sigma_{\hat{i}\hat{j}} + O(|\mathbf{X}|^3).
\end{aligned}
\tag{5.73}
$$

Here, we have introduced

$$\mathcal{P} = 1 + U, \qquad U = \mathbf{a}\cdot\mathbf{X}, \qquad \boldsymbol{\mathcal{Q}} = \boldsymbol{\Omega}\times\mathbf{X} \tag{5.74}$$

and we have used the notation

$$^F\Phi = \frac{1}{2}R_{\hat{0}\hat{i}\hat{0}\hat{j}}X^{\hat{i}}X^{\hat{j}}, \qquad ^F\mathcal{A}_{\hat{i}} = \frac{1}{3}R_{\hat{0}\hat{j}\hat{i}\hat{k}}X^{\hat{j}}X^{\hat{k}}, \qquad \Sigma_{\hat{i}\hat{j}} = \frac{1}{6}R_{\hat{i}\hat{k}\hat{j}\hat{l}}X^{\hat{k}}X^{\hat{l}}. \tag{5.75}$$

Moreover, $R_{\hat{\alpha}\hat{\beta}\hat{\gamma}\hat{\delta}}(T)$ is the projection of the Riemann curvature tensor on the orthonormal tetrad frame of the reference observer and evaluated along the reference world line; that is,

$$R_{\hat{\alpha}\hat{\beta}\hat{\gamma}\hat{\delta}}(T) := {}^0R_{\mu\nu\rho\sigma}\,\lambda^{\mu}{}_{\hat{\alpha}}\,\lambda^{\nu}{}_{\hat{\beta}}\,\lambda^{\rho}{}_{\hat{\gamma}}\,\lambda^{\sigma}{}_{\hat{\delta}}. \tag{5.76}$$

Henceforward, we will only keep terms up to second order in the metric perturbation and note that Fermi coordinates are admissible in a finite cylindrical region about the world line of the reference observer with $|\mathbf{X}| \ll r_c$, where $r_c(T)$ is the infimum of acceleration lengths ($|\mathbf{a}(T)|^{-1}, |\mathbf{\Omega}(T)|^{-1}$) as well as spacetime curvature lengths such as $|R_{\hat{\alpha}\hat{\beta}\hat{\gamma}\hat{\delta}}(T)|^{-1/2}$.

Let us now consider the class of observers that are all at *rest* in this gravitational field and carry orthonormal tetrads that have essentially the same orientation as the Fermi coordinate system. This class includes of course our reference observer at the origin of spatial Fermi coordinates. The orthonormal tetrad frame of these fundamental observers can be expressed in $(T, X^{\hat{i}})$ coordinates as

$$e^{\mu}{}_{\hat{0}} = (1 - \tilde{\Phi}, 0, 0, 0), \tag{5.77}$$

$$e^{\mu}{}_{\hat{1}} = (-2\,\tilde{\mathcal{A}}_{\hat{1}}, 1 + \tilde{\Sigma}_{\hat{1}\hat{1}}, 0, 0), \tag{5.78}$$

$$e^{\mu}{}_{\hat{2}} = (-2\,\tilde{\mathcal{A}}_{\hat{2}}, 2\,\tilde{\Sigma}_{\hat{2}\hat{1}}, 1 + \tilde{\Sigma}_{\hat{2}\hat{2}}, 0), \tag{5.79}$$

$$e^{\mu}{}_{\hat{3}} = (-2\,\tilde{\mathcal{A}}_{\hat{3}}, 2\,\tilde{\Sigma}_{\hat{3}\hat{1}}, 2\,\tilde{\Sigma}_{\hat{3}\hat{2}}, 1 + \tilde{\Sigma}_{\hat{3}\hat{3}}). \tag{5.80}$$

Here, we have defined

$$\tilde{\Phi} := {}^F\Phi + U - U^2 - \frac{1}{2}\mathcal{Q}^2, \quad \tilde{\mathcal{A}}_{\hat{i}} := {}^F\mathcal{A}_{\hat{i}} - \left(\frac{1}{2} - U\right)\mathcal{Q}_{\hat{i}}, \quad \tilde{\Sigma}_{\hat{i}\hat{j}} := \Sigma_{\hat{i}\hat{j}} - \frac{1}{2}\mathcal{Q}_{\hat{i}}\,\mathcal{Q}_{\hat{j}}. \tag{5.81}$$

As expected, $e^{\mu}{}_{\hat{\alpha}}$ reduces to $\delta^{\mu}{}_{\hat{\alpha}}$ along the reference world line, where $\mathbf{X} = 0$. It follows from $e_{\mu\hat{\alpha}} = g_{\mu\nu}\,e^{\nu}{}_{\hat{\alpha}}$ that

$$e_{\mu\hat{0}} = (-1 - U^2 - \tilde{\Phi}, -2\,\tilde{\mathcal{A}}_{\hat{1}} + U\mathcal{Q}_{\hat{1}}, -2\,\tilde{\mathcal{A}}_{\hat{2}} + U\mathcal{Q}_{\hat{2}}, -2\,\tilde{\mathcal{A}}_{\hat{3}} + U\mathcal{Q}_{\hat{3}}), \tag{5.82}$$

$$e_{\mu\hat{1}} = (0, 1 - \tilde{\Sigma}_{\hat{1}\hat{1}}, -2\,\tilde{\Sigma}_{\hat{1}\hat{2}}, -2\,\tilde{\Sigma}_{\hat{1}\hat{3}}), \tag{5.83}$$

$$e_{\mu\hat{2}} = (0, 0, 1 - \tilde{\Sigma}_{\hat{2}\hat{2}}, -2\,\tilde{\Sigma}_{\hat{2}\hat{3}}), \tag{5.84}$$

$$e_{\mu\hat{3}} = (0, 0, 0, 1 - \tilde{\Sigma}_{\hat{3}\hat{3}}). \tag{5.85}$$

Explicitly, we therefore have

$$e_{\hat{0}} = \left(1 - U + U^2 + \frac{1}{2}\mathcal{Q}^2 - {}^F\Phi\right)\partial_T,$$

$$e_{\hat{1}} = \left[-2\,{}^F\mathcal{A}_{\hat{1}} + \mathcal{Q}_{\hat{1}}(1-2U)\right]\partial_T + \left(1 - \frac{1}{2}\mathcal{Q}_{\hat{1}}^2 + \Sigma_{\hat{1}\hat{1}}\right)\partial_{X^{\hat{1}}},$$

$$e_{\hat{2}} = \left[-2\,{}^F\mathcal{A}_{\hat{2}} + \mathcal{Q}_{\hat{2}}(1-2U)\right]\partial_T + \left(2\,\Sigma_{\hat{2}\hat{1}} - \mathcal{Q}_{\hat{2}}\,\mathcal{Q}_{\hat{1}}\right)\partial_{X^{\hat{1}}} + \left(1 - \frac{1}{2}\mathcal{Q}_{\hat{2}}^2 + \Sigma_{\hat{2}\hat{2}}\right)\partial_{X^{\hat{2}}},$$

$$e_{\hat{3}} = \left[-2\,{}^F\mathcal{A}_{\hat{3}} + \mathcal{Q}_{\hat{3}}(1-2U)\right]\partial_T + \left(2\,\Sigma_{\hat{3}\hat{1}} - \mathcal{Q}_{\hat{3}}\,\mathcal{Q}_{\hat{1}}\right)\partial_{X^{\hat{1}}} + \left(2\,\Sigma_{\hat{3}\hat{2}} - \mathcal{Q}_{\hat{3}}\mathcal{Q}_{\hat{2}}\right)\partial_{X^{\hat{2}}}$$
$$+ \left(1 - \frac{1}{2}\mathcal{Q}_{\hat{3}}^2 + \Sigma_{\hat{3}\hat{3}}\right)\partial_{X^{\hat{3}}} \tag{5.86}$$

with dual frame

$$\omega^{\hat{0}} = \left(1 + U - \frac{1}{2}\mathcal{Q}^2 + {}^F\Phi\right)dT + \left[2\,{}^F\mathcal{A}_{\hat{a}} - \mathcal{Q}_{\hat{a}}(1-U)\right]dX^{\hat{a}},$$

$$\omega^{\hat{1}} = \left(1 + \frac{1}{2}\mathcal{Q}_{\hat{1}}^2 - \Sigma_{\hat{1}\hat{1}}\right)dX^{\hat{1}} + \left(\mathcal{Q}_{\hat{1}}\,\mathcal{Q}_{\hat{2}} - 2\Sigma_{\hat{1}\hat{2}}\right)dX^{\hat{2}} + \left(\mathcal{Q}_{\hat{1}}\,\mathcal{Q}_{\hat{3}} - 2\Sigma_{\hat{1}\hat{3}}\right)dX^{\hat{3}},$$

$$\omega^{\hat{2}} = \left(1 + \frac{1}{2}\mathcal{Q}_{\hat{2}}^2 - \Sigma_{\hat{2}\hat{2}}\right)dX^{\hat{2}} + \left(\mathcal{Q}_{\hat{2}}\,\mathcal{Q}_{\hat{3}} - 2\Sigma_{\hat{2}\hat{3}}\right)dX^{\hat{3}},$$

$$\omega^{\hat{3}} = \left(1 + \frac{1}{2}\mathcal{Q}_{\hat{3}}^2 - \Sigma_{\hat{3}\hat{3}}\right)dX^{\hat{3}}. \tag{5.87}$$

We can now proceed to the evaluation of the associated structure functions.

In $C_{\hat{\alpha}\hat{\beta}}{}^{\hat{\gamma}}$, for each $\hat{\gamma} = \hat{0}, \hat{1}, \hat{2}, \hat{3}$, we have an antisymmetric tensor that has "electric" and "magnetic" components in analogy with the Faraday tensor. Indeed, for $\hat{\gamma} = \hat{0}$, we have

$$C_{\hat{0}\hat{i}}{}^{\hat{0}} = -\left[\boldsymbol{\mathcal{E}} + (1-U)\,\mathbf{a} + \boldsymbol{\Omega}\times\boldsymbol{\mathcal{Q}} + \dot{\boldsymbol{\mathcal{Q}}}\right]_{\hat{i}}, \tag{5.88}$$

$$C_{\hat{i}\hat{j}}{}^{\hat{0}} = 2\,\epsilon_{\hat{i}\hat{j}\hat{k}}\left[\boldsymbol{\mathcal{B}} - (1-U)\,\boldsymbol{\Omega} + \mathbf{a}\times\boldsymbol{\mathcal{Q}}\right]^{\hat{k}}, \tag{5.89}$$

where

$$\dot{\boldsymbol{\mathcal{Q}}} = \left(\frac{d\boldsymbol{\Omega}}{dT}\right)\times\mathbf{X}, \qquad \mathbf{a}\times\boldsymbol{\mathcal{Q}} = U\boldsymbol{\Omega} - (\mathbf{a}\cdot\boldsymbol{\Omega})\mathbf{X}. \tag{5.90}$$

Furthermore, the gravitoelectric field, $\boldsymbol{\mathcal{E}} = \boldsymbol{\nabla}\,{}^F\Phi$, and the gravitomagnetic field, $\boldsymbol{\mathcal{B}} = \boldsymbol{\nabla}\times{}^F\boldsymbol{\mathcal{A}}$, are given by

$$\mathcal{E}_{\hat{i}}(T,\mathbf{X}) = R_{\hat{0}\hat{i}\hat{0}\hat{j}}(T)\,X^{\hat{j}}, \qquad \mathcal{B}_{\hat{i}}(T,\mathbf{X}) = -\frac{1}{2}\epsilon_{\hat{i}\hat{j}\hat{k}}\,R^{\hat{j}\hat{k}}{}_{\hat{0}\hat{l}}(T)\,X^{\hat{l}}. \tag{5.91}$$

Let us note here that the gravitoelectric field is directly proportional to the "electric" components of the of the Riemann curvature tensor and similarly the gravitomagnetic field is directly proportional to the "magnetic" components of the Riemann curvature tensor. It is interesting that we can couch our torsion results in the familiar language of gravitoelectromagnetism (GEM); see Matte (1953), Jantzen, Carini and Bini (1992), Mashhoon (2007d) and the references cited therein. Moreover, the spatial part of the metric perturbation away from Minkowski spacetime, $\Sigma_{ij} = \Sigma_{ji}$, is likewise proportional to the spatial components of the curvature. Next, for $\hat{\gamma} = \hat{1}, \hat{2}, \hat{3}$, the electric parts only involve terms of higher order and can be ignored, so that

$$C_{\hat{0}\hat{i}}{}^{\hat{j}} = 0. \tag{5.92}$$

However, the corresponding magnetic parts depend upon the spatial components of the curvature and we find that for $\hat{\gamma} = \hat{1}$,

$$C_{\hat{2}\hat{3}}{}^{\hat{1}} = 3\Omega_{\hat{1}}\, \mathcal{Q}_{\hat{1}} + R_{\hat{2}\hat{3}\hat{1}\hat{a}}X^{\hat{a}}, \quad C_{\hat{3}\hat{1}}{}^{\hat{1}} = 2\Omega_{\hat{2}}\, \mathcal{Q}_{\hat{1}} + \frac{2}{3}R_{\hat{3}\hat{1}\hat{1}\hat{a}}X^{\hat{a}},$$

$$C_{\hat{1}\hat{2}}{}^{\hat{1}} = 2\Omega_{\hat{3}}\, \mathcal{Q}_{\hat{1}} + \frac{2}{3}R_{\hat{1}\hat{2}\hat{1}\hat{a}}X^{\hat{a}}. \tag{5.93}$$

Similarly, for $\hat{\gamma} = \hat{2}$,

$$C_{\hat{2}\hat{3}}{}^{\hat{2}} = 2\Omega_{\hat{1}}\, \mathcal{Q}_{\hat{2}} + \frac{2}{3}R_{\hat{2}\hat{3}\hat{2}\hat{a}}X^{\hat{a}}, \quad C_{\hat{3}\hat{1}}{}^{\hat{2}} = \Omega_{\hat{2}}\, \mathcal{Q}_{\hat{2}} - \Omega_{\hat{3}}\, \mathcal{Q}_{\hat{3}} + \frac{1}{3}\left(R_{\hat{3}\hat{1}\hat{2}\hat{a}} - R_{\hat{1}\hat{2}\hat{3}\hat{a}}\right)X^{\hat{a}},$$

$$C_{\hat{1}\hat{2}}{}^{\hat{2}} = \Omega_{\hat{3}}\, \mathcal{Q}_{\hat{2}} + \frac{1}{3}R_{\hat{1}\hat{2}\hat{2}\hat{a}}X^{\hat{a}} \tag{5.94}$$

and for $\hat{\gamma} = \hat{3}$,

$$C_{\hat{2}\hat{3}}{}^{\hat{3}} = \Omega_{\hat{1}}\, \mathcal{Q}_{\hat{3}} + \frac{1}{3}R_{\hat{2}\hat{3}\hat{3}\hat{a}}X^{\hat{a}}, \quad C_{\hat{3}\hat{1}}{}^{\hat{3}} = \Omega_{\hat{2}}\, \mathcal{Q}_{\hat{3}} + \frac{1}{3}R_{\hat{3}\hat{1}\hat{3}\hat{a}}X^{\hat{a}}, \quad C_{\hat{1}\hat{2}}{}^{\hat{3}} = 0. \tag{5.95}$$

It is important to note that all of the components of $C_{\hat{\alpha}\hat{\beta}}{}^{\hat{\gamma}}$ can be obtained from eqns (5.88)–(5.95) by using the antisymmetry of $C_{\hat{\alpha}\hat{\beta}}{}^{\hat{\gamma}}$ in its first two indices. Furthermore, *all* of the components of the curvature tensor are involved in our calculation of the torsion tensor. The spatial components of the curvature tensor in eqns (5.93)–(5.95) essentially reduce to the gravitoelectric components in a Ricci-flat region of spacetime.

The torsion vector C_α, $C_\alpha := -C_{\alpha\beta}{}^\beta$, can be calculated for the static Fermi observers and turns out to be completely spatial; that is, $C_{\hat{a}} = (0, \Theta_{\hat{i}})$, where Θ is related to the gravitoelectric field as well as the spatial part of the torsion tensor. Indeed,

$$\Theta_{\hat{i}} = -\left[\boldsymbol{\mathcal{E}} + (1-U)\mathbf{a} + \boldsymbol{\Omega} \times \boldsymbol{\mathcal{Q}} + \dot{\boldsymbol{\mathcal{Q}}}\right]_{\hat{i}} - C_{\hat{i}\hat{j}}{}^{\hat{j}}. \tag{5.96}$$

On the other hand, the torsion pseudovector $\check{C}_\alpha$, $\check{C}_{\hat{a}} := -(1/6)\epsilon_{\hat{a}\hat{\beta}\hat{\gamma}\hat{\delta}}\, C^{\hat{\beta}\hat{\gamma}\hat{\delta}}$ is given by $(\check{C}_{\hat{0}}, \Gamma_{\hat{i}})$, where $\check{C}_{\hat{0}} = -(1/3)(C_{\hat{2}\hat{3}}{}^{\hat{1}} + C_{\hat{3}\hat{1}}{}^{\hat{2}})$ and $\boldsymbol{\Gamma}$ is related to the gravitomagnetic field,

$$\boldsymbol{\Gamma} = \frac{2}{3}\left[\boldsymbol{\mathcal{B}} - (1-U)\boldsymbol{\Omega} + \mathbf{a} \times \boldsymbol{\mathcal{Q}}\right]. \tag{5.97}$$

We note that in our convention $\epsilon_{0123} = 1$.

The torsion tensor vanishes along the reference world line ($\mathbf{X} = 0$) if ${}^{0}\Phi_{\hat{\alpha}\hat{\beta}} = 0$. It follows that along the reference *geodesic*, the contorsion tensor and the Weitzenböck connection both vanish. Thus, by a proper choice of coordinates and fundamental frame field, the Levi-Civita connection as well as the Weitzenböck connection can be made to vanish along a timelike geodesic. This provides a natural generalization of Fermi's original result (Levi-Civita 1926) in the context of extended GR.

The results presented here for Weitzenböck's torsion in Fermi coordinates are in qualitative agreement with the results of similar calculations that have been carried out in the standard Schwarzschild-like coordinates for the structure functions of the natural tetrad frames of the *static* observers in the *exterior* Kerr spacetime (Bini and Mashhoon 2015).

Acceleration tensor of the fundamental observers. It is interesting to compute the acceleration tensor for a congruence of fundamental observers. To this end, we have

$$\frac{{}^0D\, e^{\mu}{}_{\hat{\alpha}}}{d\bar{s}} = \Upsilon_{\hat{\alpha}}{}^{\hat{\beta}}\, e^{\mu}{}_{\hat{\beta}},\tag{5.98}$$

where $\bar{s}$ is the proper time along a fundamental observer's world line. Using eqn (5.56) and the fact that $e^{\mu}{}_{\hat{0}} = dx^{\mu}/d\bar{s}$, eqn (5.98) can be expressed as

$$e^{\rho}{}_{\hat{0}}\, e_{\sigma\hat{\alpha}} \left(\Gamma^{\sigma}{}_{\rho\mu} - {}^0\Gamma^{\sigma}{}_{\rho\mu}\right) = \Upsilon_{\hat{\alpha}\hat{\beta}}\, e_{\mu}{}^{\hat{\beta}}.\tag{5.99}$$

Next, the definition of the contorsion tensor in eqn (5.33) then implies

$$\Upsilon_{\hat{\alpha}\hat{\beta}} = K_{\hat{0}\hat{\alpha}\hat{\beta}},\tag{5.100}$$

since the contorsion tensor is antisymmetric in its last two indices. We should mention that the connection between eqns (5.98) and (5.100) is completely general and is independent of the particular coordinate system or our choice of the fundamental observers. In the particular case of Fermi coordinates and the corresponding fundamental static observers under consideration in this section, however, it follows from the decomposition of $\Upsilon_{\hat{\alpha}\hat{\beta}}$ into its "electric" and "magnetic" components and eqn (5.43) that $-C_{\hat{0}\hat{i}}{}^{\hat{0}}$ and $-\frac{1}{2}\, C_{\hat{i}\hat{j}}{}^{\hat{0}}$ are responsible for the proper acceleration and rotation of our fundamental observer family, respectively. This circumstance accounts for the nature of the terms that appear in eqns (5.88) and (5.89), such as, for instance, the centripetal and transverse (Euler) acceleration terms in the electric components.

In summary, of all possible smooth tetrad systems on the *curved* spacetime of GR, a particular frame field can be chosen that is so constructed from one event to the next via an additional Weitzenböck connection as to generate a global system of parallel axes ("teleparallelism") reminiscent of a global inertial frame in Minkowski spacetime. Indeed, there is an equivalence class of such frames that are related to each other by constant elements of the six-parameter global Lorentz group. The curvature of the Weitzenböck connection vanishes, but its torsion is in general non-zero. It is in principle possible to measure the Weitzenböck torsion tensor in a given gravitational field. For orthonormal frames that are naturally adapted to static observers in a gravitational field within the extended GR framework, the Weitzenböck torsion tensor has certain similarities with the frame components of the curvature tensor. In particular, the Weitzenböck torsion behaves like tidal acceleration and has dimensions of $(\text{length})^{-1}$, while curvature has dimensions of $(\text{length})^{-2}$. For the measured components of the torsion tensor, $C_{\hat{\alpha}\hat{\beta}}{}^{\hat{\gamma}}$, we find that $C_{\hat{0}\hat{i}}{}^{\hat{0}}$ represents what is essentially the gravitoelectric field, while $C_{\hat{i}\hat{j}}{}^{\hat{0}}$ represents what is essentially the gravitomagnetic field. Moreover, $C_{\hat{0}\hat{i}}{}^{\hat{j}}$ is related to the nonstationary character of the gravitational field and $C_{\hat{i}\hat{j}}{}^{\hat{k}}$ has in general mixed properties involving both the gravitoelectric and gravitomagnetic aspects. In extended GR, it is possible to introduce Fermi coordinates and tetrad frames in the neighborhood of an arbitrary timelike geodesic path such that the Levi-Civita and Weitzenböck connections both vanish along the geodesic world line. On the other hand, in flat spacetime, where ${}^{0}R_{\mu\nu\rho\sigma} = 0$, Weitzenböck's torsion tensor loses its gravitational significance. In arbitrary systems of admissible coordinates in Minkowski spacetime, the torsion tensor vanishes only for the family of fundamental inertial observers that are all at rest in a global inertial frame and have orthonormal tetrad axes that are all parallel to the standard Cartesian coordinate axes of the global inertial frame; otherwise, the torsion tensor is non-zero. Thus a congruence of accelerated observers is endowed with torsion; similarly, torsion is non-zero for a family of inertial observers that are all static in a global inertial frame but have fixed spatial frames that vary in space (Hayashi and Shirafuji 1979). To illustrate the latter possibility, consider, for instance, a global inertial frame and static *inertial* observers with orthonormal tetrads such that their spatial frames are all along the spherical polar coordinate axes; in this case, the torsion tensor has spatial components that do not vanish.

6

Field Equation of Nonlocal Gravity

Nonlocal gravity (NLG) is a tetrad theory established upon the frame field of a fundamental family of observers in spacetime. The main purpose of this chapter is to present the field equation of nonlocal gravity and discuss some of its main features. The gravitational field is in this case characterized by torsion, which is most directly related to the tetrad frame field of the fundamental observers. We recall that the torsion tensor is given by

$$C_{\mu\nu}{}^{\hat{\alpha}} = \partial_\mu e_\nu{}^{\hat{\alpha}} - \partial_\nu e_\mu{}^{\hat{\alpha}}, \tag{6.1}$$

which, for each $\hat{\alpha} = \hat{0}, \hat{1}, \hat{2}, \hat{3}$, is reminiscent of the electromagnetic field tensor

$$F_{\mu\nu} = \partial_\mu A_\nu - \partial_\nu A_\mu \tag{6.2}$$

expressed in terms of the vector potential A_μ. In nonlocal gravity $e_\mu{}^{\hat{\alpha}}$ is the analogous gravitational potential. Thus, using the extended GR framework, we first express the Riemannian curvature of spacetime in terms of the torsion field. Writing Einstein's field equation within this scheme leads to the teleparallel equivalent of GR, namely, GR$_{||}$. It turns out that GR$_{||}$ is the gauge theory of the Abelian group of spacetime translations (Blagojević and Hehl 2013; Aldrovandi and Pereira 2013; Maluf 2013). As such, the structure of GR$_{||}$ bears certain similarities with electrodynamics. We emphasize this analogy by writing the GR$_{||}$ field equations in terms of torsion in a form that is similar to Maxwell's equations. Maxwell's equations in a medium in an inertial frame can be expressed in terms of the field tensors $F_{\mu\nu} \mapsto (\mathbf{E}, \mathbf{B})$ and $H_{\mu\nu} \mapsto (\mathbf{D}, \mathbf{H})$ as

$$F_{[\mu\nu,\rho]} = 0, \qquad \partial_\nu H^{\mu\nu} = \frac{4\pi}{c} \bar{j}^\mu, \tag{6.3}$$

where $\bar{j}^\mu$ is the total current 4-vector associated with *free* electric charges. To complete the theory, a constitutive relation between $F_{\mu\nu}$ and $H_{\mu\nu}$ is required. If we impose the local relation $H_{\mu\nu} = F_{\mu\nu}$, we recover Maxwell's equations in vacuum. However, in a medium, the constitutive relation is in general nonlocal (Jackson 1999; Landau and Lifshitz 1960), thus leading to the nonlocal electrodynamics of media. Finally, the nonlocal version of GR$_{||}$ is presented in complete analogy with Maxwell's equations coupled with nonlocal constitutive equations. From nonlocal GR$_{||}$, we then derive the nonlocal generalization of Einstein's field equation in this extended framework. The form of the field equation of nonlocal gravity raises the possibility that nonlocal gravity may simulate dark matter.

Nonlocal Gravity. Bahram Mashhoon. © Bahram Mashhoon 2017. Published 2017 by Oxford University Press.

6.1 Riemannian Curvature of Spacetime

In this section, the curvature of the Levi-Civita connection is expressed in terms of the torsion of the Weitzenböck connection. The corresponding Ricci and scalar curvatures are then evaluated in order to express the Einstein tensor in terms of torsion.

6.1.1 Weitzenböck's connection is curvature-free

From the definition of the Weitzenböck connection (Weitzenböck 1923)

$$\Gamma^\alpha_{\nu\mu} = e^\alpha{}_{\hat\rho}\left(\partial_\nu e_\mu{}^{\hat\rho}\right), \tag{6.4}$$

we see that the derivative of this connection can be written as

$$\partial_\beta\,\Gamma^\alpha_{\nu\mu} = \left(\partial_\beta\,e^\alpha{}_{\hat\rho}\right)\left(\partial_\nu\,e_\mu{}^{\hat\rho}\right) + e^\alpha{}_{\hat\rho}\left(\partial_\beta\,\partial_\nu\,e_\mu{}^{\hat\rho}\right). \tag{6.5}$$

The symmetry of the second partial derivatives then implies

$$\partial_\beta\,\Gamma^\alpha_{\nu\mu} - \partial_\nu\,\Gamma^\alpha_{\beta\mu} = \left(\partial_\beta\,e^\alpha{}_{\hat\rho}\right)\left(\partial_\nu\,e_\mu{}^{\hat\rho}\right) - \left(\partial_\nu\,e^\alpha{}_{\hat\rho}\right)\left(\partial_\beta\,e_\mu{}^{\hat\rho}\right). \tag{6.6}$$

Next, we recall from the orthonormality of the fundamental tetrads that $e^\alpha{}_{\hat\rho}\,e_\gamma{}^{\hat\rho} = \delta^\alpha_\gamma$; hence,

$$\Gamma^\alpha_{\beta\gamma} = e^\alpha{}_{\hat\rho}\left(\partial_\beta\,e_\gamma{}^{\hat\rho}\right) = -\left(\partial_\beta\,e^\alpha{}_{\hat\rho}\right)e_\gamma{}^{\hat\rho}. \tag{6.7}$$

Therefore, we can write

$$\Gamma^\alpha_{\beta\gamma}\,\Gamma^\gamma_{\nu\mu} = \left[-\left(\partial_\beta\,e^\alpha{}_{\hat\rho}\right)e_\gamma{}^{\hat\rho}\right]\left[e^\gamma{}_{\hat\delta}\left(\partial_\nu\,e_\mu{}^{\hat\delta}\right)\right], \tag{6.8}$$

where we have used the new result in eqn (6.7) for the first part and the standard expression (6.4) for the second part. From the orthonormality of fundamental tetrads, $e_\gamma{}^{\hat\rho}\,e^\gamma{}_{\hat\delta} = \delta^{\hat\rho}_{\hat\delta}$, we find

$$\Gamma^\alpha_{\beta\gamma}\,\Gamma^\gamma_{\nu\mu} = -\left(\partial_\beta\,e^\alpha{}_{\hat\rho}\right)\left(\partial_\nu\,e_\mu{}^{\hat\rho}\right). \tag{6.9}$$

Combining this result with eqn (6.6), we see that the curvature tensor vanishes identically for the Weitzenböck connection

$$R^\alpha{}_{\mu\beta\nu} = \partial_\beta\,\Gamma^\alpha_{\nu\mu} - \partial_\nu\,\Gamma^\alpha_{\beta\mu} + \Gamma^\alpha_{\beta\gamma}\,\Gamma^\gamma_{\nu\mu} - \Gamma^\alpha_{\nu\gamma}\,\Gamma^\gamma_{\beta\mu} = 0. \tag{6.10}$$

6.1.2 Curvature of the Levi-Civita connection

Let us now calculate the Riemann curvature tensor for our spacetime manifold, namely,

$$^0R^\alpha{}_{\mu\beta\nu} = \partial_\beta\,{}^0\Gamma^\alpha_{\nu\mu} - \partial_\nu\,{}^0\Gamma^\alpha_{\beta\mu} + {}^0\Gamma^\alpha_{\beta\gamma}\,{}^0\Gamma^\gamma_{\nu\mu} - {}^0\Gamma^\alpha_{\nu\gamma}\,{}^0\Gamma^\gamma_{\beta\mu}, \tag{6.11}$$

where

$$^0\Gamma^\alpha_{\mu\nu} = \Gamma^\alpha_{\mu\nu} + K_{\mu\nu}{}^\alpha. \tag{6.12}$$

Substituting this expression in eqn (6.11) and noting that the curvature tensor vanishes for the Weitzenböck connection, we find

$$^0R^\alpha{}_{\mu\beta\nu} = \partial_\beta\,K_{\nu\mu}{}^\alpha - \partial_\nu\,K_{\beta\mu}{}^\alpha + \Gamma^\alpha_{\beta\gamma}\,K_{\nu\mu}{}^\gamma + \Gamma^\gamma_{\nu\mu}\,K_{\beta\gamma}{}^\alpha + K_{\beta\gamma}{}^\alpha\,K_{\nu\mu}{}^\gamma$$
$$-\Gamma^\alpha_{\nu\gamma}\,K_{\beta\mu}{}^\gamma - \Gamma^\gamma_{\beta\mu}\,K_{\nu\gamma}{}^\alpha - K_{\nu\gamma}{}^\alpha\,K_{\beta\mu}{}^\gamma. \tag{6.13}$$

We are interested in the Einstein tensor

$$^0G_{\mu\nu} = {}^0R_{\nu\mu} - \frac{1}{2} g_{\mu\nu} \, {}^0R, \tag{6.14}$$

where the indices on the symmetric Ricci tensor have been switched for the sake of simplicity in order to get $^0R_{\nu\mu} = {}^0R^\alpha{}_{\nu\alpha\mu}$ in a useful form as follows: using the relations

$$K_{\alpha\beta}{}^\alpha = -C_\beta \tag{6.15}$$

and

$$\Gamma^\alpha_{\alpha\beta} = C_\beta + \frac{1}{\sqrt{-g}} \frac{\partial}{\partial x^\beta} (\sqrt{-g}), \tag{6.16}$$

where C^μ is the *torsion vector* and $g := \det(g_{\mu\nu})$, $\sqrt{-g} = \det(e_\mu{}^{\hat{a}})$, we find from eqn (6.13) after some algebra that

$$^0R_{\nu\mu} = \frac{1}{\sqrt{-g}} \frac{\partial}{\partial x^\alpha} \left(\sqrt{-g}\, K_{\mu\nu}{}^\alpha \right) + \frac{\partial C_\nu}{\partial x^\mu} - C_\alpha \Gamma^\alpha_{\mu\nu}$$
$$- (\Gamma^\alpha_{\mu\beta} + K_{\mu\beta}{}^\alpha) K_{\alpha\nu}{}^\beta - \Gamma^\alpha_{\beta\nu} K_{\mu\alpha}{}^\beta. \tag{6.17}$$

Let us next compute the scalar curvature $^0R = g^{\mu\nu}\, {}^0R_{\nu\mu}$. To this end, we note that $C^\alpha = g^{\mu\nu}\, K_{\mu\nu}{}^\alpha$, so that

$$\frac{1}{\sqrt{-g}} \frac{\partial}{\partial x^\alpha} \left(\sqrt{-g}\, C^\alpha \right) = g^{\mu\nu} \frac{1}{\sqrt{-g}} \frac{\partial}{\partial x^\alpha} \left(\sqrt{-g}\, K_{\mu\nu}{}^\alpha \right) + K_{\mu\nu}{}^\alpha \partial_\alpha g^{\mu\nu}. \tag{6.18}$$

Using the metric compatibility of the Weitzenböck connection, $\nabla_\alpha g^{\mu\nu} = 0$, that is,

$$\partial_\alpha g^{\mu\nu} = -\Gamma^\mu_{\alpha\beta} g^{\beta\nu} - \Gamma^\nu_{\alpha\beta} g^{\beta\mu} \tag{6.19}$$

and the definition of the torsion tensor, we find

$$K_{\mu\nu}{}^\alpha \partial_\alpha g^{\mu\nu} = -K_\alpha{}^{\mu\beta} \Gamma^\alpha_{\mu\beta} - K^\mu{}_\alpha{}^\beta \Gamma^\alpha_{\beta\mu} + K_\alpha{}^{\beta\mu} C_{\beta\mu}{}^\alpha. \tag{6.20}$$

Moreover, from $\Gamma^\alpha_{\mu\nu} = {}^0\Gamma^\alpha_{\mu\nu} - K_{\mu\nu}{}^\alpha$, we have

$$g^{\mu\nu} \Gamma^\alpha_{\mu\nu} = -\frac{1}{\sqrt{-g}} \frac{\partial}{\partial x^\beta} \left(\sqrt{-g}\, g^{\alpha\beta} \right) - C^\alpha, \tag{6.21}$$

so that

$$g^{\mu\nu} \left(\frac{\partial C_\nu}{\partial x^\mu} - C_\alpha \Gamma^\alpha_{\mu\nu} \right) = C_\alpha C^\alpha + \frac{1}{\sqrt{-g}} \frac{\partial}{\partial x^\alpha} \left(\sqrt{-g}\, C^\alpha \right). \tag{6.22}$$

Putting these results together, we find

$$^0R = \frac{2}{\sqrt{-g}} \frac{\partial}{\partial x^\delta} \left(\sqrt{-g}\, C^\delta \right) + C_\alpha C^\alpha - K_{\alpha\beta\gamma} K^{\beta\gamma\alpha} - K_{\alpha\beta\gamma} C^{\beta\gamma\alpha}. \tag{6.23}$$

Finally, from the identities

$$K_{\alpha\beta\gamma} K^{\beta\gamma\alpha} = -\frac{1}{2} K_{\alpha\beta\gamma} C^{\beta\gamma\alpha} = -\frac{1}{2}\left(\frac{1}{2} I_1 + I_2\right), \tag{6.24}$$

where

$$I_1 = C_{\alpha\beta\gamma} C^{\alpha\beta\gamma}, \quad I_2 = C_{\alpha\beta\gamma} C^{\gamma\beta\alpha}, \quad I_3 = C_\alpha C^\alpha \tag{6.25}$$

are the three independent algebraic (Weitzenböck) invariants of the torsion tensor, we find

$$^0R = \frac{2}{\sqrt{-g}} \frac{\partial}{\partial x^\delta}\left(\sqrt{-g}\, C^\delta\right) - \frac{1}{2}\left(\frac{1}{2} I_1 + I_2 - 2\, I_3\right). \tag{6.26}$$

Let us note that

$$\mathfrak{C}_{\alpha\beta\gamma} C^{\alpha\beta\gamma} = \frac{1}{2} I_1 + I_2 - 2\, I_3, \tag{6.27}$$

where $\mathfrak{C}_{\alpha\beta\gamma}$ is the *auxiliary torsion tensor* that is also antisymmetric in its first two indices and is given by

$$\mathfrak{C}_{\alpha\beta\gamma} = \frac{1}{2}\left(C_{\gamma\beta\alpha} + C_{\alpha\beta\gamma} - C_{\gamma\alpha\beta}\right) + C_\alpha\, g_{\beta\gamma} - C_\beta\, g_{\alpha\gamma}. \tag{6.28}$$

Thus eqn (6.26) takes the form

$$^0R = -\frac{1}{2}\,\mathfrak{C}_{\alpha\beta\gamma} C^{\alpha\beta\gamma} + \frac{2}{\sqrt{-g}} \frac{\partial}{\partial x^\delta}\left(\sqrt{-g}\, C^\delta\right). \tag{6.29}$$

It is important to note here that just as $C_{\mu\nu}{}^{\hat a}$ is reminiscent of $F_{\mu\nu}$ in electrodynamics, $\mathfrak{C}^{\mu\nu}{}_{\hat a}$ turns out to be reminiscent of $H^{\mu\nu}$. The auxiliary torsion tensor $\mathfrak{C}_{\alpha\beta\gamma}(x)$ is a linear combination of the components of the torsion tensor $C_{\mu\nu\rho}(x)$; indeed, this local linear connection is in effect a constitutive relation, as will be discussed in more detail below.

The last step in the computation of the Einstein tensor (6.14) involves a proper combination of our expressions for the Ricci and scalar curvatures in eqns (6.17) and (6.29), respectively. It is useful to introduce an *auxiliary field strength* $\mathcal{H}_{\mu\nu\rho} = -\mathcal{H}_{\nu\mu\rho}$ defined by

$$\mathcal{H}_{\mu\nu\rho} := \frac{\sqrt{-g}}{\kappa}\,\mathfrak{C}_{\mu\nu\rho}, \tag{6.30}$$

where $\kappa = 8\pi G/c^4$. Next, we define the "Maxwellian" tensors

$$\mathfrak{m}_{\mu\nu} = \frac{\kappa}{\sqrt{-g}}\, e_\mu{}^{\hat\gamma}\, g_{\nu\alpha}\, \frac{\partial}{\partial x^\beta}\, \mathcal{H}^{\alpha\beta}{}_{\hat\gamma} \tag{6.31}$$

and

$$\mathfrak{t}_{\mu\nu} = C_\mu{}^{\rho\sigma}\, \mathfrak{C}_{\nu\rho\sigma} - \frac{1}{4} g_{\mu\nu}\, C^{\alpha\beta\gamma}\, \mathfrak{C}_{\alpha\beta\gamma}. \tag{6.32}$$

We wish to show that the Einstein tensor in this extended GR framework has the Maxwellian form

$$^0G_{\mu\nu} = \mathfrak{m}_{\mu\nu} - \mathfrak{t}_{\mu\nu}. \tag{6.33}$$

To this end, let us write eqn (6.31) as

$$\mathfrak{m}_{\mu\nu} = \frac{\kappa}{\sqrt{-g}} \frac{\partial}{\partial x^\beta} \left(e_\mu{}^{\hat{\gamma}} g_{\nu\alpha} \mathcal{H}^{\alpha\beta}{}_{\hat{\gamma}} \right) - \mathfrak{C}^{\alpha\beta}{}_{\hat{\gamma}} \frac{\partial}{\partial x^\beta} \left(e_\mu{}^{\hat{\gamma}} g_{\nu\alpha} \right) \tag{6.34}$$

or

$$\mathfrak{m}_{\mu\nu} = \frac{1}{\sqrt{-g}} \frac{\partial}{\partial x^\alpha} \left(\sqrt{-g}\, \mathfrak{C}_\nu{}^\alpha{}_\mu \right) - \mathfrak{C}^{\alpha\beta}{}_\mu \partial_\beta g_{\nu\alpha} - \mathfrak{C}_\nu{}^\beta{}_\alpha \Gamma^\alpha_{\beta\mu}. \tag{6.35}$$

Next, in eqn (6.35) we write the auxiliary torsion tensors in terms of the corresponding contorsion tensors; for instance,

$$\mathfrak{C}_\nu{}^\alpha{}_\mu = K_{\mu\nu}{}^\alpha + C_\nu \delta^\alpha_\mu - C^\alpha g_{\mu\nu}. \tag{6.36}$$

Furthermore, we use $\nabla_\beta g_{\nu\alpha} = 0$, or more explicitly,

$$\partial_\beta g_{\nu\alpha} = \Gamma^\rho_{\beta\nu} g_{\rho\alpha} + \Gamma^\rho_{\beta\alpha} g_{\rho\nu} \tag{6.37}$$

in eqn (6.35) in order to get an expression for $\mathfrak{m}_{\mu\nu}$ that somewhat resembles ${}^0 R_{\nu\mu}$ in eqn (6.17). Let us now use the latest expression for $\mathfrak{m}_{\mu\nu}$ and eqns (6.17), (6.29) and (6.32) to form the quantity $\mathfrak{Z}_{\mu\nu}$,

$$\mathfrak{Z}_{\mu\nu} = {}^0 R_{\nu\mu} - \frac{1}{2} g_{\mu\nu} {}^0 R - (\mathfrak{m}_{\mu\nu} - \mathfrak{t}_{\mu\nu}). \tag{6.38}$$

This tensor vanishes identically when we use in the resulting expression for $\mathfrak{Z}_{\mu\nu}$ the following identities

$$C_\mu C_\nu + C^\alpha C_{\mu\alpha\nu} + C_\mu{}^{\rho\sigma} \mathfrak{C}_{\nu\rho\sigma} = \left(\Gamma^\alpha_{\mu\beta} - \Gamma^\alpha_{\beta\mu} \right) K_{\alpha\nu}{}^\beta \tag{6.39}$$

and

$$K_{\mu\beta}{}^\alpha K_{\alpha\nu}{}^\beta = K_\mu{}^{\alpha\beta} \Gamma^\rho_{\beta\alpha} g_{\rho\nu}. \tag{6.40}$$

Therefore, the Einstein tensor in terms of the torsion field can indeed be written as in eqn (6.33).

6.2 GR$_\parallel$: Teleparallel Equivalent of GR

It is interesting to express GR in terms of the torsion field. The result is the teleparallel equivalent of GR, namely, GR$_\parallel$.

The Einstein tensor ${}^0 G_{\mu\nu}$ can be written in terms of torsion in the form of eqn (6.33),

$$ {}^0 G_{\mu\nu} = -\kappa E_{\mu\nu} + \frac{\kappa}{\sqrt{-g}} e_\mu{}^{\hat{\gamma}} g_{\nu\alpha} \frac{\partial}{\partial x^\beta} \mathcal{H}^{\alpha\beta}{}_{\hat{\gamma}}, \tag{6.41}$$

where $E_{\mu\nu}$,

$$E_{\mu\nu} := \kappa^{-1} \mathfrak{t}_{\mu\nu}, \tag{6.42}$$

turns out to be the traceless and non-symmetric energy–momentum tensor of the gravitational field in the new scheme. It resembles Minkowski's energy–momentum tensor for the electromagnetic field in a medium (Hehl 2008)

$$ {}^{(M)} T_{\mu\nu} = \frac{1}{4\pi} \left(F_{\mu\rho} H_\nu{}^\rho - \frac{1}{4} g_{\mu\nu} F_{\alpha\beta} H^{\alpha\beta} \right) \tag{6.43}$$

that is also traceless, but not symmetric. Thus, using eqn (6.30), we can write

$$\sqrt{-g}\, E_{\mu\nu} = C_{\mu\rho\sigma}\, \mathcal{H}_\nu{}^{\rho\sigma} - \frac{1}{4}\, g_{\mu\nu}\, C_{\alpha\beta\gamma}\, \mathcal{H}^{\alpha\beta\gamma}. \tag{6.44}$$

Moreover, we note that

$$^0G_{\mu\nu} = \frac{\kappa}{\sqrt{-g}}\left[g_{\nu\alpha}\, e_\mu{}^{\hat{\gamma}}\, \frac{\partial}{\partial x^\beta}\, \mathcal{H}^{\alpha\beta}{}_{\hat{\gamma}} - \left(\mathcal{H}_{\nu\rho\sigma} C_\mu{}^{\rho\sigma} - \frac{1}{4}\, g_{\nu\mu}\, \mathcal{H}_{\alpha\beta\gamma} C^{\alpha\beta\gamma} \right) \right]. \tag{6.45}$$

Einstein's gravitational field equation,

$$^0G_{\mu\nu} + {}^0\Lambda\, g_{\mu\nu} = \kappa\, T_{\mu\nu}, \tag{6.46}$$

now takes the Maxwellian form

$$\frac{\partial}{\partial x^\nu}\, \mathcal{H}^{\mu\nu}{}_{\hat{\alpha}} + \frac{\sqrt{-g}}{\kappa}\, {}^0\Lambda\, e^\mu{}_{\hat{\alpha}} = \sqrt{-g}\, (T_{\hat{\alpha}}{}^\mu + E_{\hat{\alpha}}{}^\mu). \tag{6.47}$$

Let us recall here that Maxwell's original equations in a medium in terms of $F_{\mu\nu} \mapsto (\mathbf{E}, \mathbf{B})$ and $H_{\mu\nu} \mapsto (\mathbf{D}, \mathbf{H})$ can be written in arbitrary coordinates as

$$F_{[\mu\nu,\rho]} = 0, \qquad \partial_\nu\left(\sqrt{-g}\, H^{\mu\nu} \right) = \frac{4\pi}{c}\, \sqrt{-g}\, \bar{\jmath}^\mu, \tag{6.48}$$

where $\bar{\jmath}^\mu$ is the total current of *free* electric charges. We mention that the first field equation in eqn (6.48) is already satisfied by eqn (6.2), and the torsion field naturally satisfies a similar equation due to its definition in eqn (6.1) in terms of the gravitational potentials. Moreover, eqn (6.47) is similar to the second field equation in eqn (6.48). There is therefore a close analogy between GR$_{||}$ and Maxwell's electrodynamics of media. In a medium, a constitutive relation between $H_{\mu\nu}$ and $F_{\mu\nu}$ is usually required; in the context of GR$_{||}$, the corresponding constitutive relation, eqn (6.30), is essentially the linear pointwise connection between the auxiliary torsion tensor $\mathfrak{C}_{\mu\nu\rho}$ and the torsion tensor $C_{\mu\nu\rho}$.

The universality of the gravitational interaction implies that all forms of mass–energy gravitate. This is reflected in the GR$_{||}$ field equation (6.47), since the net energy–momentum tensor includes the contribution of the gravitational field itself.

To obtain the conservation laws in the framework of GR$_{||}$, let us write eqn (6.47) in the form

$$\frac{\partial}{\partial x^\beta}\, \mathcal{H}^{\alpha\beta}{}_{\hat{\gamma}} = \sqrt{-g}\, e^\mu{}_{\hat{\gamma}}\, g^{\alpha\nu} \left(T_{\mu\nu} + E_{\mu\nu} - \frac{{}^0\Lambda}{\kappa} g_{\mu\nu} \right). \tag{6.49}$$

It follows from Eq. (6.49) and the antisymmetry of $\mathcal{H}^{\alpha\beta}{}_{\hat{\gamma}}$ in its first two indices that

$$\frac{\partial}{\partial x^\mu}\left[\sqrt{-g}\, \left(T_{\hat{\alpha}}{}^\mu + E_{\hat{\alpha}}{}^\mu - \frac{{}^0\Lambda}{\kappa}\, e^\mu{}_{\hat{\alpha}} \right) \right] = 0 \tag{6.50}$$

or

$$\frac{\partial}{\partial x^\nu}\left[\sqrt{-g}\, e_\mu{}^{\hat{\alpha}}\, \left(T^{\mu\nu} + E^{\mu\nu} - \frac{{}^0\Lambda}{\kappa} g^{\mu\nu} \right) \right] = 0. \tag{6.51}$$

This is the general law of conservation of total energy–momentum tensor that consists of contributions due to matter, the gravitational field and the cosmological constant.

Equation (6.51) is of the general form of the current conservation law discussed in Section 5.1.

In summary, within the context of GR, we have chosen a fundamental frame field which together with the corresponding Weitzenböck connection has generated a $GR_{||}$ framework that is the teleparallel equivalent of GR. We emphasize that *any* smooth orthonormal frame field can be chosen to act as the fundamental frame field; the field equation of $GR_{||}$ must then determine this frame field. In GR, the ten gravitational field equations can be used in principle to determine the ten components of the spacetime metric tensor. A tetrad frame field has, however, sixteen components, which are subject to ten orthonormality relations that, in effect, determine the metric in terms of the tetrad frame. The field equation of $GR_{||}$, the teleparallel *equivalent* of GR, can only determine the metric tensor. This circumstance points to the fundamental sixfold degeneracy of $GR_{||}$. In fact, the extra six degrees of freedom are elements of the *local* Lorentz group; that is, the boosts and rotations that locally characterize one system of observers with respect to a fiducial system. This basic degeneracy of $GR_{||}$ will be removed in the nonlocal generalization of this theory later in this chapter.

We recall from Chapter 1 that $\mathcal{L}_g = (2\kappa)^{-1}\sqrt{-g}\,{}^0R$ is the Lagrangian density of the gravitational field in GR. In terms of the torsion field, we have from eqn (6.29) that

$$\mathcal{L}_g = -\frac{1}{4\kappa}\sqrt{-g}\,\mathcal{C}_{\alpha\beta\gamma}C^{\alpha\beta\gamma} + \frac{1}{\kappa}\partial_\mu\left(\sqrt{-g}\,C^\mu\right). \tag{6.52}$$

Neglecting the total divergence in the Lagrangian density (6.52), we find that the Lagrangian density for $GR_{||}$ is given by

$$\mathcal{L}_{g||} = -\frac{1}{4}\mathcal{H}_{\alpha\beta\gamma}C^{\alpha\beta\gamma}, \tag{6.53}$$

which is quadratic in the gravitational field strength in analogy with the similar situation in electrodynamics. This Lagrangian density is proportional to a particular combination of the three Weitzenböck invariants, cf. eqn (6.27); more generally, one can generate a three-parameter class of such teleparallel theories of gravity. It is interesting to note that

$$\mathcal{H}^{\mu\nu}{}_{\hat{\alpha}} = -2\,\frac{\partial\,\mathcal{L}_{g||}}{\partial\,C_{\mu\nu}{}^{\hat{\alpha}}}, \tag{6.54}$$

which explains the original motivation for the introduction of the auxiliary field strength in eqn (6.30). For further discussions of $GR_{||}$ along these lines, see Hehl and Mashhoon (2009b) and Maluf (2013).

The tetrad formulation of GR goes back to Einstein's attempt at a classical unified field theory of gravitation and electromagnetism (Einstein 1930). Later, in a purely gravitational context, Møller pointed out that the fundamental problem of gravitational energy in GR can be solved in the tetrad framework (Møller 1961; Pellegrini and Plebanski 1963; Hayashi and Shirafuji 1979). It turns out that general teleparallel theories of gravitation can be constructed on the basis of the dynamics of a fundamental parallel tetrad frame field; see, for instance, Blagojević and Hehl (2013), Aldrovandi and Pereira (2013), Maluf (2013) and the references cited therein. If in such a framework the resulting gravitational field equation reduces to that of GR, then we have the teleparallel equivalent of GR, namely, GR$_{||}$. Let $g_{\mu\nu}(x)$ be a solution of GR field equation; then, in this gravitational field *any* smooth frame field $\lambda_\mu{}^{\hat{a}}(x)$ that is orthonormal, namely,

$$g_{\mu\nu}(x) = \lambda_\mu{}^{\hat{\alpha}} \lambda_\nu{}^{\hat{\beta}} \eta_{\hat{\alpha}\hat{\beta}}, \tag{6.55}$$

is a solution of the GR$_{||}$ field equation. Equation (6.55) is invariant under the *local* Lorentz group. The Weitzenböck connection is simply auxiliary in GR$_{||}$ and helps in the resolution of the energy problem in GR. On the other hand, the extra structure can also allow the introduction of nonlocality into the theory, but then the corresponding fundamental frame field acquires basic significance; that is, given a solution $g_{\mu\nu}$ of Einstein's gravitational field equation, any smooth orthonormal tetrad field turns out to be a possible fundamental frame field in GR$_{||}$. However, this degeneracy is removed by nonlocality. Thus in nonlocal gravity, there is a unique distribution of gravitational energy in spacetime up to invariance under the *global* Lorentz group. A detailed treatment of the approach to GR$_{||}$ adopted here can be found in the review by Maluf (2013). This concludes our brief presentation of the salient features of GR$_{||}$, the teleparallel equivalent of GR.

In nonlocal gravity, the gravitational field is directly dependent upon the past history of the torsion field. To render general relativity history-dependent, we envision a nonlocal extension of the constitutive relation of GR$_{||}$, eqn (6.30), involving an average of torsion field with a weight function given, for simplicity, by a certain *scalar* constitutive kernel $\mathcal{K}(x, x')$. This kernel is assumed to be *causal*, so that event x is in the future of event x'. Indeed, we assume that x can be connected to x' by a *unique* timelike or null geodesic. To perform the integration over the past state of the gravitational field, here represented by the torsion field, we need the *world function* $\Omega(x, x')$, which provides the causal connection between the past and the present via bitensors $\Omega_{\mu\mu'}(x, x')$. We therefore turn to a discussion of the world function (Hadamard 1952; Ruse 1931; Synge 1931, 1971).

6.3   World Function

Consider a geodesic path $x^\mu(\zeta)$ in spacetime. The geodesic equation of motion can be derived from the variational principle $\delta \int L_g \, d\zeta = 0$, where (Bini and Mashhoon 2016)

$$L_g = -\frac{1}{2} \left(\frac{ds}{d\zeta} \right)^2 = -\frac{1}{2} g_{\alpha\beta} \frac{dx^\alpha}{d\zeta} \frac{dx^\beta}{d\zeta}. \tag{6.56}$$

The momentum corresponding to the 4-velocity vector $dx^\alpha/d\zeta$ is given by

$$p_\alpha = \frac{\partial L_g}{\partial (dx^\alpha/d\zeta)} = -g_{\alpha\beta} \frac{dx^\beta}{d\zeta}. \tag{6.57}$$

It is straightforward to show that the Euler–Lagrange equation in this case leads to

$$\frac{d^2 x^\mu}{d\zeta^2} + {}^0\Gamma^\mu{}_{\alpha\beta} \frac{dx^\alpha}{d\zeta} \frac{dx^\beta}{d\zeta} = 0. \tag{6.58}$$

Moreover, the Hamiltonian, $-L_g + p_\alpha \, dx^\alpha/d\zeta$, in this case amounts to $-\frac{1}{2} p^\alpha p_\alpha$, which is conserved, since the dynamical system does not explicitly depend upon ζ. The magnitude of the Hamiltonian is the same as that of the Lagrangian along the path; therefore, $ds/d\zeta$ is a constant of the motion. This constant vanishes for a null geodesic path; otherwise, the proper spacetime distance along the path is linearly related to ζ. The path is a timelike (spacelike) geodesic if the Hamiltonian is positive (negative). The geodesic equation (6.58) is independent of a constant *linear* rescaling of parameter ζ; hence, ζ is an *affine* parameter along the geodesic path.

Imagine now that a unique geodesic path connects event $P' : x' = \xi(\zeta_0)$ to event $P : x = \xi(\zeta_1)$ along the geodesic path $x^\mu = \xi^\mu(\zeta)$; see Fig. 6.1a. It proves useful to employ the world function Ω, which denotes half the square of the proper distance from P' to P; that is, we define (Synge 1971)

$$\Omega(x, x') = \frac{1}{2}(\zeta_1 - \zeta_0) \int_{\zeta_0}^{\zeta_1} g_{\alpha\beta} \frac{d\xi^\alpha}{d\zeta} \frac{d\xi^\beta}{d\zeta} \, d\zeta. \tag{6.59}$$

It turns out that Ω is independent of the affine parameter ζ; that is, Ω is invariant under a linear rescaling of ζ. Moreover, the integrand in eqn (6.59) is constant by virtue of the geodesic equation; therefore, $\Omega = 0$ for a null geodesic, $\Omega = -\frac{1}{2}\tau^2$ for a timelike geodesic of length τ and $\Omega = \frac{1}{2}\sigma^2$ for a spacelike geodesic of length σ. To proceed, indices $\mu', \nu', \rho', \dots$ refer to event x', while indices $\mu, \nu, \rho, \dots$ refer to event x; moreover, we use indices $\alpha, \beta, \dots$ to refer to an arbitrary event on the geodesic between P' and P.

To illustrate the main properties of $\Omega(x, x')$, we consider a variation of eqn (6.59) that changes the endpoints; that is, in eqn (6.59) we replace ξ by $\xi + \delta\xi$, where $\delta\xi$ is non-zero only at the endpoints. Then,

$$\delta\Omega(x, x') = (\zeta_1 - \zeta_0) \left[g_{\alpha\beta} \frac{d\xi^\beta}{d\zeta} \delta\xi^\alpha \right]_{\zeta_0}^{\zeta_1}. \tag{6.60}$$

On the other hand,

$$\delta\Omega = \frac{\partial\Omega}{\partial x^\mu} \delta x^\mu + \frac{\partial\Omega}{\partial x'^{\mu'}} \delta x'^{\mu'}, \tag{6.61}$$

so that

$$\frac{\partial\Omega}{\partial x^\mu} = (\zeta_1 - \zeta_0) g_{\mu\nu}(x) \frac{dx^\nu}{d\zeta},$$

$$\frac{\partial\Omega}{\partial x'^{\mu'}} = -(\zeta_1 - \zeta_0) g_{\mu'\nu'}(x') \frac{dx'^{\nu'}}{d\zeta}. \tag{6.62}$$

Thus from the viewpoint of an observer at P, $-g^{\mu\nu} \partial\Omega/\partial x^\nu := -\Omega^\mu$ is the natural generalization of the position vector of P' with respect to P, while from the viewpoint of an observer at P', $-g^{\mu'\nu'} \partial\Omega/\partial x'^{\nu'} := -\Omega^{\mu'}$ is the natural generalization of the position vector of P with respect to P'; see Fig. 6.1b.

It is possible to see from the geodesic equation that the integrand in eqn (6.59) is indeed constant; therefore,

$$\begin{aligned} \Omega(x, x') &= \frac{1}{2}(\zeta_1 - \zeta_0)^2 g_{\mu\nu}(x) \frac{dx^\mu}{d\zeta} \frac{dx^\nu}{d\zeta} \\ &= \frac{1}{2}(\zeta_1 - \zeta_0)^2 g_{\mu'\nu'}(x') \frac{dx'^{\mu'}}{d\zeta} \frac{dx'^{\nu'}}{d\zeta}. \end{aligned} \tag{6.63}$$

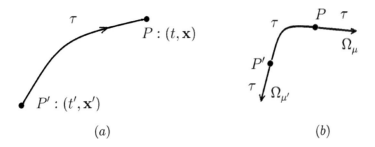

(a)   (b)

Fig. 6.1 (a) A timelike geodesic segment with proper time τ that starts at point P' and ends at point P. (b) The derivatives of the world function $\Omega_{\mu'}$ and Ω_μ are tangents to the path at P' and P, respectively, and have the same length τ as the geodesic segment. Republished from Bini, D. and Mashhoon, B., 2016, "Nonlocal Gravity: Conformally Flat Spacetimes", *Int. J. Geom. Methods Mod. Phys.* **13**, 1650081 (17 pages), with the permission of World Scientific Publishing Co., Inc.; permission conveyed through Copyright Clearance Center, Inc. DOI: 10.1142/S021988781650081X

We define
$$\Omega_\mu(x, x') := \frac{\partial \Omega}{\partial x^\mu}, \qquad \Omega_{\mu'}(x, x') := \frac{\partial \Omega}{\partial x'^{\mu'}}. \tag{6.64}$$

Then, it follows from eqns (6.62)–(6.63) that
$$2\Omega = g^{\mu\nu}\Omega_\mu\Omega_\nu = g^{\mu'\nu'}\Omega_{\mu'}\Omega_{\nu'}, \tag{6.65}$$

which are the fundamental partial differential equations for $\Omega(x, x')$. They have the interpretation that the proper length of each tangent vector at P and P' is equal to the proper length of the geodesic from P' to P; see Fig. 6.1b.

It can be shown that covariant derivatives at x and x' commute for any bitensor (Synge 1971). Thus $\Omega_{\mu\mu'}(x, x') = \Omega_{\mu'\mu}(x, x')$ is a smooth dimensionless bitensor. From the basic equation $2\Omega = \Omega_\mu\,\Omega^\mu$ we obtain, via differentiation,
$$\Omega_{\mu'} = \Omega_{\mu\mu'}\,\Omega^\mu. \tag{6.66}$$

It follows from this result that as $x' \to x$, we have
$$\lim_{x' \to x} \Omega_{\mu\mu'}(x, x') = -g_{\mu\mu'}(x), \tag{6.67}$$

cf. Fig. 6.1b.

Let k^α be a Killing vector field; then, $k_\alpha\, dx^\alpha/d\tau$ is a constant along a geodesic path. It follows from this fact and Fig. 6.1b that
$$k^\mu(x)\,\Omega_\mu(x, x') + k^{\mu'}(x')\,\Omega_{\mu'}(x, x') = 0. \tag{6.68}$$

In the case of Minkowski spacetime, $\Omega \to {}^M\Omega$, where

$$
{}^M\Omega = \frac{1}{2}\eta_{\alpha\beta}\left(x^\alpha - x'^\alpha\right)\left(x^\beta - x'^\beta\right). \tag{6.69}
$$

According to our convention, $\eta_{\alpha\beta}$ is given by $\mathrm{diag}(-1, 1, 1, 1)$; hence,

$$
{}^M\Omega(x, x') = -\frac{1}{2}\left[(t' - t)^2 - |\mathbf{x}' - \mathbf{x}|^2\right]. \tag{6.70}
$$

It is straightforward to see that

$$
{}^M\Omega_{\mu\mu'} = \frac{\partial^2\,{}^M\Omega(x, x')}{\partial x^\mu \partial x'^{\mu'}} = -\eta_{\mu\mu'}, \tag{6.71}
$$

while

$$
{}^M\Omega_{\mu\nu} = \frac{\partial^2\,{}^M\Omega}{\partial x^\mu \partial x^\nu} = \eta_{\mu\nu}, \qquad {}^M\Omega_{\mu'\nu'} = \frac{\partial^2\,{}^M\Omega}{\partial x'^{\mu'} \partial x'^{\nu'}} = \eta_{\mu'\nu'}. \tag{6.72}
$$

Beyond Minkowski spacetime, it is in general not possible to obtain an explicit expression for the world function, except perhaps in the framework of certain approximation schemes (Poisson, Pound and Vega 2011). To illustrate the problem, we consider here the world function in certain simple conformally flat spacetimes (Bini and Mashhoon 2016).

6.3.1 Conformally flat spacetimes

Consider the metric of a conformally flat spacetime of the form

$$
ds^2 = e^{2\mathbb{U}}\eta_{\mu\nu}\,dx^\mu\,dx^\nu, \tag{6.73}
$$

where $\mathbb{U}(x)$ is a scalar under general coordinate transformations. To simplify matters even further, we will assume that $\mathbb{U} = \mathbb{U}(t)$, where $t \geq 0$ in our convention. As is well known, the geodesics of this spacetime can be obtained from the Lagrangian

$$
L_g = -\frac{1}{2}\left(\frac{ds}{d\tau}\right)^2, \tag{6.74}
$$

where τ is initially just an affine parameter but will turn out to be the proper time along a timelike geodesic, once we set $L_g = 1/2$ along the geodesic,

$$
\frac{d^2x^\mu}{d\tau^2} + {}^0\Gamma^\mu_{\alpha\beta}\frac{dx^\alpha}{d\tau}\frac{dx^\beta}{d\tau} = 0. \tag{6.75}
$$

Thus

$$
L_g = \frac{1}{2}e^{2\mathbb{U}}\left(\dot{t}^2 - \delta_{ij}\dot{x}^i\,\dot{x}^j\right), \tag{6.76}
$$

where an overdot indicates differentiation with respect to τ. Hence,

$$
p_0 = \frac{\partial L_g}{\partial \dot{t}} = e^{2\mathbb{U}}\dot{t}, \qquad p_i = \frac{\partial L_g}{\partial \dot{x}^i} = -e^{2\mathbb{U}}\,\delta_{ij}\,\dot{x}^j \tag{6.77}
$$

and the Euler–Lagrange equations are

$$\frac{d}{d\tau}\left(e^{2U}\dot{t}\right) = U_{,t}\,e^{2U}(\dot{t}^2 - \delta_{ij}\dot{x}^i\,\dot{x}^j),\tag{6.78}$$

$$\frac{d}{d\tau}\left(e^{2U}\,\delta_{ij}\,\dot{x}^j\right) = 0.\tag{6.79}$$

Equations (6.78) and (6.79) can be written as

$$\ddot{t} + U_{,t}\,(\dot{t}^2 + \delta_{ij}\dot{x}^i\,\dot{x}^j) = 0,\tag{6.80}$$

$$\delta_{ij}\,\ddot{x}^j + 2U_{,t}\,\dot{t}\,\delta_{ij}\,\dot{x}^j = 0.\tag{6.81}$$

Introducing the Euclidean spatial interval $d\ell$, $d\ell^2 = \delta_{ij}\,dx^i\,dx^j$, we have from the integration of these equations for a *timelike geodesic* that

$$e^{2U}\left[\dot{t}^2 - \left(\frac{d\ell}{d\tau}\right)^2\right] = 1,\tag{6.82}$$

$$\frac{d\ell}{d\tau} = \eta\,e^{-2U},\tag{6.83}$$

where η is a constant of integration.

Assuming that the timelike geodesic starts at P' and moves forward to P as in Fig. 6.1a, we have $\eta \ge 0$ and

$$\tau = \int_{t'}^{t} \frac{e^{2U(\theta)}}{\sqrt{\eta^2 + e^{2U(\theta)}}}\,d\theta,\tag{6.84}$$

$$|\mathbf{x} - \mathbf{x}'| = \eta \int_{t'}^{t} \frac{1}{\sqrt{\eta^2 + e^{2U(\theta)}}}\,d\theta.\tag{6.85}$$

Moreover, it follows from these results that

$$\tau + \eta\,|\mathbf{x} - \mathbf{x}'| = \int_{t'}^{t} \sqrt{\eta^2 + e^{2U(\theta)}}\,d\theta.\tag{6.86}$$

The parameter $\eta : 0 \to \infty$ must be eliminated between eqns (6.84) and (6.85) to give us the world function $\Omega(x, x') = -\tau^2/2$. We note that $\eta = 0$ for no movement in space at all, while $\eta = \infty$ corresponds to null motion. Null geodesics are invariant under a conformal transformation; therefore, we see that for $\eta = \infty$, $|\mathbf{x} - \mathbf{x}'| = t - t'$ and $\tau = 0$, as expected.

In many interesting situations, such as the curved spacetimes examined below, it turns out that $\exp[U(\theta)] = \bar{C}\,\theta^\nu$ for $\nu > 0$ and constant $\bar{C} > 0$. In this case, eqns (6.84) and (6.85) become

$$\tau = \frac{\eta}{\omega_\nu}\left[\mathbb{X}_\nu\left(\omega_\nu t\right) - \mathbb{X}_\nu\left(\omega_\nu t'\right)\right]\tag{6.87}$$

and

$$|\mathbf{x} - \mathbf{x}'| = \frac{1}{\omega_\nu}\left[\mathbb{Y}_\nu\left(\omega_\nu t\right) - \mathbb{Y}_\nu\left(\omega_\nu t'\right)\right],\tag{6.88}$$

where

$$\omega_\nu(\eta) := \left(\frac{\bar{C}}{\eta}\right)^{1/\nu} \tag{6.89}$$

and

$$\mathbb{X}_\nu(x) = \int_0^x \frac{\psi^{2\nu}}{\sqrt{1 + \psi^{2\nu}}} d\psi, \qquad \mathbb{Y}_\nu(x) = \int_0^x \frac{1}{\sqrt{1 + \psi^{2\nu}}} d\psi. \tag{6.90}$$

Formally, we have for $\nu \neq -1$,

$$\mathbb{X}_\nu(x) = \frac{1}{\nu + 1}\left[x\sqrt{1 + x^{2\nu}} - \mathbb{Y}_\nu(x)\right]. \tag{6.91}$$

The $\nu = -1$ case, excluded here, corresponds in fact to the metric of de Sitter spacetime discussed later in this chapter. From the integral representation of the hypergeometric function (Abramowitz and Stegun 1964)

$$F(a, b; \bar{c}; \zeta) = \frac{\Gamma(\bar{c})}{\Gamma(b)\,\Gamma(\bar{c} - b)} \int_0^1 t^{b-1}\,(1 - t)^{\bar{c}-b-1}\,(1 - t\,\zeta)^{-a}\,dt, \tag{6.92}$$

valid for $\mathrm{Re}(\bar{c}) > \mathrm{Re}(b) > 0$, with $t^{1/(2\nu)}\,x = \psi$, we find

$$\mathbb{Y}_\nu(x) = x\,F\left(\frac{1}{2}, \frac{1}{2\nu}; 1 + \frac{1}{2\nu}; -x^{2\nu}\right). \tag{6.93}$$

It is not in general possible to eliminate η between eqns (6.84) and (6.85), or in their new forms (6.87) and (6.88), to obtain an *explicit* expression for the world function. To illustrate this situation, we consider some special cases.

Minkowski spacetime.   Let $\mathbb{U} = 0$, as in Minkowski spacetime. Then,

$$\tau = \frac{t - t'}{\sqrt{\eta^2 + 1}}, \qquad |\mathbf{x} - \mathbf{x}'| = \frac{\eta\,(t - t')}{\sqrt{\eta^2 + 1}}. \tag{6.94}$$

Here $\eta = v/\sqrt{1 - v^2}$, where v is the speed of uniform motion along the straight line from P' to P. Thus, eliminating η, we find, as expected, eqn (6.70) for $^M\Omega$ of the Minkowski spacetime.

de Sitter spacetime.   Let $\exp \mathbb{U} = 1/(\bar{\lambda}\,t)$, where $\bar{\lambda} := \sqrt{^0\Lambda/3}$ and $t \geq 0$, as in the de Sitter spacetime (Ruse 1930). Then, eqns (6.84) and (6.85) imply that for $\bar{u} := \eta\,\bar{\lambda}\,t$ and $\bar{u}' := \eta\,\bar{\lambda}\,t'$,

$$e^{\bar{\lambda}\tau} = \frac{\bar{u}}{\bar{u}'}\,\frac{1 + \sqrt{1 + \bar{u}'^2}}{1 + \sqrt{1 + \bar{u}^2}}, \tag{6.95}$$

$$\eta\,\bar{\lambda}\,|\mathbf{x} - \mathbf{x}'| = \sqrt{1 + \bar{u}^2} - \sqrt{1 + \bar{u}'^2}. \tag{6.96}$$

Eliminating η in this case leads to $\Omega(x, x') = -\tau^2/2$, where

$$\mathbb{Q} = \cosh(\bar{\lambda}\,\tau), \qquad \mathbb{Q} = \frac{t^2 + t'^2 - |\mathbf{x} - \mathbf{x}'|^2}{2t\,t'}. \tag{6.97}$$

Here, $\mathbb{Q}$ satisfies

$$g^{\mu\nu}\,\mathbb{Q}_{,\mu}\,\mathbb{Q}_{,\nu} = \bar{\lambda}^2\,(1 - \mathbb{Q}^2). \tag{6.98}$$

More specifically, we have $\mathbb{Q} \geq 1$,

$$\bar{\lambda}\,\tau = \ln(\mathbb{Q} + \sqrt{\mathbb{Q}^2 - 1}) \tag{6.99}$$

and

$$\Omega = -(2\,\bar{\lambda}^2)^{-1}\,\ln^2(\mathbb{Q} + \sqrt{\mathbb{Q}^2 - 1}). \tag{6.100}$$

Using relations (6.97) and (6.99), it is straightforward to check that over an infinitesimal interval with $t - t' = dt > 0$ and $|\mathbf{x} - \mathbf{x}'| = |d\mathbf{x}|$, we have, as expected,

$$d\tau = (\bar{\lambda}\,t)^{-1}\,\sqrt{dt^2 - d\mathbf{x}^2} \tag{6.101}$$

for the de Sitter spacetime.

FLRW spacetime with stiff equation of state. Next, we consider the case of a spatially flat Friedmann–Lemaître–Robertson–Walker (FLRW) universe with a stiff equation of state (Roberts 1993). Here, $\exp U = \sqrt{\beta_0}\,t$, where $\beta_0 > 0$ is a constant parameter. The density (ρ) and pressure (p) are equal in this universe model and are given by $\rho = p = 3/(32\pi G \beta_0 t^3)$. From eqn (6.85), we find

$$\beta_0\,|\mathbf{x} - \mathbf{x}'| = 2\eta\,(\mathbb{Z} - \mathbb{Z}'), \tag{6.102}$$

where $\mathbb{Z} := \sqrt{\eta^2 + \beta_0\,t}$. Squaring relation (6.102) twice and defining

$$\tau_M^2 := (t - t')^2 - |\mathbf{x} - \mathbf{x}'|^2, \tag{6.103}$$

we find, after some algebra, that

$$\eta^2 = \frac{\beta_0}{4\,\tau_M^2}\,|\mathbf{x} - \mathbf{x}'|^2\,\left(t + t' + \sqrt{4\,t\,t' + |\mathbf{x} - \mathbf{x}'|^2}\right), \tag{6.104}$$

which properly diverges for null motion $(\tau_M = 0)$. Next, eqn (6.84) implies that

$$\frac{3}{2}\,\tau + \eta\,|\mathbf{x} - \mathbf{x}'| = t\,\mathbb{Z} - t'\,\mathbb{Z}'. \tag{6.105}$$

Substituting for η in this equation, we eventually find the world function in this case (Roberts 1993).

Einstein–de Sitter universe. Finally, we consider the case of a spatially flat FLRW universe with $p = 0$ and $\rho = 3/(2\pi G\,b_0^4\,t^6)$, where $b_0 = 1/(3\,\bar{t}_0)$ and $3\,\bar{t}_0$ is the age of the universe in this model. Thus the present energy density of matter ρ_0 is given by $6\pi G \rho_0\,\bar{t}_0^2 = 1$. In this case, $\exp U = b_0^2\,t^2$ and only an implicit form of the world function is possible.

6.4   Nonlocal Gravity (NLG)

In his fruitful approach to GR, Einstein interpreted the experimentally well-established principle of equivalence of inertial and gravitational masses to mean that there is an intimate connection between *inertia* and *gravitation* (Einstein 1950). This notion eventually led to Einstein's extremely *local* principle of equivalence and GR. Following Einstein, we wish to employ the general connection between inertia and gravitation as a guiding principle to render GR (or, equivalently, $GR_{||}$) *nonlocal* in just the same way that accelerated observers in Minkowski spacetime are nonlocal. In field measurements of accelerated observers, the memory of past acceleration appears as an integral over the past that is linear in the field. To implement the same idea in the theory of gravitation, we note that Einstein's field equation, represented by eqn (6.49) in our tetrad framework, has the same general form as in electromagnetism, and eqn (6.30) is a *local* constitutive relation, where the auxiliary torsion tensor $\mathfrak{C}_{\mu\nu\rho}$ is linearly related to the torsion tensor $C_{\mu\nu\rho}$, cf. eqn (6.28). In the electrodynamics of media, the constitutive relation between $H_{\mu\nu} \mapsto (\mathbf{D}, \mathbf{H})$ and $F_{\mu\nu} \mapsto (\mathbf{E}, \mathbf{B})$ could be nonlocal (Jackson 1999; Landau and Lifshitz 1960). Therefore, in the nonlocal electrodynamics of media, Maxwell's original equations remain unchanged, but the constitutive relation now involves the past history of the electromagnetic field. We wish to construct here a nonlocal theory of gravitation in analogy with the nonlocal electrodynamics of media. To render observers nonlocal in a gravitational field in the same sense as in nonlocal special relativity, we simply change eqn (6.30) to

$$^{NLG}\mathcal{H}_{\mu\nu\rho} := \frac{\sqrt{-g}}{\kappa}\left(\mathfrak{C}_{\mu\nu\rho} + N_{\mu\nu\rho}\right), \tag{6.106}$$

where $N_{\mu\nu\rho} = -N_{\nu\mu\rho}$ is a tensor involving the past history of the gravitational field. We emphasize that in order to preserve the invariance of the theory under arbitrary coordinate transformations, $N_{\mu\nu\rho}$ and hence the resulting *nonlocal* auxiliary field strength $^{NLG}\mathcal{H}_{\mu\nu\rho}$ should be antisymmetric in their first two indices. The simplest expression for the nonlocality tensor $N_{\mu\nu\rho}$ would involve a *scalar* kernel; that is,

$$N_{\mu\nu\rho} = -\int \Omega_{\mu\mu'}\Omega_{\nu\nu'}\Omega_{\rho\rho'}\,\mathcal{K}(x,x')\,X^{\mu'\nu'\rho'}(x')\sqrt{-g(x')}\,d^4x', \tag{6.107}$$

where $\mathcal{K}$ is the scalar *causal* kernel of the nonlocal theory (Hehl and Mashhoon 2009a, 2009b) and $X_{\mu\nu\rho}(x)$ is a tensor that is antisymmetric in its first two indices and involves a linear combination of the components of the torsion tensor. We note that there is no physical connection between kernel $\mathcal{K}$ and the nonlocal kernel of accelerated observers in Minkowski spacetime due to the extreme locality of Einstein's principle of equivalence. In eqn (6.107), $\Omega(x, x')$ is Synge's *world function* (Synge 1971), which involves a unique future directed timelike or null geodesic of $g_{\mu\nu}$ that connects event x' to event x and the square of its proper length is $2\,\Omega$. It is important to emphasize that in eqn (6.107) the integrand as a function of x' is a scalar invariant, which can therefore be integrated over the curved spacetime manifold, while the result is properly a tensor in x. Event x' is in the past of event x; otherwise, $\mathcal{K} = 0$. In other words, eqn (6.107) represents a certain average over the past history of the gravitational field.

Let us observe here that the constitutive ansatz (6.106) involves a *linear* nonlocal relation between the two field strengths involving $^{NLG}\mathcal{H}_{\mu\nu\rho}$ and $C_{\mu\nu\rho}$ via $\mathfrak{C}_{\mu\nu\rho}$; however, as in electrodynamics, such a nonlocal relation could well become nonlinear when the field strengths are sufficiently high. We will not have occasion here to discuss such nonlinearities since at this early stage in the development of nonlocal gravity (NLG) the relation between $X_{\mu\nu\rho}$ and torsion is assumed to be linear for the sake of simplicity.

In electrodynamics, the local constitutive relation between $H_{\mu\nu}$ and $F_{\mu\nu}$, considered as 6-vectors, can be described via a 6×6 matrix. One can similarly envision the local linear relationship between $X_{\mu\nu\rho} = -X_{\nu\mu\rho}$ and $\mathfrak{C}_{\mu\nu\rho}$ in eqn (6.107) in a rather general context, namely,

$$X_{\mu\nu\rho} = \chi_{\mu\nu\rho}{}^{\alpha\beta\gamma} \mathfrak{C}_{\alpha\beta\gamma}. \tag{6.108}$$

This relation is reminiscent of the local constitutive relation between $H_{\mu\nu}$ and $F_{\mu\nu}$ in electrodynamics (Hehl and Obukhov 2003). Various forms of eqn (6.108) have been explored in Mashhoon (2014) and the relation that has been adopted for NLG is

$$X_{\mu\nu\rho} = \mathfrak{C}_{\mu\nu\rho} + \check{p}\left(\check{C}_\mu\, g_{\nu\rho} - \check{C}_\nu\, g_{\mu\rho}\right). \tag{6.109}$$

Here, $\check{p} \neq 0$ is a constant dimensionless parameter and $\check{C}^\mu$ is the torsion pseudovector given by

$$\check{C}_\mu = \frac{1}{3!} C^{\alpha\beta\gamma}\, E_{\alpha\beta\gamma\mu}, \qquad \check{C}_\alpha = \frac{1}{3} E_{\alpha\beta\gamma\delta}\, \mathfrak{C}^{\beta\gamma\delta}, \tag{6.110}$$

where $E_{\alpha\beta\gamma\delta}$ is the Levi-Civita tensor.

The constitutive kernel $\mathcal{K}(x,x')$ could in general depend upon scalars at x and x' that can be formed from the gravitational potentials, the world function $\Omega(x,x')$ and their derivatives. For instance, we can tentatively assume that $\mathcal{K}(x,x')$ is simply a function of spacetime scalars such as

$$\Omega_\mu(x,x')\, e^{\mu}{}_{\hat{a}}(x), \qquad \Omega_{\mu'}(x,x')\, e^{\mu'}{}_{\hat{a}}(x'), \tag{6.111}$$

where the Lorentz freedom in the choice of the fundamental frame has been fixed relative to the rest frame of the gravitational source, as discussed in detail in the following chapter, where the consequences of this form for $\mathcal{K}(x,x')$ are worked out in detail within the framework of the linearized theory of nonlocal gravity.

It is not known at present whether the field equation of nonlocal gravity can be derived from an action principle. There are in general problems with action principles for nonlocal theories if the kernel is not symmetric—and causal kernels cannot be symmetric in time; in fact, this issue has been discussed in detail in Hehl and Mashhoon (2009b). For instance, nonlocality can arise from integrating out certain physical degrees of freedom (Galley 2013). Other than the implicit connection between inertia and gravitation, elucidated by Einstein, that has led us from acceleration-induced nonlocality in Minkowski spacetime to a nonlocal generalization of GR, the explicit physical origin of nonlocality in the case of the gravitational field is thus far unknown. Therefore, the theory is incomplete without a thorough examination of the physical origin of the nonlocal kernel $\mathcal{K}$ and constant parameter $\check{p}$. The constitutive kernel in the classical nonlocal electrodynamics of media is ultimately obtained from the underlying atomic and molecular physics of the material medium involved (Jackson 1999).

The corresponding underlying physics of the constitutive kernel $\mathcal{K}$ and parameter $\check{p}$ for gravity is not known at present; therefore, as discussed in detail in the next chapter, we take the view that $\mathcal{K}$ and $\check{p}$ can be determined instead from the comparison of nonlocal gravity with experiment. Perhaps someday $\mathcal{K}$ and $\check{p}$ will be ascertained from a more complete future theory.

6.5   Nonlocal GR$_{||}$

The field equation of nonlocal gravity is obtained from eqns (6.49) and (6.44) by substituting $^{NLG}\mathcal{H}_{\mu\nu\rho}$ for $\mathcal{H}_{\mu\nu\rho}$. In close analogy with the nonlocal electrodynamics of media, the main field equation of the teleparallel equivalent of general relativity is maintained, that is,

$$\frac{\partial}{\partial x^\beta}\,^{NLG}\mathcal{H}^{\alpha\beta}{}_{\hat{\gamma}} = \sqrt{-g}\,e^\mu{}_{\hat{\gamma}}\,g^{\alpha\nu}\left(T_{\mu\nu} + {}^{NLG}E_{\mu\nu} - \frac{{}^0\Lambda}{\kappa}g_{\mu\nu}\right), \tag{6.112}$$

where $^{NLG}E_{\mu\nu}$ is given by

$$\sqrt{-g}\,^{NLG}E_{\mu\nu} = C_{\mu\rho\sigma}\,^{NLG}\mathcal{H}_\nu{}^{\rho\sigma} - \frac{1}{4}g_{\mu\nu}\,C_{\alpha\beta\gamma}\,^{NLG}\mathcal{H}^{\alpha\beta\gamma}, \tag{6.113}$$

except that the constitutive relation is now $^{NLG}\mathcal{H}_{\mu\nu\rho} := (\sqrt{-g}/\kappa)\,(\mathfrak{C}_{\mu\nu\rho} + N_{\mu\nu\rho})$, where $N_{\mu\nu\rho}$ represents the past history of the gravitational field, namely,

$$\begin{aligned}^{NLG}\mathcal{H}_{\mu\nu\rho} = \frac{\sqrt{-g}}{\kappa}\Big[\mathfrak{C}_{\mu\nu\rho} \\ - \int \Omega_{\mu\mu'}\Omega_{\nu\nu'}\Omega_{\rho\rho'}\,\mathcal{K}(x,x')\,X^{\mu'\nu'\rho'}(x')\sqrt{-g(x')}\,d^4x'\Big].\end{aligned} \tag{6.114}$$

Here,

$$X_{\mu\nu\rho} = \mathfrak{C}_{\mu\nu\rho} + \check{p}\,(\check{C}_\mu\,g_{\nu\rho} - \check{C}_\nu\,g_{\mu\rho}) \tag{6.115}$$

and $\check{p}$ is a non-zero constant; see eqns (6.107) and (6.109).

Let us define

$$\mathcal{E}_{\mu\nu} := {}^{NLG}E_{\mu\nu}; \tag{6.116}$$

then, the field equation of nonlocal gravity can be expressed as

$$\frac{\partial}{\partial x^\beta}\left[\frac{\sqrt{-g}}{\kappa}\left(\mathfrak{C}^{\alpha\beta}{}_{\hat{\gamma}} + N^{\alpha\beta}{}_{\hat{\gamma}}\right)\right] = \sqrt{-g}\,e^\mu{}_{\hat{\gamma}}\,g^{\alpha\nu}\left(T_{\mu\nu} + \mathcal{E}_{\mu\nu} - \frac{{}^0\Lambda}{\kappa}g_{\mu\nu}\right), \tag{6.117}$$

where

$$\kappa\,\mathcal{E}_{\mu\nu} = C_{\mu\rho\sigma}(\mathfrak{C}_\nu{}^{\rho\sigma} + N_\nu{}^{\rho\sigma}) - \frac{1}{4}g_{\mu\nu}\,C_{\alpha\beta\gamma}(\mathfrak{C}^{\alpha\beta\gamma} + N^{\alpha\beta\gamma}). \tag{6.118}$$

The main new element here is of course the nonlocality tensor $N_{\mu\nu\rho}$, which is an average over the past history of the gravitational field. As mentioned before, the law of conservation of total energy–momentum tensor takes the form

$$\frac{\partial}{\partial x^\mu}\left[\sqrt{-g}\,(T_{\hat{\alpha}}{}^\mu + \mathcal{E}_{\hat{\alpha}}{}^\mu - \frac{{}^0\Lambda}{\kappa}e^\mu{}_{\hat{\alpha}})\right] = 0. \tag{6.119}$$

6.6 Nonlocal GR

It is possible to express the nonlocal gravitational field equation as modified Einstein's equation. To this end, we separate out in eqn (6.117) the partial derivative term involving $(\sqrt{-g}/\kappa)\,\mathfrak{C}^{\mu\nu}{}_{\hat{a}}$ and insert it into the expression (6.45) for the Einstein tensor ${}^{0}G_{\mu\nu}$ to get the *nonlocal generalization of Einstein's field equation*, namely,

$$
{}^{0}G_{\mu\nu} + \mathcal{N}_{\mu\nu} = \kappa\,T_{\mu\nu} - {}^{0}\Lambda\,g_{\mu\nu} + Q_{\mu\nu}. \tag{6.120}
$$

Here, $\mathcal{N}_{\mu\nu}$, defined by

$$
\mathcal{N}_{\mu\nu} := g_{\nu\alpha}\,e_{\mu}{}^{\hat{\gamma}}\,\frac{1}{\sqrt{-g}}\,\frac{\partial}{\partial x^{\beta}}\left(\sqrt{-g}\,N^{\alpha\beta}{}_{\hat{\gamma}}\right), \tag{6.121}
$$

is indeed a proper tensor, since $N_{\alpha\beta\gamma} = -N_{\beta\alpha\gamma}$ by assumption; moreover, the quantity $Q_{\mu\nu} := \kappa\,(\mathcal{E}_{\mu\nu} - E_{\mu\nu})$ is a traceless tensor given by

$$
Q_{\mu\nu} := C_{\mu\rho\sigma}N_{\nu}{}^{\rho\sigma} - \frac{1}{4}\,g_{\mu\nu}\,C_{\delta\rho\sigma}N^{\delta\rho\sigma}. \tag{6.122}
$$

It is clear that Einstein's gravitational field equation is recovered when the nonlocal kernel vanishes, $\mathcal{K} = 0$, and hence $N_{\mu\nu\rho} = 0$. In GR, the ten components of the metric tensor $g_{\mu\nu}$ can be determined, in principle, from the ten gravitational field equations. Here, however, the sixteen components of the fundamental observers' frame field $e^{\mu}{}_{\hat{a}}$ can be obtained, in principle, from the sixteen gravitational field eqns (6.120)–(6.122) of nonlocal general relativity. In other words, nonlocality removes the essential degeneracy of $\mathrm{GR}_{\parallel}$; moreover, as expected, nonlocal gravity is invariant under the *global* Lorentz group. The integro-differential field equations of nonlocal gravity in general contain Fredholm integral relations that turn into Volterra integral relations whenever causal kernels are involved (Lovitt 1950; Tricomi 1957).

To compare and contrast further the field equation of nonlocal gravity with the Einstein field equation of GR, one can separate out eqn (6.120) into its symmetric and antisymmetric components. In this way, we get the ten nonlocally modified Einstein equations given by

$$
{}^{0}G_{\mu\nu} + \mathcal{N}_{(\mu\nu)} = \kappa\,T_{\mu\nu} - {}^{0}\Lambda\,g_{\mu\nu} + Q_{(\mu\nu)} \tag{6.123}
$$

as well as the six integral constraint equations involving the nonlocality tensor $N_{\mu\nu\rho}$, namely,

$$
\mathcal{N}_{[\mu\nu]} = Q_{[\mu\nu]} = \frac{1}{2}\left(C_{\mu\rho\sigma}N_{\nu}{}^{\rho\sigma} - C_{\nu\rho\sigma}N_{\mu}{}^{\rho\sigma}\right), \tag{6.124}
$$

that are dominated by averaging over past events and vanish for $\mathcal{K} = 0$. The energy–momentum tensor is symmetric; therefore, there is no contribution from $T_{[\mu\nu]} = 0$ to eqn (6.124). Let us recall here that these sixteen field equations are required to determine the sixteen components of $e^{\mu}{}_{\hat{a}}(x)$, of which ten are fixed by the spacetime metric $g_{\mu\nu}$ via orthonormality and the other six are Lorentz degrees of freedom (i.e. boosts and rotations). This division is reflected in eqns (6.123) and (6.124), respectively. The general mathematical investigation of the existence and uniqueness of the solutions of the partial integro-differential eqn (6.123) with integral constraints (6.124) is beyond the scope of the present book.

Nonlocality—in the sense of an influence ("memory") from the past that endures—could be a natural feature of the universal gravitational interaction. Some of the consequences of our nonlocal gravity model in the linear weak-field regime are considered in the rest of this book. This involves detailed studies of the nonlocal modifications of Newtonian gravity and linearized gravitational waves. As explained in the following chapter, the notion that nonlocal gravity (NLG) can simulate dark matter is completely consistent with causality; moreover, the theoretical results appear to be consistent with experiment at the linear level. The nonlinear regime of NLG has not yet been studied; therefore, exact cosmological models or issues involving the influence of nonlocality on the formation and evolution of black holes are beyond the scope of our present considerations. Indeed, the investigation of the nonlinear regime of NLG remains a task for the future.

6.7 Effective Dark Matter

Let us now return to eqn (6.120) and define $\mathfrak{T}_{\mu\nu}$ in terms of the nonlocal parts of the field equation of NLG, namely,

$$\mathfrak{T}_{\mu\nu} = \kappa^{-1}\left(Q_{\mu\nu} - \mathcal{N}_{\mu\nu}\right), \tag{6.125}$$

so that eqn (6.120) can now be written as

$$^{0}G_{\mu\nu} + {}^{0}\Lambda\, g_{\mu\nu} = \kappa\left(T_{\mu\nu} + \mathfrak{T}_{\mu\nu}\right). \tag{6.126}$$

Here $\mathfrak{T}_{(\mu\nu)}$ has the interpretation of the symmetric energy–momentum tensor of the *effective dark matter*, while

$$\mathfrak{T}_{[\mu\nu]} = 0 \tag{6.127}$$

are the six constraint equations that are necessary in order to determine the sixteen components of the tetrad frame field of the fundamental observers; that is, the field equation of NLG consists of the ten nonlocally modified Einstein equations

$$^{0}G_{\mu\nu} + {}^{0}\Lambda\, g_{\mu\nu} = \kappa\left[T_{\mu\nu} + \mathfrak{T}_{(\mu\nu)}\right] \tag{6.128}$$

together with the six constraint equations (6.127). Furthermore, it follows from the reduced Bianchi identity, $^{0}\nabla_{\nu}\, {}^{0}G^{\mu\nu} = 0$, that the *total* matter energy–momentum tensor is conserved, namely,

$$^{0}\nabla_{\nu}\left[T^{\mu\nu} + \mathfrak{T}^{(\mu\nu)}\right] = 0. \tag{6.129}$$

Could the effective dark matter of nonlocal gravity correspond to *dark matter*? Is it possible that what appears in astrophysics and cosmology as dark matter may in fact turn out to be the nonlocal aspect of the gravitational interaction? This important issue will be discussed at length in the remaining chapters of this book.

It is interesting to investigate gravitational systems that may consist entirely of effective dark matter with ($T_{\mu\nu} = 0$). Such systems would be expected to lack proper

Newtonian limits and hence could exist only in highly relativistic situations. Furthermore, there could be systems for which the nonlocal contribution to the gravitational field equation vanishes. In other words, the fundamental observers' tetrad frames in eqn (6.126) could be such that the corresponding metric tensor $g_{\mu\nu}$ satisfies this equation with $\mathfrak{T}_{\mu\nu} = 0$. In this case, $\mathcal{N}_{\mu\nu} = Q_{\mu\nu}$. An immediate consequence of this equality is that $\mathcal{N}_{\mu\nu}$ must be traceless, since $Q_{\mu\nu}$ is traceless. It follows from eqn (6.121) and $g^{\mu\nu} \mathcal{N}_{\mu\nu} = 0$ that

$$e_\alpha{}^{\hat{\gamma}} \frac{\partial}{\partial x^\beta} \left(\sqrt{-g}\, N^{\alpha\beta}{}_{\hat{\gamma}} \right) = 0. \tag{6.130}$$

These special gravitational systems require further investigation.

To find a solution of NLG, we must ultimately determine the sixteen components of the tetrad frame field $e^\mu{}_{\hat{\alpha}}$ of the fundamental observers. Of these spacetime functions, ten would then specify the metric tensor in accordance with the orthonormality condition. Let us note that with $e^\mu{}_{\hat{\alpha}} = \delta^\mu_{\hat{\alpha}}$, $g_{\mu\nu} = \eta_{\mu\nu}$, ${}^0\Lambda = 0$ and $T_{\mu\nu} = 0$, we find that the tetrad frame field of the fundamental inertial observers in Minkowski spacetime is an exact solution of NLG. *No other exact solution of NLG is known*, because of the complicated structure of nonlocal gravity theory. Thus far, the main observational consequences of the NLG theory *linearized about the fundamental inertial observers in Minkowski spacetime* have been investigated; see Chapter 7. In these studies, it is sufficient to employ the world function of Minkowski spacetime, which enormously simplifies the task of finding solutions of the theory. Otherwise, the world function for the timelike geodesics of curved spacetime would be required in the nonlocal ansatz (6.107). At the present stage of the development of NLG, the causal scalar kernel $\mathcal{K}(x, x')$ should in any case be ultimately determined from the comparison of the theory with observation.

7
Linearized Nonlocal Gravity

The fundamental observers' frame field in source-free Minkowski spacetime is the only known exact solution of the field equation of nonlocal gravity (NLG) at present. In the absence of an exact solution involving a non-zero gravitational field, we look for an approximate solution of NLG that is a first-order perturbation about Minkowski spacetime. The purpose of this chapter is to develop and study the general linear weak-field approximation of NLG beyond Minkowski spacetime (Mashhoon 2014; Chicone and Mashhoon 2016a).

7.1 Linear Approximation of Nonlocal Gravity

Imagine a finite source of mass energy in a compact region of space. We suppose that the gravitational field is everywhere weak and falls off to zero far away from the source. We also set the cosmological constant equal to zero, $^0\Lambda = 0$, and assume that in the absence of gravity, we are in the "rest" frame of the source in Minkowski spacetime with the fundamental tetrad frame $e^\mu{}_{\hat\alpha} = \delta^\mu_\alpha$. In the presence of gravity, the fundamental frame field of nonlocal gravity is then assumed to be

$$e_\mu{}^{\hat\alpha} = \delta^\alpha_\mu + \psi^\alpha{}_\mu, \quad e^\mu{}_{\hat\alpha} = \delta^\mu_\alpha - \psi^\mu{}_\alpha, \tag{7.1}$$

where $\psi_{\mu\nu}$ is treated to linear order in perturbation away from Minkowski spacetime and hence the distinction between spacetime and tetrad indices disappears at this level of approximation. Let us note that in eqn (7.1), the invariance of the theory under global Lorentz transformations has been broken, since the unperturbed fundamental frame field coincides with the "rest" frame of the gravitational source. It is useful to decompose $\psi_{\mu\nu}$ into its symmetric and antisymmetric components; that is, we define,

$$h_{\mu\nu} := 2\psi_{(\mu\nu)}, \quad \phi_{\mu\nu} := 2\psi_{[\mu\nu]}. \tag{7.2}$$

It then follows from orthonormality, $g_{\mu\nu}(x) = e_\mu{}^{\hat\alpha} e_\nu{}^{\hat\beta} \eta_{\hat\alpha\hat\beta}$, that

$$g_{\mu\nu} = \eta_{\mu\nu} + h_{\mu\nu}. \tag{7.3}$$

Moreover, it is convenient to employ the trace-reversed potentials

$$\bar{h}_{\mu\nu} = h_{\mu\nu} - \frac{1}{2}\eta_{\mu\nu}h, \quad h := \eta_{\mu\nu}h^{\mu\nu}, \tag{7.4}$$

just as in GR. Here $\bar{h} = -h$ and we have

Nonlocal Gravity. Bahram Mashhoon. © Bahram Mashhoon 2017. Published 2017 by Oxford University Press.

$$\psi_{\mu\nu} = \frac{1}{2}\bar{h}_{\mu\nu} + \frac{1}{2}\phi_{\mu\nu} - \frac{1}{4}\eta_{\mu\nu}\bar{h}. \tag{7.5}$$

It is now straightforward to work out the field components in terms of $\psi_{\mu\nu}$. The torsion tensor is then,

$$C_{\mu\nu\sigma} = \partial_\mu \psi_{\sigma\nu} - \partial_\nu \psi_{\sigma\mu} \tag{7.6}$$

and the auxiliary torsion tensor is given by

$$\mathfrak{C}_{\mu\sigma\nu} = -\bar{h}_{\nu[\mu,\sigma]} - \eta_{\nu[\mu}\bar{h}_{\sigma]\rho,}{}^{\rho} + \frac{1}{2}\phi_{\mu\sigma,\nu} + \eta_{\nu[\mu}\phi_{\sigma]\rho,}{}^{\rho}, \tag{7.7}$$

in terms of which the Einstein tensor can be expressed as

$$^{0}G_{\mu\nu} = \partial_\sigma \mathfrak{C}_{\mu}{}^{\sigma}{}_{\nu} = -\frac{1}{2}\Box\bar{h}_{\mu\nu} + \bar{h}^{\rho}{}_{(\mu,\nu)\rho} - \frac{1}{2}\eta_{\mu\nu}\bar{h}^{\rho\sigma}{}_{,\rho\sigma}, \tag{7.8}$$

where $\Box := \eta^{\alpha\beta}\partial_\alpha\partial_\beta$. Moreover, the torsion vector and pseudovector are given by

$$C_\mu = \frac{1}{4}\partial_\mu\bar{h} + \frac{1}{2}\partial_\nu(\bar{h}^\nu{}_\mu + \phi^\nu{}_\mu), \qquad \check{C}^\mu = \frac{1}{6}\epsilon^{\mu\nu\rho\sigma}\phi_{\nu\rho,\sigma}. \tag{7.9}$$

In the linear regime, eqn (6.107) reduces to

$$N_\mu{}^\sigma{}_\nu = \int \mathcal{K}(x,y)X_\mu{}^\sigma{}_\nu(y)\,d^4y. \tag{7.10}$$

Moreover, in the nonlocal generalization of Einstein's field equation in eqn (6.120), $Q_{\mu\nu}$ is of second order in torsion and can therefore be neglected; see eqn (6.122). Thus the linearized forms of the field eqns (6.123) and (6.124) of nonlocal gravity are given by

$$^{0}G_{\mu\nu} + \frac{1}{2}\partial_\sigma(N_\mu{}^\sigma{}_\nu + N_\nu{}^\sigma{}_\mu) = \kappa T_{\mu\nu} \tag{7.11}$$

and

$$\partial_\sigma N_\mu{}^\sigma{}_\nu = \partial_\sigma N_\nu{}^\sigma{}_\mu, \tag{7.12}$$

respectively. It follows immediately from the antisymmetry of the auxiliary torsion tensor in its first two indices in eqn (7.8) and the symmetry of Einstein's tensor that $\partial_\nu {}^{0}G^{\mu\nu} = 0$, as expected. Furthermore, eqns (7.11)–(7.12) imply that

$$\partial_\nu T^{\mu\nu} = 0, \tag{7.13}$$

since $N^{\mu\sigma\nu} = -N^{\sigma\mu\nu}$. We thus recover the energy–momentum conservation law for mass–energy, just as in linearized GR.

Let us next discuss the gauge freedom of the gravitational potentials. An infinitesimal coordinate transformation, $x^\mu \mapsto x'^\mu = x^\mu - \epsilon^\mu(x)$, leads to $\psi_{\mu\nu} \mapsto \psi'_{\mu\nu} = \psi_{\mu\nu} + \epsilon_{\mu,\nu}$ that is valid to linear order in ϵ^μ. Thus under a gauge transformation,

$$\bar{h}'_{\mu\nu} = \bar{h}_{\mu\nu} + \epsilon_{\mu,\nu} + \epsilon_{\nu,\mu} - \eta_{\mu\nu}\epsilon^\alpha{}_{,\alpha}, \qquad \phi'_{\mu\nu} = \phi_{\mu\nu} + \epsilon_{\mu,\nu} - \epsilon_{\nu,\mu} \tag{7.14}$$

and $\bar{h}' = \bar{h} - 2\epsilon^\alpha{}_{,\alpha}$; however, as expected, the gravitational field tensors $C_{\mu\nu\sigma}$ and $\mathfrak{C}_{\mu\sigma\nu}$ are left unchanged. It follows that the linearized gravitational field equation of NLG is gauge-invariant.

To proceed further, we must discuss the nature of the nonlocal kernel in the linearized theory. The kernel that appears in eqn (7.10) is the nonlocal kernel in the Minkowski spacetime limit. In Minkowski spacetime, the world function is given by

$$^M\Omega(x, x') = \frac{1}{2}\eta_{\alpha\beta}(x^\alpha - x'^\alpha)(x^\beta - x'^\beta); \tag{7.15}$$

see eqn (6.69). Hence, to lowest order in the perturbation, we find

$$^M\Omega_\mu(x, x')\, e^\mu{}_{\hat{\alpha}}(x) = -^M\Omega_{\mu'}(x, x')\, e^{\mu'}{}_{\hat{\alpha}}(x') = \eta_{\alpha\beta}(x^\beta - x'^\beta). \tag{7.16}$$

It follows from this result and our brief discussion of the kernel in the previous chapter, cf. eqn (6.111), that we have a *convolution* kernel in the linearized theory; that is, we can tentatively assume that the nonlocal kernel $\mathcal{K}(x, y)$ is a *universal* function of $x^\alpha - y^\alpha$, so that

$$\mathcal{K}(x, y) := K(x - y). \tag{7.17}$$

Moreover, to ensure causality, we assume that the convolution kernel K is non-zero only when $x^\alpha - y^\alpha$ is a future-directed timelike or null vector in Minkowski spacetime, which means that event y must be within or on the past light cone of event x, or equivalently, that event x must be within or on the future light cone of event y. In other words, $x^0 \geq y^0$ and

$$\eta_{\alpha\beta}(x^\alpha - y^\alpha)(x^\beta - y^\beta) \leq 0. \tag{7.18}$$

It follows that causality is ensured whenever

$$x^0 - y^0 \geq |\mathbf{x} - \mathbf{y}|. \tag{7.19}$$

Hence, $K(x-y)$ must be proportional to $u(x^0-y^0-|\mathbf{x}-\mathbf{y}|)$, where $u(t)$ is the Heaviside unit step function such that $u(t) = 0$ for $t < 0$ and $u(t) = 1$ for $t \geq 0$; that is,

$$K(x - y) \propto u(x^0 - y^0 - |\mathbf{x} - \mathbf{y}|). \tag{7.20}$$

Returning to field eqns (7.11) and (7.12), let us now write them more explicitly as follows

$$^0G_{\mu\nu}(x) + \partial_\sigma \int K(x - y)\, X_{(\mu}{}^\sigma{}_{\nu)}(y)\, d^4y = \kappa\, T_{\mu\nu}(x) \tag{7.21}$$

and

$$\partial_\sigma \int K(x - y)\, X_{[\mu}{}^\sigma{}_{\nu]}(y)\, d^4y = 0. \tag{7.22}$$

The consequences of these equations for various choices of $X_{\mu\sigma\nu}$ are briefly discussed in Appendix A. As in Chapter 6, we choose eqn (6.109), namely, $X_{\mu\sigma\nu} = \mathfrak{C}_{\mu\sigma\nu} + \check{p}(\check{C}_\mu\, g_{\sigma\nu} - \check{C}_\sigma\, g_{\mu\nu})$ with $\check{p} \neq 0$. Then, in the *linear* regime we have

$$X_{(\mu}{}^\sigma{}_{\nu)} = \mathfrak{C}_{(\mu}{}^\sigma{}_{\nu)} + \check{p}\left[\check{C}_{(\mu}\delta^\sigma_{\nu)} - \check{C}^\sigma\eta_{\mu\nu}\right], \qquad X_{[\mu}{}^\sigma{}_{\nu]} = \mathfrak{C}_{[\mu}{}^\sigma{}_{\nu]} + \check{p}\check{C}_{[\mu}\delta^\sigma_{\nu]}. \tag{7.23}$$

Let us recall here the fact that the torsion pseudovector $\check{C}^\sigma$ is the dual of $C_{[\mu\nu\rho]}$, which in the linear approximation is given by $C_{[\mu\nu\rho]} = -\phi_{[\mu\nu,\rho]}$. Moreover, in the linear approximation, $\check{C}^\sigma{}_{,\sigma} = 0$. Thus the part of the constitutive relation proportional to $\check{p}$ is given exclusively by the derivatives of the antisymmetric tetrad potentials $\phi_{\mu\nu}$ and vanishes for $\phi_{\mu\nu} = 0$.

We are interested in the general case of time-dependent gravitational fields. The implications of the gravitational field equation for steady-state configurations are considered in Section 7.6. The present section is devoted to time-varying situations. In the calculation of the nonlocal terms in eqns (7.21) and (7.22), $\partial K/\partial x^\sigma = -\partial K/\partial y^\sigma$ and

$$X_\mu{}^\sigma{}_\nu = \mathfrak{C}_\mu{}^\sigma{}_\nu + \check{p}\,(\check{C}_\mu\,\delta_\nu^\sigma - \check{C}^\sigma\,\eta_{\mu\nu}). \tag{7.24}$$

It then follows via integration by parts that

$$\partial_\sigma \int K(x-y)\,\mathfrak{C}_\mu{}^\sigma{}_\nu(y)\,d^4y = -S_{\mu\nu} + \int K(x-y)\,{}^0G_{\mu\nu}(y)\,d^4y, \tag{7.25}$$

where we have used eqn (7.8) and $S_{\mu\nu}$ is given by

$$S_{\mu\nu} := \int \frac{\partial}{\partial y^\sigma}\Big[K(x-y)\mathfrak{C}_\mu{}^\sigma{}_\nu(y)\Big]\,d^4y. \tag{7.26}$$

Gauss's theorem then implies that

$$S_{\mu\nu} = \oint K(x-y)\mathfrak{C}_\mu{}^\alpha{}_\nu(y)\,d^3\Sigma_\alpha(y), \tag{7.27}$$

where the only contribution to the integral comes from the boundary hypersurface at the light cone given by $y^0 = x^0 - |\mathbf{x} - \mathbf{y}|$. Therefore,

$$S_{\mu\nu}(x) = \int K(|\mathbf{x}-\mathbf{y}|, \mathbf{x}-\mathbf{y})\,\mathfrak{C}_\mu{}^0{}_\nu(x^0 - |\mathbf{x}-\mathbf{y}|, \mathbf{y})\,d^3y, \tag{7.28}$$

where $\mathfrak{C}_\mu{}^0{}_\nu = \mathfrak{C}_{(\mu}{}^0{}_{\nu)} + \mathfrak{C}_{[\mu}{}^0{}_{\nu]}$ is given by eqn (7.7), namely,

$$\mathfrak{C}_{(\mu}{}^0{}_{\nu)} = \frac{1}{2}\left(\bar{h}_{\mu\nu,0} - \bar{h}_{0(\mu,\nu)} + \eta_{\mu\nu}\,\bar{h}_{0\rho}{}^{,\rho} - \eta_{0(\mu}\,\bar{h}_{\nu)\rho}{}^{,\rho} + \phi_{0(\mu,\nu)} - \eta_{\mu\nu}\,\phi_{0\rho}{}^{,\rho} + \eta_{0(\mu}\,\phi_{\nu)\rho}{}^{,\rho}\right) \tag{7.29}$$

and

$$\mathfrak{C}_{[\mu}{}^0{}_{\nu]} = \frac{1}{2}\left(\bar{h}_{0[\mu,\nu]} + \phi_{0[\mu,\nu]} + \eta_{0[\mu}\,\bar{h}_{\nu]\rho}{}^{,\rho} - \eta_{0[\mu}\,\phi_{\nu]\rho}{}^{,\rho}\right). \tag{7.30}$$

In a similar way, we find that

$$U_{\mu\nu} := \partial_\sigma \int K(x-y)\,(\check{C}_\mu\,\delta_\nu^\sigma - \check{C}^\sigma\,\eta_{\mu\nu})(y)\,d^4y \tag{7.31}$$

can be written as

$$U_{\mu\nu} = -\int K(|\mathbf{x}-\mathbf{y}|, \mathbf{x}-\mathbf{y})\,(\check{C}_\mu\,\delta_\nu^0 - \check{C}^0\,\eta_{\mu\nu})(x^0 - |\mathbf{x}-\mathbf{y}|, \mathbf{y})\,d^3y$$
$$+ \int K(x-y)\check{C}_{\mu,\nu}(y)\,d^4y. \tag{7.32}$$

We recall here that $U_{\mu\nu}$ depends only upon the derivatives of $\phi_{\mu\nu}$ and vanishes for $\phi_{\mu\nu} = 0$.

It follows from these results that in the linear regime with $^0\Lambda = 0$, eqn (6.120), which is the nonlocal extension of Einstein's field equation, can be written as

$$^0G_{\mu\nu}(x) + \int K(x-y)\,^0G_{\mu\nu}(y)\,d^4y = \kappa T_{\mu\nu}(x) + S_{\mu\nu}(x) - \check{p}\,U_{\mu\nu}(x). \tag{7.33}$$

This is the main field equation of linearized nonlocal gravity and can be split into its symmetric and antisymmetric components, namely,

$$^0G_{\mu\nu}(x) + \int K(x-y)\,^0G_{\mu\nu}(y)\,d^4y = \kappa T_{\mu\nu}(x) + S_{(\mu\nu)}(x) - \check{p}\,U_{(\mu\nu)}(x) \tag{7.34}$$

and

$$S_{[\mu\nu]}(x) = \check{p}\,U_{[\mu\nu]}(x). \tag{7.35}$$

Let us first note here that $S_{0\nu}(x) = 0$ due to the antisymmetry of $\mathfrak{C}_{\mu\sigma\nu}$ in its first two indices. Moreover, it proves useful to define the quantity

$$\mathcal{W}_i := -\bar{h}_{00,i} + \bar{h}_{ij,}{}^j - \phi_{ij,}{}^j \tag{7.36}$$

and introduce the *light-cone kernel* $K_c(x-y)$, that is,

$$K_c(x-y) := K(x-y)\,\delta(x^0 - y^0 - |\mathbf{x} - \mathbf{y}|). \tag{7.37}$$

Then, the purely nonlocal source-free integral constraints (7.35) consist of six equations given by

$$\int K_c(x-y)\,\mathcal{W}_i(y)\,d^4y = 4\,\check{p}\,U_{[i\,0]}(x) \tag{7.38}$$

and

$$\int K_c(x-y)\left(\bar{h}_{0i,j} + \phi_{0i,j} - \bar{h}_{0j,i} - \phi_{0j,i}\right)(y)\,d^4y = 4\,\check{p}\,U_{[i\,j]}(x). \tag{7.39}$$

Furthermore, from $S_{0\nu} = 0$ and eqn (7.33), we have that

$$^0G_{0\nu}(x) + \int K(x-y)\,^0G_{0\nu}(y)\,d^4y = \kappa T_{0\nu}(x) - \check{p}\,U_{0\nu}(x), \tag{7.40}$$

where $U_{0\nu}$ can be determined from eqn (7.32), namely,

$$U_{0\nu}(x) = \int K(x-y)\check{C}_{0,\nu}(y)\,d^4y. \tag{7.41}$$

In Appendix A, we show that $\check{C}_0$ can be determined in principle in terms of T_{00}; see eqn (7.226). Finally, the source term for the field equation involving $^0G_{ij}$ contains $S_{(ij)}$ and $U_{(ij)}$, where

$$S_{(ij)}(x) = \frac{1}{2}\int K_c(x-y)\left[\bar{h}_{ij,0} - \bar{h}_{0(i,j)} + \phi_{0(i,j)} + \delta_{ij}\left(\bar{h}_{0\rho,}{}^\rho - \phi_{0k,}{}^k\right)\right](y)\,d^4y \tag{7.42}$$

and $U_{(ij)}$ can be simply determined from eqn (7.32).

It is clear from these results that in our decomposition of the linear gravitational potentials $\psi_{\mu\nu}$ in eqn (7.2), the symmetric *metric* part $\bar{h}_{\mu\nu}$ that satisfies eqn (7.34) has primary dynamical content, while the antisymmetric *tetrad* part $\phi_{\mu\nu}$ plays a secondary role and is constrained via eqn (7.35). In general, $\bar{h}_{\mu\nu}$ and $\phi_{\mu\nu}$ are *inextricably connected* in both sets of equations and cannot be simply disentangled. In the case of $X_{\mu\nu\rho} = \mathfrak{C}_{\mu\nu\rho} + \check{p}\,(\check{C}_\mu\,g_{\nu\rho} - \check{C}_\nu\,g_{\mu\rho})$ under consideration here, certain simplifications occur that are discussed in Section 7.3.

Before discussing the solution of the linearized field equations, we must digress here and point out a significant consequence of gravitational dynamics given by eqn (7.33). Working in the space of continuous functions on spacetime that are absolutely integrable (L^1) as well as square integrable (L^2), it is possible to write eqn (7.33) in the form

$$^0G_{\mu\nu} = \kappa\,T_{\mu\nu} + S_{\mu\nu} - \check{p}\,U_{\mu\nu} + \int R(x - y)\,[\kappa\,T_{\mu\nu} + S_{\mu\nu} - \check{p}\,U_{\mu\nu}](y)\,d^4y, \qquad (7.43)$$

where $R(x-y)$ is a kernel that is *reciprocal* to $K(x-y)$; see Appendix B. The reciprocal kernel is of the *convolution* type and is *causal* as well. Aside from nonlocal terms involving $S_{\mu\nu}$ and $U_{\mu\nu}$, eqn (7.43) exhibits an important feature that must be stressed. That the linearized gravitational field equation can be expressed as in eqn (7.43) is a crucial result, since it means that nonlocal gravity in the linear regime is essentially equivalent to general relativity, except that in addition to the usual gravitational source, there is an additional effective "dark" source that is given by the convolution of the usual source with the *causal* reciprocal kernel. Most of the matter in the universe is currently thought to be in the form of certain elusive particles that have not been directly detected (Aprile *et al.* 2012; Agnese *et al.* 2014; Akerib *et al.* 2014; Baudis 2016); indeed, the existence and properties of this *dark matter* have thus far been deduced only through its gravity. Could the additional source in nonlocal gravity be identified as the main component of what appears as *dark matter* in astrophysics? Would observational data support the notion that the effective dark matter of NLG corresponds to astrophysical dark matter? It is necessary to investigate the interesting possibility that nonlocality could simulate dark matter in this linearized theory, since the "dark" source is simply the manifestation of the nonlocal aspect of the gravitational interaction.

Nonlocality has been introduced into the theory of gravitation via the constitutive kernel K. In effect, nonlocal gravity can be expressed as GR with an additional "dark" source term. This effective dark matter source is the convolution of the standard source with the reciprocal kernel R. To investigate the nature of the effective dark matter in nonlocal gravity, we must first study the reciprocal kernel R.

7.2 Causal Reciprocal Kernel

Due to the importance of eqn (7.43) for the physical interpretation of NLG, this section is devoted to a brief description of the mathematical steps that lead to this result. Appendix B contains important background information regarding the transition from eqn (7.33) to eqn (7.43). It turns out that the convolution property of the kernels under consideration is independent of their crucial causality properties. Therefore, we first

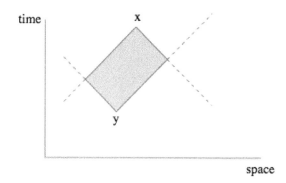

Fig. 7.1 Schematic plot indicating the *finite* shaded domain $\mathcal{D}(x, y)$ in spacetime. It is the region common to the light cone that has its vertex at event x and the light cone that has its vertex at event y. Reprinted with permission from Chicone, C. and Mashhoon, B., 2013, *Phys. Rev. D* **87**, 064015. DOI: 10.1103/PhysRevD.87.064015

consider a kernel $K(x, y)$ that is causal, so that $K(x, y)$ vanishes unless eqn (7.19) is satisfied in this case.

A *Volterra kernel* is defined to be a *causal* kernel function $K(x, y)$ that is continuous over causally ordered sets in Minkowski spacetime. The product of two Volterra kernels K and K' is defined to be

$$\bar{V}(x, y) = \int_{\mathcal{D}(x,y)} K(x, z) K'(z, y) \, d^4 z, \tag{7.44}$$

which is a Volterra kernel, since the above integrand is non-zero only when z is simultaneously in the past light cone of x and in the future light cone of y, so that y is in the past light cone of x. Thus the integration domain $\mathcal{D}(x, y)$ in eqn (7.44) is the *finite* region in Minkowski spacetime bounded by the past light cone of event x and the future light cone of event y, as depicted schematically in Fig. 7.1. Alternatively, consider the causality conditions for K and K', namely,

$$x^0 - z^0 \geq |\mathbf{x} - \mathbf{z}|, \qquad z^0 - y^0 \geq |\mathbf{z} - \mathbf{y}|, \tag{7.45}$$

respectively. These imply, via addition, that $\bar{V}$ is causal, since

$$x^0 - y^0 \geq |\mathbf{x} - \mathbf{z}| + |\mathbf{z} - \mathbf{y}| \geq |\mathbf{x} - \mathbf{y}|, \tag{7.46}$$

by the triangle inequality. Volterra kernels thus form an *algebra* over the causally ordered events in Minkowski spacetime.

Consider next the generalized Volterra integral equation of the second kind given by

$$B(x, y) + \int_{\mathcal{D}(x,y)} K(x, z) \, B(z, y) \, d^4 z = A(x, y), \tag{7.47}$$

where $A(x, y)$ and $K(x, y)$ are given Volterra kernels and we wish to find a Volterra kernel $B(x, y)$ that satisfies this equation. According to a general theorem due to M. Riesz, there is a *unique solution* given by (Riesz 1949; Faraut and Viano 1986)

$$A(x, y) + \int_{\mathcal{D}(x,y)} R(x, z)\, A(z, y)\, d^4 z = B(x, y), \tag{7.48}$$

where the reciprocal Volterra kernel $R(x, y)$ can be expressed as

$$R(x, y) = \sum_{n=1}^{\infty} K_n(x, y). \tag{7.49}$$

Here the iterated Volterra kernels $K_n(x, y)$ for $n = 1, 2, 3, \ldots$ are defined such that $K_1(x, y) := -K(x, y)$ and

$$K_{n+1}(x, y) := \int_{\mathcal{D}(x,y)} K_n(x, z)\, K_1(z, y)\, d^4 z. \tag{7.50}$$

The Neumann series (7.49) converges uniformly on *bounded* domains and the reciprocal kernel R is indeed a Volterra kernel. This is proved in the paper of Faraut and Viano (1986) using generalized Riemann–Liouville kernels. The work of M. Riesz (1949) employed a wider context; here, we have followed the treatment of Faraut and Viano (1986).

It is simple to demonstrate that this significant mathematical result holds just as well if Volterra kernels are all of the convolution type; that is, we can replace $K(x, y)$ by $K(x - y)$, etc. For instance, a simple change of variable in the corresponding integral in eqn (7.44) is enough to show that $\bar{V}$, the product of Volterra kernels K and K' of convolution type, is also of convolution type and that, furthermore, $\bar{V}$ is also the product of K' and K. Therefore, *convolution Volterra kernels* form a *commutative subalgebra* of the Volterra algebra.

Henceforth, we limit our considerations to Volterra *convolution* kernels that are L^1 and L^2 functions on spacetime. We wish to reduce the generalized Volterra integral eqns (7.47) and (7.48) to the following Volterra integral equations:

$$\mathcal{G}(x) + \int K(x - y)\, \mathcal{G}(y)\, d^4 y = \mathcal{F}(x) \tag{7.51}$$

and

$$\mathcal{F}(x) + \int R(x - y)\, \mathcal{F}(y)\, d^4 y = \mathcal{G}(x). \tag{7.52}$$

To this end, consider any continuous L^1 function $f(x)$ over spacetime and define

$$\mathcal{F}(x) := \int A(x - y) f(y)\, d^4 y, \qquad \mathcal{G}(x) := \int B(x - y) f(y)\, d^4 y, \tag{7.53}$$

where A and B are closely related to the Volterra kernels defined in eqns (7.47) and (7.48). In other words, replacing the kernels in eqns (7.47) and (7.48) by L^1 and

L^2 *convolution* kernels, multiplying the resulting equations by $f(y)$ and integrating over spacetime, we obtain eqns (7.51) and (7.52).

Let us now express the operation of folding ("Faltung") in eqn (7.53) as $\mathcal{F} = A * f$ and $\mathcal{G} = B * f$, and suppose that $\| \bar{g} \|_n$ is the norm of the function $\bar{g}$ in L^n for $n \geq 1$. It is possible to show that if $\bar{f} \in L^1$ and $\bar{g} \in L^n$, then $\bar{g} * \bar{f}$ is Lebesgue measurable and

$$\| \bar{g} * \bar{f} \|_n \leq \| \bar{g} \|_n \| \bar{f} \|_1 . \tag{7.54}$$

This inequality can be proved using either Minkowski's integral inequality or Young's inequality for convolutions (Bogachev 2007; Hardy, Littlewood and Pólya 1988; Rudin 1966). We need inequality (7.54) here for $n = 1, 2$. It follows that if f is L^1 and A is L^n, then their convolution $\mathcal{F}$ is L^n. Similarly, if f is L^1 and B is L^n, then their convolution $\mathcal{G}$ is L^n. Therefore, $\mathcal{F}(x)$ and $\mathcal{G}(x)$ are L^1 and L^2 functions over spacetime as well.

The substitution of eqn (7.51) into eqn (7.52), or vice versa, results in the basic *reciprocity* integral equation

$$K(x - y) + R(x - y) + \int K(x - z)R(z - y) \, d^4z = 0. \tag{7.55}$$

By changing the integration variable in eqn (7.55) from z to z', where $z' := x + y - z$, it is simple to see that the convolution Volterra kernels K and R can be interchanged in this reciprocity relation.

Writing $\mathcal{G}$ for $^0G_{\mu\nu}$ and $\mathcal{F}$ for $\kappa\, T_{\mu\nu} + S_{\mu\nu} - \check{p}\, U_{\mu\nu}$ in eqn (7.33), we recover eqn (7.51), which means that eqn (7.52) is then equivalent to eqn (7.43); in particular, we have the remarkable result that *in the space of continuous and absolutely integrable as well as square integrable functions on spacetime, the reciprocal kernel exists and is causal,* so that

$$R(x - y) \propto u(x^0 - y^0 - |\mathbf{x} - \mathbf{y}|). \tag{7.56}$$

Furthermore, it is possible to express eqns (7.51) and (7.52) in the Fourier domain.

Thus, let

$$\widehat{f}(\xi) = \int f(x)e^{-i\xi \cdot x} \, d^4x \tag{7.57}$$

be the *spacetime* Fourier transform of f, where $\xi \cdot x := \eta_{\alpha\beta}\xi^\alpha x^\beta$. Then,

$$f(x) = \frac{1}{(2\pi)^4} \int \widehat{f}(\xi)e^{i\xi \cdot x} \, d^4\xi. \tag{7.58}$$

It follows from the convolution theorem for Fourier transforms that in the Fourier domain one can write eqns (7.51) and (7.52) as $\widehat{\mathcal{F}} = \widehat{\mathcal{G}}(1 + \widehat{K})$ and $\widehat{\mathcal{G}} = \widehat{\mathcal{F}}(1 + \widehat{R})$, respectively. Therefore,

$$(1 + \widehat{K})(1 + \widehat{R}) = 1, \tag{7.59}$$

which can also be obtained directly via Fourier transformation from eqn (7.55) and is an expression of the complete reciprocity between K and R. The reciprocity between the nonlocal kernels K and R implies that it is in principle sufficient to determine only one of them. It turns out that the reciprocal kernel is more directly connected to observational data. Therefore, suppose that $R(x - y)$ can be estimated from the comparison of the nonlocal theory with experiment, then the kernel of nonlocal gravity $K(x - y)$ can be determined from the Fourier transform of

$$\widehat{K}(\xi) = -\frac{\widehat{R}(\xi)}{1 + \widehat{R}(\xi)}, \tag{7.60}$$

provided $1 + \widehat{R}(\xi) \neq 0$.

7.3   Linearized Field Equation with $\bar{h}^{\mu\nu}{}_{,\nu} = 0$

Let us now return to eqns (7.33)–(7.43) that characterize linearized nonlocal gravity and discuss the general structure and the formal solution of the nonlocal field equation for the gravitational field of an isolated source. For $K = R = 0$ in these equations, nonlocality disappears and the field equation reduces to the familiar second-order partial differential equation of linearized GR. We assume, for the present discussion, that kernels K and R are known; in fact, their determination is the subject of the next section.

In connection with eqn (7.43), it is useful to define the *total* matter energy–momentum tensor $\mathcal{T}_{\mu\nu}$,

$$\mathcal{T}_{\mu\nu} := T_{\mu\nu} + T^D_{\mu\nu}, \tag{7.61}$$

where $T^D_{\mu\nu}$, the convolution of $T_{\mu\nu}$ and R, is the "dark" counterpart of the matter energy–momentum tensor $T_{\mu\nu}$. That is,

$$T^D_{\mu\nu}(x) = \int R(x - y) \, T_{\mu\nu}(y) \, d^4y. \tag{7.62}$$

Similarly, we define

$$\mathcal{S}_{\mu\nu}(x) := S_{\mu\nu}(x) + \int R(x - y) \, S_{\mu\nu}(y) \, d^4y \tag{7.63}$$

and

$$\mathcal{U}_{\mu\nu}(x) := U_{\mu\nu}(x) + \int R(x-y)\, U_{\mu\nu}(y)\, d^4y. \tag{7.64}$$

It is possible to write these equations as

$$\mathcal{S}_{\mu\nu}(x) = \int W(x-y)\, \mathfrak{C}_\mu{}^0{}_\nu(y)\, d^4y, \tag{7.65}$$

where $\mathfrak{C}_\mu{}^0{}_\nu$ is given by eqns (7.29)–(7.30), and

$$\mathcal{U}_{\mu\nu}(x) = -\int W(x-y)\,(\check{C}_\mu\,\delta_\nu^0 - \check{C}^0\,\eta_{\mu\nu})(y)\, d^4y - \int R(x-y)\,\check{C}_{\mu,\nu}(y)\, d^4y. \tag{7.66}$$

Here, we have introduced a new convolution kernel W,

$$W(x-y) := K_c(x-y) + \int R(x-z) K_c(z-y)\, d^4z, \tag{7.67}$$

where in the integrand R and K_c can be interchanged. Moreover, in deriving eqn (7.66), we have used the reciprocity relation (7.55).

As in GR, the gauge freedom of the gravitational potentials may be used to impose the transverse gauge condition

$$\bar{h}^{\mu\nu}{}_{,\nu} = 0. \tag{7.68}$$

The remaining gauge degrees of freedom involve four functions $\epsilon^\mu(x)$ such that $\Box \epsilon^\mu = 0$. With the imposition of the transverse gauge condition, we find from eqn (7.8) that

$$^0G_{\mu\nu} = -\frac{1}{2}\,\Box \bar{h}_{\mu\nu}. \tag{7.69}$$

Hence, our main dynamical result, eqn (7.43), can be expressed as

$$\Box \bar{h}_{\mu\nu} + 2\,\mathcal{S}_{\mu\nu} = -2\,\kappa\,\mathcal{T}_{\mu\nu} + 2\,\check{p}\,\mathcal{U}_{\mu\nu}. \tag{7.70}$$

In other words,

$$\Box \bar{h}_{0\mu} = -2\,\kappa\,\mathcal{T}_{0\mu} - 2\,\check{p}\int R(x-y)\,\check{C}_{0,\mu}(y)\, d^4y, \tag{7.71}$$

since $\mathcal{S}_{0\mu} = 0$ and hence $\mathcal{S}_{0\mu} = 0$ as well. Furthermore,

$$\Box \bar{h}_{ij} + \int W(x-y)\big[\bar{h}_{ij,0} - \bar{h}_{0(i,j)} + \phi_{0(i,j)} - \delta_{ij}\,\phi_{0k,}{}^k\big](y)\, d^4y = -2\,\kappa\,\mathcal{T}_{ij} + 2\,\check{p}\,\mathcal{U}_{(ij)}, \tag{7.72}$$

where

$$\mathcal{U}_{(ij)}(x) = -\delta_{ij}\int W(x-y)\,\check{C}_0(y)\, d^4y - \int R(x-y)\,\check{C}_{(i,j)}(y)\, d^4y. \tag{7.73}$$

We must solve these dynamic field equations subject to the six integral constraints given by eqns (7.38) and (7.39). Once the ten components of $\bar{h}_{\mu\nu}$ have been determined, one can find the metric perturbation

$$h_{\mu\nu} = \bar{h}_{\mu\nu} - \frac{1}{2}\,\eta_{\mu\nu}\,\bar{h}. \tag{7.74}$$

On the other hand, the constraints appear to be dominated by $\phi_{\mu\nu} = -\phi_{\nu\mu}$. Let us recall that the gravitational potentials of linearized nonlocal gravity, $\psi_{\mu\nu} = \psi_{(\mu\nu)} +$

$\psi_{[\mu\nu]}$, consist of ten metric variables $\psi_{(\mu\nu)} = \frac{1}{2} h_{\mu\nu}$ and six tetrad variables $\psi_{[\mu\nu]} = \frac{1}{2} \phi_{\mu\nu}$. These variables are all intertwined in the linearized field equations of NLG.

It is shown in Appendix A that the field equation for $\bar{h}_{00}$ can be combined with constraint (7.38) to derive eqn (7.226) for $\check{C}_0 = O(c^{-2})$. Assuming that $\check{C}_0$ can be determined in terms of T_{00} from eqn (7.226), we can then calculate $\mathcal{U}_{0\mu}$ via

$$\mathcal{U}_{0\mu} = -\int R(x-y)\check{C}_{0,\mu}(y)\, d^4y. \tag{7.75}$$

The general solution of eqn (7.71) involves the superposition of a particular solution of the inhomogeneous equation plus a general solution of the wave equation. Assuming the absence of incoming gravitational waves, we are interested in the special retarded solution

$$\bar{h}_{0\mu}(x^0,\mathbf{x}) = \frac{\kappa}{2\pi}\int \frac{\left[T_{0\mu} - (\check{p}/\kappa)\mathcal{U}_{0\mu}\right](x^0 - |\mathbf{x}-\mathbf{y}|,\mathbf{y})}{|\mathbf{x}-\mathbf{y}|}\, d^3y. \tag{7.76}$$

In general, the other variables cannot be simply decoupled.

In connection with the propagation of gravitational waves, let us note an aspect of eqn (7.72) that leads to a *nonlocal damping* feature discussed in Chapter 9. Thinking about eqn (7.72) in terms of a simple analogy with the mechanics of a linear damped oscillator, we note that the term $\partial \bar{h}_{ij}/\partial t$ in eqn (7.72) is reminiscent of the "velocity" of the corresponding oscillator. It is interesting that such a nonlocal damping is completely absent in eqn (7.71), which for $\bar{h}_{00}$ is the physical basis for the modified Poisson equation in the Newtonian regime of nonlocal gravity.

The general solution of the linearized field equation of NLG is not known at present; however, some special cases of particular physical interest are treated later in this chapter after we have a more explicit knowledge of kernels K and R. The first step in this direction involves the Newtonian limit of nonlocal gravity, which can be used to determine R in the Newtonian regime from the gravitational physics of the Solar System and the comparison of the theory with observational data regarding spiral galaxies as well as clusters of galaxies.

7.4 Newtonian Limit

The Newtonian regime is marked by instantaneous action at a distance; that is, we formally let $c \to \infty$. Therefore, it is natural to assume that as $c \to \infty$, gravitational memory becomes purely spatial and all retardation effects vanish. It follows that in the Newtonian limit

$$K(x-y) = \delta(x^0 - y^0)\,\chi(\mathbf{x}-\mathbf{y}), \tag{7.77}$$

then reciprocity requires that

$$R(x-y) = \delta(x^0 - y^0)\,q(\mathbf{x}-\mathbf{y}). \tag{7.78}$$

The spatial kernels χ and q are *universal* functions that must be determined from observation. To arrive at the Newtonian limit of nonlocal gravity, we insert eqns (7.77) and (7.78) for the Newtonian kernels in our basic relations (7.51) and (7.52), respectively, where $\mathcal{G}$ stands for $^0G_{\mu\nu}$ and $\mathcal{F}$ stands for $\kappa T_{\mu\nu} + S_{\mu\nu} - \check{p}U_{\mu\nu}$. Moreover, as

$c \to \infty$, the dominant contribution to $\mathcal{G}$ is due to $^{0}G_{00} = -(\Box \bar{h}_{00})/2$, using eqn (7.69), where $\bar{h}_{00} = -4\Phi/c^2$ and Φ is the gravitational potential in the Newtonian limit of nonlocal gravity. It follows that in this case $h_{\mu\nu} = -(2\Phi/c^2)\,\mathrm{diag}(1,1,1,1)$. Similarly, the dominant term in the matter energy–momentum tensor is given by $T_{00} = \rho c^2$, where ρ is the density of matter. Furthermore, eqn (7.28) implies that $S_{00} = 0$, while eqn (7.41) implies that $U_{00} = O(c^{-3})$, since it follows from eqn (7.226) of Appendix A that $\check{C}_0 = O(c^{-2})$. In this way, we find that in the Newtonian limit, the basic equations of nonlocal gravity reduce to

$$\nabla^2 \Phi(\mathbf{x}) + \int \chi(\mathbf{x} - \mathbf{y}) \nabla^2 \Phi(\mathbf{y})\, d^3 y = 4\pi G\, \rho(\mathbf{x}), \tag{7.79}$$

and

$$\nabla^2 \Phi = 4\pi G\, (\rho + \rho_D), \qquad \rho_D(\mathbf{x}) = \int q(\mathbf{x} - \mathbf{y}) \rho(\mathbf{y})\, d^3 y, \tag{7.80}$$

where ρ_D is the density of effective dark matter and we have suppressed the dependence of Φ, ρ and ρ_D upon time t for the sake of simplicity. The nonlocal aspect of gravity appears in eqn (7.80) as an extra "dark" matter source whose density is the convolution of the reciprocal kernel q with the density of matter ρ. In this sense, nonlocality appears to simulate dark matter. Moreover, no such effective dark matter exists in the complete absence of matter; that is, $\rho_D = 0$ if $\rho = 0$.

The nonlocal memory reduces in the Newtonian limit to an instantaneous average over space, since retardation effects vanish as $c \to \infty$. Equation (7.79) is a Fredholm integral equation of the second kind that can be solved in principle using the Fourier transform method described below or via the Liouville–Neumann method of successive substitutions described in Appendix B. In the latter approach, for instance, if the Neumann series converges uniformly, we obtain a unique solution involving the reciprocal kernel. The unique solution of the Fredholm equation can then be expressed as eqn (7.80). The Fourier transform method is described in detail in Chicone and Mashhoon (2012).

It is possible to derive eqn (7.80) from a variational principle if we assume that $\chi(\mathbf{x} - \mathbf{y})$ is only a function of $|\mathbf{x} - \mathbf{y}|$ and therefore *symmetric*. Indeed, it turns out that χ is invariant under the exchange of $\mathbf{x}$ and $\mathbf{y}$ for all nonlocal Newtonian kernels of interest in this book. In this case, the variation of action S_P,

$$S_P = \int \mathcal{L}_P\, d^3 x \tag{7.81}$$

with

$$\mathcal{L}_P = \frac{1}{8\pi G}\left[(\nabla_{\mathbf{x}} \Phi)^2 + \int \chi(\mathbf{x} - \mathbf{y})(\nabla_{\mathbf{x}} \Phi) \cdot (\nabla_{\mathbf{y}} \Phi)\, d^3 y\right] + \rho\, \Phi \tag{7.82}$$

results in eqn (7.79).

7.4.1 Fourier transform method

Equation (7.80) represents the nonlocal generalization of Poisson's equation of Newtonian gravity. The reciprocal kernel q can be determined in principle from the comparison of eqn (7.80) with experiment. To find χ, the kernel of nonlocal gravity in the Newtonian regime, we use the Fourier transform method (Chicone and Mashhoon 2012). To this end, we work in the space of functions that are absolutely integrable (L^1) as well as square integrable (L^2) over all space.

It follows from combining eqns (7.79) and (7.80) that the spatial kernels χ and q are indeed reciprocal to each other; that is, two reciprocity relations can in general be deduced in this way that for convolution kernels reduce to

$$\chi(\mathbf{x} - \mathbf{y}) + q(\mathbf{x} - \mathbf{y}) + \int \chi(\mathbf{x} - \mathbf{z})\, q(\mathbf{z} - \mathbf{y})\, d^3 z = 0. \tag{7.83}$$

Indeed, in the integrand of eqn (7.83), the change of variable $\mathbf{z}$ to $\mathbf{z}'$, via $\mathbf{z} - \mathbf{y} = \mathbf{x} - \mathbf{z}'$, leads to the result that eqn (7.83) is completely symmetric with respect to the interchange of χ and q.

Let $\hat{s}(\boldsymbol{\xi})$ be the Fourier integral transform of a function $s(\mathbf{x})$ that is both L^1 and L^2; then,

$$\hat{s}(\boldsymbol{\xi}) = \int s(\mathbf{x})\, e^{-i\boldsymbol{\xi}\cdot\mathbf{x}}\, d^3 x, \qquad s(\mathbf{x}) = \frac{1}{(2\pi)^3} \int \hat{s}(\boldsymbol{\xi})\, e^{i\boldsymbol{\xi}\cdot\mathbf{x}}\, d^3 \xi. \tag{7.84}$$

It follows from the convolution theorem for Fourier integral transforms and eqn (7.83) that

$$(1 + \hat{\chi})(1 + \hat{q}) = 1. \tag{7.85}$$

The spatial kernels χ and q turn out to be symmetric in the sense that $\chi(\mathbf{x} - \mathbf{y})$ is only a function of $|\mathbf{x} - \mathbf{y}|$, etc. Thus in the Fourier domain, we have

$$\hat{\chi}(|\boldsymbol{\xi}|) + \hat{q}(|\boldsymbol{\xi}|) + \hat{\chi}(|\boldsymbol{\xi}|)\, \hat{q}(|\boldsymbol{\xi}|) = 0. \tag{7.86}$$

It follows that if $q(\mathbf{x})$ is given by experimental data and subsequently $\hat{q}(|\boldsymbol{\xi}|)$ is calculated from the Fourier integral transform of $q(\mathbf{x})$, then the kernel of nonlocal gravity $\chi(\mathbf{x})$ can be determined from the Fourier transform of $\hat{\chi}(|\boldsymbol{\xi}|)$ that is given by eqn (7.86), namely,

$$\hat{\chi}(|\boldsymbol{\xi}|) = -\frac{\hat{q}(|\boldsymbol{\xi}|)}{1 + \hat{q}(|\boldsymbol{\xi}|)}, \tag{7.87}$$

provided

$$1 + \hat{q}(|\boldsymbol{\xi}|) \neq 0. \tag{7.88}$$

Thus an acceptable reciprocal kernel $q(\mathbf{x})$ should be a smooth function that is L^1, L^2 and satisfies requirement (7.88). We now proceed to the determination of $q(\mathbf{x})$.

7.4.2   Kuhn kernel q_K

The nonlocal Poisson eqn (7.80) is in a form that can be compared with observational data regarding, for instance, the rotation curves of spiral galaxies. Imagine, for instance, the circular motion of stars (or gas clouds) in the disk of a spiral galaxy about the galactic bulge. According to the Newtonian laws of motion, such a star (or gas cloud) has a centripetal acceleration of v_c^2/r, where v_c is its constant speed; moreover, this centripetal acceleration must be equal to the gravitational acceleration of the star.

Observational data indicate that v_c is nearly the same for all stars (and gas clouds) in the galactic disk, thus leading to the nearly *flat* rotation curves of spiral galaxies (Rubin and Ford 1970; Roberts and Whitehurst 1975; Sofue and Rubin 2001).

If we adopt the standard Newtonian theory of gravitation, we find that $v_c^2/r = GM_{gc}/r^2$, where M_{gc} is the effective mass of the *galactic core* from the viewpoint of the star (or gas cloud). As v_c is essentially independent of radial distance r, M_{gc} must increase linearly with r, that is, $M_{gc} = v_c^2 r/G$. Assuming spherical symmetry, we conclude that there must exist *dark matter* of density $v_c^2/(4\pi Gr^2)$ in the spiral galaxy beyond the galactic bulge.

Suppose, however, that there is no such dark matter, but the gravitational interaction is described instead by nonlocal gravity. This means that the nonlocal force of gravity varies essentially as $1/r$ beyond the galactic bulge. Attributing this circumstance to an *effective density of dark matter* and assuming spherical symmetry, we find from

$$\nabla \cdot [D(r)\,\hat{\mathbf{r}}] = \frac{1}{r^2}\frac{d}{dr}[r^2\,D(r)] \tag{7.89}$$

that for $D = 1/r$, we get from the nonlocal generalization of Poisson's equation that the corresponding *effective* density of dark matter ρ_D must be $v_c^2/(4\pi Gr^2)$, which is the same as the density of actual dark matter in the standard treatment. Thus in this case, the nonlocal aspect of the gravitational interaction simulates dark matter. Using eqn (7.80) with $\rho(\mathbf{x}) = M_{gc}\,\delta(\mathbf{x})$, where M_{gc} is now the *constant* effective mass of the galactic bulge, we find for kernel q,

$$q_K(\mathbf{x} - \mathbf{y}) = \frac{1}{4\pi\lambda_{TK}}\frac{1}{|\mathbf{x} - \mathbf{y}|^2}, \tag{7.90}$$

where $\lambda_{TK} = GM_{gc}/v_c^2$ is a constant galactic length of the order of 1 kpc.

It is remarkable that a modified Poisson equation of the form (7.80) with kernel (7.90) was suggested by Kuhn about 30 years ago; in fact, it is interesting to digress briefly here and mention the phenomenological Tohline–Kuhn modified gravity approach to the problem of dark matter (Tohline 1983, 1984; Kuhn, Burns and Schorr 1986; Kuhn and Kruglyak 1987). According to this scheme, the "flat" rotation curves of spiral galaxies lead to a Tohline–Kuhn extension of the Newtonian inverse square law of gravity for point masses m_1 and m_2, namely,

$$F_{TK}(r) = \frac{Gm_1m_2}{r^2} + \frac{Gm_1m_2}{\lambda_{TK}\,r}, \tag{7.91}$$

where the relative deviation from Newton's law due to the long-range ("galactic") contribution is given by r/λ_{TK}. In 1983, Tohline showed that this modification leads to the stability of the galactic disk (Tohline 1983). The gravitational potential for a point mass M corresponding to this modified force law can be written as (Tohline 1983)

$$\Phi_T(\mathbf{x}) = -\frac{GM}{|\mathbf{x}|} + \frac{GM}{\lambda_{TK}}\ln\left(\frac{|\mathbf{x}|}{\lambda_{TK}}\right). \tag{7.92}$$

This Tohline potential satisfies eqn (7.80) with Kuhn kernel (7.90) when $\rho(\mathbf{x}) = M\,\delta(\mathbf{x})$.

The work of Kuhn and his collaborators contained a significant generalization of Tohline's original suggestion (Kuhn, Burns and Schorr 1986; Kuhn and Kruglyak 1987); the Tohline–Kuhn scheme has been admirably reviewed by Bekenstein (1988).

7.4.3   Derivation of reciprocal kernel q

The reciprocal kernel $q(\mathbf{r})$ of nonlocal gravity must satisfy the mathematical requirements discussed before; namely, it must be a smooth function that is L^1, L^2 and satisfies condition (7.88). Moreover, it should reduce to the Kuhn kernel in appropriate limits in order to recover the observational data related to the nearly flat rotation curves of spiral galaxies. However, these conditions are not sufficient to specify a unique functional form for q.

Our physical considerations thus far involved the motion of stars and gas clouds in circular orbits around the galactic core. The radii of such orbits extend from the core radius to the outer reaches of the spiral galaxy. The resulting Kuhn kernel q_K captures important physical aspects of the problem, but it is not mathematically suitable as it is not L^1 and L^2. In fact, q_K integrated over all space leads to an infinite amount of effective dark matter for any point mass. From the standpoint of nonlocal gravity, the Tohline–Kuhn approach reflects the appropriate generalization of Newtonian gravity in the *intermediate* galactic regime from the bulge to the outer limits of a spiral galaxy; however, the $r \to 0$ and $r \to \infty$ regimes are not taken into account. It follows from these considerations that q must be constructed out of q_K by moderating its short and long distance behaviors (Chicone and Mashhoon 2016a).

To proceed, let us start from the Kuhn kernel (7.90) and recall that it leads to flat rotation curves in the intermediate distance regime extending from the core radius to the outer limits of a spiral galaxy. The $r \to \infty$ behavior of q is related to the fading of spatial memory with distance. If the decay rate of a quantity is proportional to itself, then the quantity dies out exponentially. We therefore adopt the simple rule that q behaves as $\exp\left(-\mu_0\, r\right)$ for $r \to \infty$, where μ_0^{-1} is a new length parameter that characterizes the rate of spatial decay of gravitational memory. For $r \ll \mu_0^{-1}$, where we expect to recover the nearly flat rotation curve of a spiral galaxy, the modified Kuhn kernel becomes

$$\frac{1}{4\pi\lambda_{TK}}\,\frac{1}{r^2}\,e^{-\mu_0\, r} = \frac{1}{4\pi\lambda_{TK}}\,\frac{1}{r^2}\left(1 - \mu_0\, r + \frac{1}{2}\mu_0^2\, r^2 - \cdots\right), \qquad (7.93)$$

where the dominant correction is of linear order in $\mu_0\, r \ll 1$. To cancel the linear correction in eqn (7.93) and hence provide a better approximation to the Kuhn kernel for $\mu_0\, r \ll 1$, we consider instead

$$\frac{1}{4\pi\lambda_{TK}}\,\frac{1}{r^2}\,(1 + \mu_0\, r)\,e^{-\mu_0\, r} = \frac{1}{4\pi\lambda_{TK}}\,\frac{1}{r^2}\left[1 - \frac{1}{2}\,(\mu_0\, r)^2 + \frac{1}{3}\,(\mu_0\, r)^3 - \cdots\right]. \qquad (7.94)$$

Kernel (7.94) is integrable over all space, but it is not square integrable. We must therefore modify the $r \to 0$ behavior of kernel (7.94) to make it square integrable by essentially replacing r with $a_0 + r$, where $a_0 > 0$ is a new constant length parameter. We note that two simple square-integrable possibilities exist

$$\frac{1}{4\pi\lambda_{TK}} \frac{1+\mu_0(a_0+r)}{r(a_0+r)} e^{-\mu_0(a_0+r)} \tag{7.95}$$

and

$$\frac{1}{4\pi\lambda_{TK}} \frac{1+\mu_0(a_0+r)}{(a_0+r)^2} e^{-\mu_0(a_0+r)}. \tag{7.96}$$

Moreover, we can define

$$\frac{1}{\lambda_0} := \frac{1}{\lambda_{TK}} e^{-\mu_0 a_0}, \tag{7.97}$$

so that the Tohline–Kuhn parameter λ_{TK} is modified and is henceforth replaced by λ_0. In this way, we find from eqns (7.95) and (7.96) two possible solutions for q, namely, q_1 and q_2 given by (Chicone and Mashhoon 2012)

$$q_1 = \frac{1}{4\pi\lambda_0} \frac{1+\mu_0(a_0+r)}{r(a_0+r)} e^{-\mu_0 r} \tag{7.98}$$

and

$$q_2 = \frac{1}{4\pi\lambda_0} \frac{1+\mu_0(a_0+r)}{(a_0+r)^2} e^{-\mu_0 r}, \tag{7.99}$$

where $r = |\mathbf{x} - \mathbf{y}|$ and q_1 and q_2 are symmetric functions of $\mathbf{x}$ and $\mathbf{y}$. Here, λ_0, a_0 and μ_0 are three positive constant parameters that must be determined via observational data. The fundamental length scale of nonlocal gravity is λ_0, which is expected to be of the order of 1 kpc and is reminiscent of the parameter λ_{TK} of the Kuhn kernel. We note that for $i = 1, 2$, $q_i \to 0$ and nonlocality disappears as $\lambda_0 \to \infty$. Furthermore, a_0 moderates the $r \to 0$ behavior of the reciprocal kernel, while the kernel decays exponentially for $r \gg \mu_0^{-1}$, as the spatial gravitational memory fades. Henceforth, we will refer to a_0 and μ_0 as the short-distance and the large-distance parameters of the reciprocal kernel, respectively.

In agreement with the requirements of the Fourier transform method, kernels q_1 and q_2 are continuous positive functions that are integrable as well as square integrable over all space. The Fourier transform of q_1 is always real and positive and hence satisfies eqn (7.88) regardless of the value of a_0/λ_0. On the other hand, the Fourier transform of q_2 is such that eqn (7.88) is satisfied if $a_0 < \lambda_0$; see Appendix C. In any case, it is natural to expect on physical grounds that $a_0 < \lambda_0 < \mu_0^{-1}$; that is, the (intermediate) nonlocality parameter is expected to be smaller than the large-distance parameter and larger than the short-distance parameter. It then follows from the Fourier transform method that the corresponding kernels χ_1 and χ_2 exist, are symmetric and have other desirable physical properties; see Appendix C.

It is important to emphasize that q_1 and q_2 are by no means unique. More complicated expressions that include more parameters are certainly possible. Kernels q_1 and q_2 appear to be the simplest functions that satisfy the requirements of nonlocal gravity theory (Chicone and Mashhoon 2012).

The reciprocal kernels q_1 and q_2 thus depend upon three parameters: the nonlocality parameter λ_0, the large-distance parameter μ_0 and the short-distance parameter a_0. We expect that these three parameters will be determined via observational data, which will, in addition, point to a unique function (i.e. either q_1 or q_2) for q.

It is interesting to note that for $a_0 = 0$, q_1 and q_2 both reduce to q_0,

$$q_0 = \frac{1}{4\pi\lambda_0} \frac{(1+\mu_0 r)}{r^2} e^{-\mu_0 r}, \qquad (7.100)$$

where for any finite $r : 0 \to \infty$, we have for $i = 1, 2$,

$$q_0(r) > q_i(r). \qquad (7.101)$$

Moreover, q_0 is *not* square integrable over all space and the behavior of q_0 for $r \to 0$ is precisely the same as that of the Kuhn kernel; for instance, in the Solar System, we recover the Tohline–Kuhn force (7.91). For observational data related to the rotation curves of spiral galaxies as well as the internal gravitational physics of clusters of galaxies, we expect that the short-distance behavior of the kernel would be unimportant and hence q_0 may be employed to fit the data. This has indeed been done in Rahvar and Mashhoon (2014) and parameters λ_0 and μ_0 have thus been determined; see Chapter 8. In this connection, it is useful to introduce the dimensionless parameter α_0,

$$\alpha_0 := \int q_0(|\mathbf{x}|) \, d^3 x, \qquad \alpha_0 = \frac{2}{\lambda_0 \mu_0}. \qquad (7.102)$$

Then, it follows from observational data that (Rahvar and Mashhoon 2014)

$$\alpha_0 = 10.94 \pm 2.56, \qquad \mu_0 = 0.059 \pm 0.028 \text{ kpc}^{-1}. \qquad (7.103)$$

Hence, $\lambda_0 = 2/(\alpha_0 \mu_0)$ turns out to be $\lambda_0 \approx 3 \pm 2$ kpc. Regarding the short-distance parameter a_0, it is useful to introduce a new dimensionless parameter ς,

$$\varsigma := \mu_0 \, a_0, \qquad (7.104)$$

and provisionally assume, on the basis of $a_0 < \lambda_0$ and eqn (7.103), that

$$0 < \varsigma < \frac{1}{5} \qquad (7.105)$$

for the sake of simplicity. Preliminary lower limits can be placed on a_0 on the basis of current data regarding planetary orbits in the Solar System; see Chapter 9. For instance, using the data for the orbit of Saturn, a preliminary lower limit of $a_0 \gtrsim 2 \times 10^{15}$ cm can be established if we use q_1, while $a_0 \gtrsim 5.5 \times 10^{14}$ cm if we use q_2 (Chicone and Mashhoon 2016a).

It is abundantly clear from our considerations here that the choice of the kernel is not unique. In the absence of a physical principle that could uniquely lead to the appropriate kernel, we must adopt simple functional forms that satisfy the mathematical requirements discussed earlier and are based on agreement with observation. Let us recall that the relativistic framework of Einstein's field theory of gravitation has properly generalized Newton's inverse square force law, which is ultimately based on Solar System observations that originally led to Kepler's laws of planetary motion. That is, an acceptable theory of gravitation must agree with Newton's theory in some form. How did Newton come up with the inverse square law? As explained in his *Principia*, he explored various functional forms such as r and r^{-3} in addition to r^{-2} and concluded that only r^{-2} agreed with Kepler's empirical laws of planetary motion (Cohen 1960). In short, the inverse square force law was not derived from a physical principle; rather, it was chosen to agree with observation. Moreover, observational data never have infinite accuracy; therefore, to Newton's r^{-2}, for example, one can add other functional forms with sufficiently small coefficients such that agreement with experimental results can be maintained. The same is true, of course, in Einstein's general theory of relativity.

7.4.4 Modified force law

We next proceed to the determination of the short-distance behaviors of the modified force laws associated with q_1 and q_2 (Chicone and Mashhoon 2016a).

The gravitational force acting on a point particle of mass m in a gravitational field with potential Φ is $\mathbf{F} = -m\nabla\Phi$ and the geodesic equation reduces in the Newtonian limit to Newton's equation of motion

$$\frac{d^2\mathbf{r}}{dt^2} = -\nabla\Phi(\mathbf{r}). \tag{7.106}$$

Let us now imagine that potential Φ is due to a point mass M at the origin of spatial coordinates with mass density $\rho(\mathbf{r}) = M\,\delta(\mathbf{r})$. Thus we find from eqn (7.80) that

$$\nabla^2\Phi_i(\mathbf{r}) = 4\pi GM\,[\delta(\mathbf{r}) + q_i(r)], \tag{7.107}$$

where $i = 1, 2$, depending upon which reciprocal kernel is employed, since experiment must ultimately decide between q_1 and q_2. Assuming that the force on a point mass m at $\mathbf{r}$ due to M is radial, namely, $\mathbf{F} = -m\,(d\Phi/dr)\,\hat{\mathbf{r}}$, where $\hat{\mathbf{r}}$ is the radial unit vector, we have

$$\frac{d\Phi}{dr} = G\,M\,f(r), \tag{7.108}$$

so that the gravitational force between the two point masses is $\mathbf{F} = -G\,m\,M\,f(r)\,\hat{\mathbf{r}}$.

The solution of eqn (7.107) is the sum of the Newtonian potential plus $F_i(r)$, which is the contribution from the reciprocal kernel; that is,

$$\Phi_i(r) = GM\left[-\frac{1}{r} + F_i(r)\right]. \tag{7.109}$$

It follows from

$$\nabla^2\left(\frac{1}{r}\right) = -4\pi\delta(\mathbf{r}) \tag{7.110}$$

that

$$\nabla^2 F_i = 4\pi\,q_i. \tag{7.111}$$

It then proves useful to write

$$f_i(r) = \frac{1}{r^2} + N_i(r), \tag{7.112}$$

where $N_i(r) := dF_i/dr$ and we have again separated the Newtonian contribution from the nonlocal contribution. Thus we find from eqns (7.89) and (7.111) that

$$\frac{1}{r^2}\frac{d}{dr}[r^2 N_i(r)] = 4\pi\, q_i(r). \tag{7.113}$$

The solution of this equation can be expressed as

$$N_i(r) = \frac{4\pi}{r^2}\int_0^r s^2\, q_i(s)\,ds, \tag{7.114}$$

where we have assumed that as $r \to 0$, $r^2 N_i(r) \to 0$, so that in the limit of $r \to 0$, the force on m due to M is given by the Newtonian inverse square law. This important assumption is based on the results of experiments that have verified the gravitational inverse square force law down to a radius of $r \approx 50\,\mu$m (Adelberger, Heckel and Nelson 2003; Hoyle *et al.* 2004; Adelberger *et al.* 2007; Kapner *et al.* 2007). Furthermore, no significant deviation from Newton's law of gravitation has been detected thus far in laboratory experiments (Meyer *et al.* 2012; Little and Little 2014).

It proves interesting to define

$$N_0(r) := \frac{4\pi}{r^2}\int_0^r s^2\, q_0(s)\,ds = \frac{\alpha_0}{r^2}\left[1 - \left(1 + \frac{1}{2}\mu_0\, r\right)e^{-\mu_0\, r}\right], \tag{7.115}$$

where q_0 is given by eqn (7.100), so that we can write

$$N_i(r) = -\frac{1}{r^2}\,\mathcal{E}_i(r) + N_0(r). \tag{7.116}$$

Here, we have defined

$$\mathcal{E}_i(r) := 4\pi\int_0^r s^2\,[q_0(s) - q_i(s)]\,ds, \tag{7.117}$$

such that $\mathcal{E}_i(r) = 0$ for $a_0 = 0$ and $\mathcal{E}_i(r) > 0$ for $r > 0$. It follows from eqn (7.114) and the fact that q_1 and q_2 are positive functions that $N_i(r) \geq 0$; therefore, $f_i(r) > 0$ by eqn (7.112). Putting eqns (7.112), (7.115) and (7.116) together, we find

$$f_i(r) = \frac{1}{r^2}[1 - \mathcal{E}_i(r) + \alpha_0] - \frac{\alpha_0}{r^2}\left(1 + \frac{1}{2}\mu_0\, r\right)e^{-\mu_0\, r}. \tag{7.118}$$

Thus, we finally have the force of gravity on point mass m due to point mass M, namely,

$$\mathbf{F}_i(\mathbf{r}) = -GmM\frac{\hat{\mathbf{r}}}{r^2}\left\{[1 - \mathcal{E}_i(r) + \alpha_0] - \alpha_0\left(1 + \frac{1}{2}\mu_0\, r\right)e^{-\mu_0\, r}\right\}, \tag{7.119}$$

which, except for the $\mathcal{E}_i(r)$ term, is due to kernel q_0. This force is conservative, satisfies Newton's third law of motion and is *always attractive*. The gravitational force of attraction in eqn (7.119) consists of two parts: an enhanced attractive "Newtonian" part involving $\alpha_0 \approx 11$ and a repulsive "Yukawa" part with an exponential decay length of $\mu_0^{-1} \approx 17$ kpc. The exponential decay in the Yukawa term originates from the fading of spatial *memory*.

Imagine a uniform thin spherical shell of matter and a point mass m inside the hollow shell. As is well known, Newton's inverse square law of gravity implies that there is no net force on m, regardless of the location of m within the shell. However, Newton's shell theorem does not hold in nonlocal gravity, so that m would in general be subject to a gravitational force that is along the diameter that connects m to the center of the shell and can be calculated by suitably integrating $\mathbf{F}_i(\mathbf{r}) + GmM\,(1 + \alpha_0)\,r^{-2}\,\hat{\mathbf{r}}$ over the shell, where $\mathbf{F}_i(\mathbf{r})$ is given by eqn (7.119).

The short-distance parameter a_0 appears only in $\mathcal{E}_i(r)$; therefore, we now turn to the study of $\mathcal{E}_i(r)$. To this end, let us first define the *exponential integral function* (Abramowitz and Stegun 1964)

$$E_1(x) := \int_x^\infty \frac{e^{-t}}{t}\,dt. \tag{7.120}$$

For $x : 0 \to \infty$, $E_1(x)$ is a positive function that monotonically decreases from infinity to zero. Indeed, $E_1(x)$ behaves like $-\ln x$ near $x = 0$ and vanishes exponentially as $x \to \infty$. Moreover,

$$E_1(x) = -C - \ln x - \sum_{n=1}^\infty \frac{(-x)^n}{n\,n!}, \tag{7.121}$$

where $C = 0.577\ldots$ is Euler's constant. It is useful to note that

$$\frac{e^{-x}}{x+1} < E_1(x) \le \frac{e^{-x}}{x}; \tag{7.122}$$

see formula 5.1.19 in Abramowitz and Stegun (1964).

From eqn (7.117), we find by straightforward integration that

$$\mathcal{E}_1(r) = \frac{a_0}{\lambda_0} e^\varsigma \left[E_1(\varsigma) - E_1(\varsigma + \mu_0 r) \right] \tag{7.123}$$

and

$$\mathcal{E}_2(r) = \frac{a_0}{\lambda_0} \left\{ -\frac{r}{r + a_0} e^{-\mu_0 r} + 2e^\varsigma \left[E_1(\varsigma) - E_1(\varsigma + \mu_0 r) \right] \right\}, \tag{7.124}$$

where ς has been defined in eqns (7.104) and (7.105). Furthermore, it follows from eqn (7.117) that

$$\frac{d\mathcal{E}_i}{dr} = 4\pi\,r^2 \left[q_0(r) - q_i(r) \right], \tag{7.125}$$

where the right-hand side is positive by eqn (7.101). More explicitly,

$$\frac{d\mathcal{E}_1}{dr} = \frac{a_0}{\lambda_0} \frac{1}{a_0 + r} e^{-\mu_0\,r} \tag{7.126}$$

and

$$\frac{d\mathcal{E}_2}{dr} = \frac{a_0}{\lambda_0} \left[\mu_0 + \frac{2 - \varsigma}{a_0 + r} - \frac{a_0}{(a_0 + r)^2} \right] e^{-\mu_0\,r}. \tag{7.127}$$

Thus $\mathcal{E}_1(r)$ and $\mathcal{E}_2(r)$ are positive, monotonically increasing functions of r that start from zero at $r = 0$ and asymptotically approach, for $r \to \infty$, $\mathcal{E}_1(\infty) = \mathcal{E}_\infty$ and $\mathcal{E}_2(\infty) = 2\,\mathcal{E}_\infty$, respectively. Here,

$$\mathcal{E}_\infty = \frac{1}{2}\,\alpha_0\,\varsigma\,e^{\varsigma}\,E_1(\varsigma). \tag{7.128}$$

It then follows from eqn (7.122) that

$$\mathcal{E}_\infty < \frac{\alpha_0}{2}, \tag{7.129}$$

so that in formula (7.119) for the gravitational force,

$$\alpha_0 - \mathcal{E}_i(r) > 0. \tag{7.130}$$

Thus for $r \gg \mu_0^{-1}$, the Yukawa part of eqn (7.119) can be neglected and

$$\mathbf{F}_i(\mathbf{r}) \approx -\frac{GmM\left[1 + \alpha_0 - \mathcal{E}_i(\infty)\right]}{r^2}\,\hat{\mathbf{r}}, \tag{7.131}$$

so that $M\left[\alpha_0 - \mathcal{E}_i(\infty)\right]$ has the interpretation of the total effective dark mass associated with M.

For $a_0 = 0$, the net effective dark mass associated with point mass M is simply $\alpha_0\,M$, where $\alpha_0 \approx 11$. On the other hand, for $a_0 \neq 0$, the corresponding result is $\alpha_0\,\epsilon_i(\varsigma)\,M$, where

$$\epsilon_1(\varsigma) = 1 - \frac{1}{2}\,\varsigma\,e^{\varsigma}\,E_1(\varsigma), \qquad \epsilon_2(\varsigma) = 1 - \varsigma\,e^{\varsigma}\,E_1(\varsigma). \tag{7.132}$$

These functions are plotted in Fig. 7.2 for $\varsigma : 0 \to 0.2$ in accordance with eqn (7.105).

Finally, let us note that the solution of eqn (7.108) for the gravitational potential Φ_i due to a point mass M at $r = 0$ is given by

$$\Phi_i(r) = GM \int_\infty^r f_i(r')\,dr', \tag{7.133}$$

where, as expected, we have assumed that $\Phi_i(r) \to 0$, when $r \to \infty$. It follows from a detailed but straightforward calculation that for $i = 1, 2$, corresponding to q_1 and q_2, respectively,

$$\Phi_1(r) = -\frac{GM}{r}\left(1 + \alpha_0 - \mathcal{E}_\infty - \alpha_0\,e^{-\mu_0 r}\right) - \frac{GM}{\lambda_0}\left(1 + \frac{a_0}{r}\right)e^{\varsigma}\,E_1(\varsigma + \mu_0 r) \tag{7.134}$$

and

$$\Phi_2(r) = -\frac{GM}{r}\left(1 + \alpha_0 - 2\,\mathcal{E}_\infty - \alpha_0\,e^{-\mu_0 r}\right) - \frac{GM}{\lambda_0}\left(1 + 2\,\frac{a_0}{r}\right)e^{\varsigma}\,E_1(\varsigma + \mu_0 r). \tag{7.135}$$

In these expressions, we can use the Taylor expansion of $E_1(\varsigma + \mu_0 r)$ about $\varsigma = \mu_0\,a_0$ to write

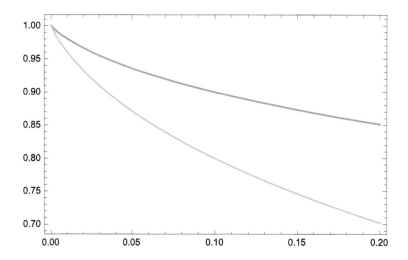

Fig. 7.2 The figure depicts the graph of the function $\epsilon_1(\varsigma)$ that lies above the graph of $\epsilon_2(\varsigma)$ for $0 < \varsigma < 1/5$. Reproduced from Chicone, C. and Mashhoon, B., 2016, "Nonlocal Gravity in the Solar System", *Class. Quantum Grav.* **33**, 075005 (21 pages), with the permission of IOP Publishing. DOI: 10.1088/0264-9381/33/7/075005

$$e^\varsigma E_1(\varsigma + \mu_0 r) = \frac{\lambda_0}{a_0}\mathcal{E}_\infty - \frac{r}{a_0} + \frac{1}{2}(1+\varsigma)\frac{r^2}{a_0^2} - \cdots . \tag{7.136}$$

In this way, we see that $\Phi_i(r) \to -GM/r$ for $r \to 0$. It follows from eqns (7.134) and (7.135) that in the limiting case where $a_0 = 0$, we have $\Phi_1 = \Phi_2 = \Phi_0$,

$$\Phi_0(r) = -\frac{GM}{r}\left(1 + \alpha_0 - \alpha_0\, e^{-\mu_0\, r}\right) - \frac{GM}{\lambda_0} E_1(\mu_0\, r), \tag{7.137}$$

which is the gravitational potential corresponding to kernel q_0.

The implications of these results for various astronomical systems are considered in Chapter 8.

7.5 Beyond the Newtonian Limit

Memory generally dies out; therefore, we expect nonlocal kernels K and R to decay exponentially in space and time. The exponential decay term in q already indicates that the lengthscale associated with spatial gravitational memory is $\mu_0^{-1} \approx 17$ kpc. We should therefore expect a similar temporal behavior in K and R; moreover, causality requires that these kernels be proportional to the Heaviside unit step function as in eqns (7.20) and (7.56). Thus the Dirac delta function $\delta(x^0 - y^0)$ that appears in eqns (7.77) and (7.78) should be suitably generalized for *finite* c to satisfy these requirements.

Consider the set of functions $\delta_n(s)$ for $n = 1, 2, 3, \ldots$ given by

$$\delta_n(s) := \nu\, n\, e^{-\nu n\, (s - \frac{r}{n})}\, u\left(s - \frac{r}{n}\right), \tag{7.138}$$

where $\nu > 0$ and $r \geq 0$ are constants. These functions are normalized,

$$\int_{-\infty}^{\infty} \delta_n(s)\, ds = 1, \tag{7.139}$$

and form a *Dirac sequence*, since it can be shown that for any smooth function $f(s)$,

$$\lim_{n \to \infty} \int_{-\infty}^{\infty} \delta_n(s)\, f(s)\, ds = f(0^+). \tag{7.140}$$

Therefore, the Dirac delta function $\delta(s)$ may be regarded as a certain distributional limit of the sequence of normalized functions $\delta_n(s)$ as $n \to \infty$. Moreover, we note that the singularity of this Dirac delta function occurs at 0^+, the positive side of the origin.

In eqn (7.138), let us now formally replace s by $t_\mathbf{x} - t_\mathbf{y}$, r by $|\mathbf{x} - \mathbf{y}|$ and n by the speed of light c; then, it is straightforward to check that in the limit as $c \to \infty$, we have

$$\nu\, c\, e^{-\nu c \left(t_\mathbf{x} - t_\mathbf{y} - \frac{|\mathbf{x} - \mathbf{y}|}{c} \right)}\, u\!\left(t_\mathbf{x} - t_\mathbf{y} - \frac{|\mathbf{x} - \mathbf{y}|}{c} \right) \to \delta(t_\mathbf{x} - t_\mathbf{y}) \tag{7.141}$$

in the distributional sense of eqn (7.140). It follows from these considerations that when the finite magnitude of the speed of light is taken into account, $\delta(x^0 - y^0)$ in eqn (7.78) can be replaced by

$$\nu\, e^{-\nu\, (x^0 - y^0 - |\mathbf{x} - \mathbf{y}|)}\, u\!\left(x^0 - y^0 - |\mathbf{x} - \mathbf{y}| \right), \tag{7.142}$$

where we recall that $x^0 = c\, t_\mathbf{x}$, $y^0 = c\, t_\mathbf{y}$ and $\delta(t_\mathbf{x} - t_\mathbf{y}) = c\, \delta(x^0 - y^0)$. Here, ν^{-1} is a constant length that should ultimately be determined on the basis of observational data. We speculate that $\nu^{-1} > \mu_0^{-1}$ is a galactic length that is comparable with μ_0^{-1}.

Henceforward, we assume that

$$R(x - y) = \nu\, e^{-\nu\, (x^0 - y^0 - |\mathbf{x} - \mathbf{y}|)}\, u\!\left(x^0 - y^0 - |\mathbf{x} - \mathbf{y}| \right)\, q(\mathbf{x} - \mathbf{y}). \tag{7.143}$$

This reciprocal kernel R is consistent with our physical requirements and depends only upon $x^0 - y^0$ and $|\mathbf{x} - \mathbf{y}|$. An important consequence of the normalization property of eqn (7.142), namely,

$$\int_{-\infty}^{\infty} \nu\, e^{-\nu\, (x^0 - y^0 - |\mathbf{x} - \mathbf{y}|)}\, u\!\left(x^0 - y^0 - |\mathbf{x} - \mathbf{y}| \right) dy^0 = 1, \tag{7.144}$$

is that

$$\int R(x - y)\, Z(\mathbf{y})\, d^4 y = \int q(\mathbf{x} - \mathbf{y})\, Z(\mathbf{y})\, d^3 y \tag{7.145}$$

for any smooth purely *spatial* function $Z(\mathbf{x})$. In the Fourier domain, this relation amounts to

$$\widehat{R}(0, \boldsymbol{\xi}) = \hat{q}(|\boldsymbol{\xi}|), \tag{7.146}$$

which implies, via eqn (7.59) for $\xi^0 = 0$ and eqn (7.86), that

$$\widehat{K}(0, \boldsymbol{\xi}) = \hat{\chi}(|\boldsymbol{\xi}|), \tag{7.147}$$

or, in the spacetime domain,

$$\int K(x-y)\, Z(\mathbf{y})\, d^4 y = \int \chi(\mathbf{x}-\mathbf{y})\, Z(\mathbf{y})\, d^3 y. \qquad (7.148)$$

A detailed discussion of the functions $\hat{q}$ and χ is contained in Appendix C.

Finally, it is interesting to note that for $Z = 1$ in eqn (7.145), the integral of the reciprocal kernel R over the whole spacetime is given by

$$\int R(x)\, d^4 x = \int q(\mathbf{x})\, d^3 x = \hat{q}(0), \qquad (7.149)$$

which can be easily computed using eqn (7.117) for q_i, $i = 1, 2$. The result is

$$\hat{q}_i(0) = \alpha_0 - \mathcal{E}_i(\infty) = \alpha_0\, \epsilon_i(\varsigma); \qquad (7.150)$$

see Fig. 7.2. Similarly, it follows from eqn (7.148) that

$$\int K(x)\, d^4 x = \int \chi(\mathbf{x})\, d^3 x = \hat{\chi}(0), \qquad (7.151)$$

where $\hat{\chi}(0)$ is related to $\hat{q}(0)$ via eqn (7.87).

7.5.1   Kernel K of linearized NLG

The procedure followed above for the determination of kernel R cannot be simply repeated for kernel K, since it turns out that the fundamental reciprocity relation (7.55) cannot be satisfied in this way. It is therefore necessary to determine K via the Fourier transform method, cf. eqn (7.60). Let us note that our basic expression for R in eqn (7.143) implies that

$$\widehat{R}(\xi) = \frac{\nu}{\nu - i\,\xi^0} \int e^{i\,\xi^0\,|\mathbf{x}|}\, q(\mathbf{x})\, e^{-i\boldsymbol{\xi}\cdot\mathbf{x}}\, d^3 x. \qquad (7.152)$$

Substituting for q in this expression either q_1 given by eqn (7.98) or q_2 given by eqn (7.99), we can calculate $\widehat{R}(\xi)$ explicitly in terms of the exponential integral function. Then, $\widehat{K}(\xi)$ is given by eqn (7.60) and $K(x)$ can, in principle, be determined by inverse Fourier transformation.

For a more tractable result, we can employ an approximation scheme that involves neglecting certain retardation effects in eqn (7.143). This means in practice that we replace $x^0 - y^0 - |\mathbf{x} - \mathbf{y}|$ in eqn (7.143) by $x^0 - y^0$; that is, instead of eqn (7.143), we consider

$$R(x-y) \approx \nu\, e^{-\nu(x^0 - y^0)}\, u(x^0 - y^0)\, q(\mathbf{x} - \mathbf{y}). \qquad (7.153)$$

The Fourier transform of this approximate kernel is

$$\widehat{R}(\xi) \approx \frac{\nu}{\nu - i\xi^0}\, \hat{q}(|\boldsymbol{\xi}|). \qquad (7.154)$$

If we use for q the simplified kernel q_0 defined by eqn (7.100), we get

$$\hat{q}_0(|\boldsymbol{\xi}|) = \frac{\mu_0}{\lambda_0(\mu_0^2 + |\boldsymbol{\xi}|^2)} + \frac{1}{\lambda_0|\boldsymbol{\xi}|}\, \arctan\left(\frac{|\boldsymbol{\xi}|}{\mu_0}\right); \qquad (7.155)$$

see eqn (7.261) of Appendix C. We note that relation (7.146) is satisfied by both eqns (7.152) and (7.154).

In eqn (7.154), $1 + \widehat{R} \neq 0$; hence, $K(x)$ can be obtained from

$$\widehat{K}(\xi) \approx -\frac{\nu\, \hat{q}(|\boldsymbol{\xi}|)}{\nu\,[1 + \hat{q}(|\boldsymbol{\xi}|)] - i\xi^0}. \tag{7.156}$$

Let us note that in this case, eqn (7.147) is satisfied. It can be shown, by means of contour integration and Jordan's lemma, that

$$K(x) \approx -\frac{\nu}{(2\pi)^3}\, u(x^0) \int \hat{q}(|\boldsymbol{\xi}|) e^{i\boldsymbol{\xi}\cdot\mathbf{x}} e^{-\nu(1+\hat{q})x^0}\, d^3\xi. \tag{7.157}$$

Moreover, it is straightforward to verify, by integrating this expression for $K(x)$ over all spacetime, that eqn (7.151) is satisfied in this case. Our approximation method has thus led to a manageable expression for kernel K; the nature and limitations of this simplification have been studied in Chicone and Mashhoon (2013) and Mashhoon (2013b).

Following the determination of the reciprocal kernel R in eqn (7.143) and the approximate determination of kernel K, it is now possible to treat more explicitly the gravitational field of an isolated source in the linear post-Newtonian approximation of nonlocal gravity. We begin with the treatment of the time-independent field of a stationary source in the next section, which amounts to a nonlocal extension of steady-state gravitoelectromagnetism (GEM) of GR. A dynamic nonlocal generalization of the standard GEM appears to be intractable.

7.6 Gravitational Field of a Stationary Source

The purpose of this section is to study the implications of the linearized nonlocal field equation in the *transverse gauge* ($\bar{h}^{\mu\nu}{}_{,\nu} = 0$) for the weak time-independent gravitational field of an isolated *stationary* source. To this end, let us note that in the field eqns (7.71)–(7.72),

$$\mathcal{T}_{\mu\nu}(\mathbf{x}) = T_{\mu\nu}(\mathbf{x}) + \int q(\mathbf{x} - \mathbf{y})\, T_{\mu\nu}(\mathbf{y})\, d^3y, \tag{7.158}$$

as a result of eqn (7.145). In a similar way, we can show that $S_{\mu\nu} = 0$, since $S_{\mu\nu} = 0$ in this case. To see this, let us consider eqn (7.25) that defines $S_{\mu\nu}$; for a time-independent torsion field, eqn (7.25) takes the form

$$\partial_i \int \chi(\mathbf{x} - \mathbf{y})\, \mathcal{C}_\mu{}^i{}_\nu(\mathbf{y})\, d^3y = -S_{\mu\nu} + \int \chi(\mathbf{x} - \mathbf{y})\, {}^0G_{\mu\nu}(\mathbf{y})\, d^3y, \tag{7.159}$$

as a consequence of eqn (7.148). Following essentially the same steps as in our discussion of eqn (7.25), we find that $S_{\mu\nu} = 0$, since the boundary surface in this case is at spatial infinity. Here, the seeming disappearance of the light cone is consistent with the complete temporal independence of the gravitational field. It follows from $S_{\mu\nu} = 0$ and eqn (7.35) that the integral constraints in the stationary case reduce to $U_{[\mu\nu]} = 0$, which contain only $\phi_{\mu\nu}$ and the constraints vanish for $\phi_{\mu\nu} = 0$. We can therefore set $\phi_{\mu\nu} = 0$ in the gravitational potentials of a stationary source. In the transverse gauge, the linearized field equation (7.70) of nonlocal gravity thus reduces in the stationary case to the ten field equations

$$\nabla^2 \bar{h}_{\mu\nu}(\mathbf{x}) = -2\kappa \left[T_{\mu\nu}(\mathbf{x}) + \int q(\mathbf{x} - \mathbf{y}) \, T_{\mu\nu}(\mathbf{y}) \, d^3y \right]. \tag{7.160}$$

The spatial reciprocal kernel q is independent of the speed of light; therefore, the standard static GEM approach can be adopted in this nonlocal case. Let us write the energy–momentum tensor for a slowly rotating source with $|\mathbf{v}| \ll c$ as $T^{00} = \rho c^2$ and $T^{0i} = c j^i$, where $\mathbf{j} = \rho \mathbf{v}$ is the matter current; moreover, the matter stresses are assumed to be independent of c and of the form $T_{ij} \sim \rho v_i v_j + p \, \delta_{ij}$, where p is the pressure. Then, with $\bar{h}_{00} = -4\Phi/c^2$, we have a static gravitoelectric potential $\Phi(\mathbf{x})$ that satisfies eqn (7.80) of the Newtonian limit of nonlocal gravity. Next, $\bar{h}_{0i} = -2\,{}^g A_i/c^2$, where ${}^g\mathbf{A}(\mathbf{x})$ is the static gravitomagnetic vector potential that satisfies

$$\nabla^2 \, {}^g\mathbf{A}(\mathbf{x}) = -\frac{8\pi G}{c} \left[\mathbf{j}(\mathbf{x}) + \int q(\mathbf{x} - \mathbf{y}) \, \mathbf{j}(\mathbf{y}) \, d^3y \right]. \tag{7.161}$$

It is interesting to note here the contribution of the "dark" current, $\mathbf{j}_D(\mathbf{x})$, which is the convolution of the regular current with the reciprocal spatial kernel q, to the gravitomagnetic vector potential. The solution of eqn (7.161) is thus given by

$$\frac{1}{2} \, {}^g\mathbf{A}(\mathbf{x}) = \frac{G}{c} \int \frac{\mathbf{j}(\mathbf{y}) + \mathbf{j}_D(\mathbf{y})}{|\mathbf{x} - \mathbf{y}|} \, d^3y. \tag{7.162}$$

Finally, eqn (7.160) implies that $\bar{h}_{ij} = O(c^{-4})$ and is therefore neglected. Indeed, all terms of $O(c^{-4})$ are neglected in the standard linear GEM analysis (Mashhoon 2007d).

It is simple to check that the energy–momentum conservation law, eqn (7.13), reduces in our nonlocal steady-state GEM treatment to $\nabla \cdot \mathbf{j} = 0$, which leads to $\nabla \cdot \mathbf{j}_D = 0$ as well, and is consistent with the transverse gauge condition $\nabla \cdot {}^g\mathbf{A} = 0$. With these conditions, one can develop a nonlocal version of the steady-state GEM for any suitable stationary source (Teyssandier 1977, 1978). In fact, with $\mathbf{E}_g = \nabla\Phi$ and $\mathbf{B}_g = \nabla \times {}^g\mathbf{A}$, we have GEM fields with dimensions of acceleration such that

$$\nabla \cdot \mathbf{E}_g = 4\pi G \left(\rho + \rho_D \right), \qquad \nabla \times \mathbf{E}_g = 0, \tag{7.163}$$

$$\nabla \cdot \left(\frac{1}{2} \mathbf{B}_g \right) = 0, \qquad \nabla \times \left(\frac{1}{2} \mathbf{B}_g \right) = \frac{4\pi G}{c} (\mathbf{j} + \mathbf{j}_D). \tag{7.164}$$

These are the steady-state field equations of nonlocal GEM.

The GEM spacetime metric in this nonlocal case has the usual form

$$ds^2 = -c^2 \left(1 + 2\frac{\Phi}{c^2} \right) dt^2 - \frac{4}{c} ({}^g\mathbf{A} \cdot d\mathbf{x}) dt + \left(1 - 2\frac{\Phi}{c^2} \right) \delta_{ij} \, dx^i dx^j. \tag{7.165}$$

Here, $\Phi(\mathbf{x})$ is the gravitoelectric potential of nonlocal gravity in the Newtonian regime given by eqn (7.80) and ${}^g\mathbf{A}(\mathbf{x}) = O(c^{-1})$ is the gravitomagnetic vector potential given by eqn (7.162). It is now possible to discuss the motion of test particles and null rays that follow geodesics associated with this metric. For instance, for the motion of test particles, we recover the gravitational analog of the Lorentz force law (Mashhoon 2007d).

In view of possible astrophysical applications, it is convenient, for the sake of simplicity, to assume that the reciprocal kernel is q_0 given by eqn (7.100); then, Φ and $^g\mathbf{A}$ are given by

$$\Phi(\mathbf{x}) = -G \int \left[1 + \alpha_0(1 - e^{-\mu_0 r}) + \frac{r}{\lambda_0} E_1(\mu_0 r) \right] \frac{\rho(\mathbf{y})}{|\mathbf{x} - \mathbf{y}|} d^3 y \qquad (7.166)$$

and

$$\frac{1}{2} {}^g\mathbf{A}(\mathbf{x}) = \frac{G}{c} \int \left[1 + \alpha_0(1 - e^{-\mu_0 r}) + \frac{r}{\lambda_0} E_1(\mu_0 r) \right] \frac{\mathbf{j}(\mathbf{y})}{|\mathbf{x} - \mathbf{y}|} d^3 y, \qquad (7.167)$$

where $r = |\mathbf{x} - \mathbf{y}|$ and E_1 is the exponential integral function defined in eqn (7.120). Moreover, we note that (Abramowitz and Stegun 1964)

$$\frac{\alpha_0}{2} \frac{\mu_0 r}{\mu_0 r + 1} e^{-\mu_0 r} < \frac{r}{\lambda_0} E_1(\mu_0 r) \leq \frac{\alpha_0}{2} e^{-\mu_0 r}. \qquad (7.168)$$

These potentials can be explicitly calculated in any given situation involving an isolated material source using general methods familiar from classical electrodynamics (Jackson 1999). We are particularly interested in the propagation of light rays in this gravitational field. This is necessary in order to explain astrophysical phenomena associated with gravitational lensing without invoking dark matter. In linearized nonlocal gravity, just as in linearized GR, the effects due to gravitoelectric and gravitomagnetic fields could be treated separately and then linearly superposed. Thus, as is well known, the bending of light rays due to the gravitoelectric potential Φ is given by twice the Newtonian expectation; see Chapter 8. The influence of the gravitomagnetic field on the propagation of light in GR has been discussed in Mashhoon (1993b); according to GR, the gravitomagnetic bending of light rays passing near a slowly rotating source is generally smaller in magnitude than the gravitoelectric deflection by a factor of the order of $|\mathbf{v}|/c \ll 1$. It is therefore usually ignored in the discussion of gravitational lensing (Schneider, Ehlers and Falco 1999; Petters, Levine and Wambsganss 2001; Perlick 2004). The situation regarding the gravitomagnetic deflection of light in nonlocal gravity is, however, somewhat more complicated. For instance, if the integration in eqns (7.166) and (7.167) extends over a structure such as a cluster of galaxies for which $\mu_0 r \gg 1$, then the quantity in square brackets in these equations essentially reduces to $1 + \alpha_0$. Therefore, we are in effect working in the domain of linearized GR, but with enhanced gravity, namely, with $G \to G(1 + \alpha_0)$.

Imagine the propagation of light in the gravitational field of an isolated static source that moves uniformly with speed $c\beta$, $-1 < \beta < 1$ in the background Minkowski spacetime. This case is of interest in connection with the Bullet Cluster (Clowe *et al.* 2006; Clowe, Randall and Markevitch 2007) and is treated in the next section; however, the general case of a time-dependent source requires a more extensive investigation.

7.7 Light Deflection due to a Uniformly Moving Mass

Consider the stationary case treated in Section 7.6 with no matter current. In the rest frame of such a static gravitational source, it is convenient to think of this body in terms of a collection of fixed mass elements $m_j, j = 0, 1, 2, \ldots, N$. Then in eqn (7.166), we can write

$$\rho(\mathbf{x}) = \sum_j m_j \, \delta(\mathbf{x} - \mathbf{x}_j), \qquad \Phi(\mathbf{x}) = \sum_j m_j \, \phi(|\mathbf{x} - \mathbf{x}_j|), \qquad (7.169)$$

where,

$$\phi(r) = -\frac{G}{r} \left[1 + \alpha_0(1 - e^{-\mu_0 r}) + \frac{r}{\lambda_0} E_1(\mu_0 r) \right]. \qquad (7.170)$$

The spacetime metric in the rest frame of the source is given by eqn (7.165) with $^g\mathbf{A} = 0$. Let us remark here that for $\mu_0 r \gg 1$, $\phi(r) \approx -(1 + \alpha_0)G/r$ in NLG, which is $1 + \alpha_0$ times the Newtonian gravitational potential per unit mass. To return to GR, we can formally set $\lambda_0 = \infty$ and $\alpha_0 = 0$ in NLG.

In the background global inertial frame with coordinates $x^\mu = (t, x, y, z)$, the gravitational source under consideration here moves uniformly with speed β, $|\beta| < 1$, along the x axis. The moving source acts as a gravitational lens in deflecting a ray of light that, in its unperturbed state, is parallel to the z axis, pierces the (x, y) plane at the point (a, b) and passes over the body. We assume that the lens is relatively thin and its matter is mostly distributed in and near the (x, y) plane. We are interested in the deflection of the ray by the lens when the point (a, b) and the lens are in a definite geometric configuration as recorded by the static inertial observers at spatial infinity. It will turn out that the end result is independent of such a configuration. Let us assume that the desired configuration—that is, the observationally preferred position of the source relative to the unperturbed ray of light—occurs at time $t = t_0$, when, for instance, mass element m_j of the lens is at $\mathbf{x}_j$. The source is then completely at rest in a comoving frame with coordinates $x'^\mu = (t', x', y', z')$. To write the Lorentz transformation that connects the two frames, let us choose mass point m_0 to be the origin of the comoving system; then,

$$t' = \gamma[(t - t_0) - \beta(x - x_0)],$$
$$x' = \gamma[(x - x_0) - \beta(t - t_0)], \quad y' = y - y_0, \quad z' = z - z_0. \qquad (7.171)$$

Here, γ is the Lorentz factor corresponding to β. Thus m_0 with coordinates $x_0^\mu = (t_0, x_0, y_0, z_0)$ is at the origin of coordinates in the rest frame of the source, namely, $x_0'^\mu = (0, 0, 0, 0)$. As the whole static source is at rest in the comoving frame at t_0, eqn (7.171) can be written with respect to any other mass point m_j as

$$t' - t'_j = \gamma[(t - t_0) - \beta(x - x_j)],$$
$$x' - x'_j = \gamma[(x - x_j) - \beta(t - t_0)], \quad y' - y'_j = y - y_j,$$
$$z' - z'_j = z - z_j, \qquad (7.172)$$

where $t'_j = -\gamma\beta(x_j - x_0)$, etc. The result of the Lorentz transformation is that the invariant spacetime interval (7.165) can be written in the observers' rest frame as

$$ds^2 = (\eta_{\mu\nu} + h_{\mu\nu}) \, dx^\mu \, dx^\nu, \qquad (7.173)$$

where the non-zero components of $h_{\mu\nu}$ are given by

$$h_{00} = h_{11} = -2\gamma^2(1+\beta^2)\,\Phi, \tag{7.174}$$

$$h_{01} = h_{10} = 4\beta\gamma^2\,\Phi, \qquad h_{22} = h_{33} = -2\,\Phi. \tag{7.175}$$

Here, Φ depends upon time and is given by

$$\Phi = \sum_j m_j\,\phi(u_j), \tag{7.176}$$

where $u_j := |\mathbf{x}' - \mathbf{x}'_j|$ is the *positive* square root of

$$u_j^2 = \gamma^2\,[(x-x_j) - \beta(t-t_0)]^2 + (y-y_j)^2 + (z-z_j)^2, \tag{7.177}$$

in accordance with eqn (7.172). In practice, $|\beta| \ll 1$; nevertheless, we perform the calculations in this section for arbitrary β, but then we set $|\beta| \ll 1$ in the end result. To preserve our linear weak-field approximation scheme, however, β^2 cannot be too close to unity. Moreover, as in Section 7.6, the antisymmetric tetrad potentials vanish (i.e. $\phi_{\mu\nu} = 0$) and the transverse gauge condition is satisfied, as these are maintained under Lorentz transformation.

In the geometric optics approximation, a light ray propagates along a null geodesic

$$\frac{dk^\mu}{d\varkappa} + {}^0\Gamma^\mu_{\alpha\beta}\,k^\alpha k^\beta = 0, \tag{7.178}$$

where the spacetime propagation vector $k^\mu = dx^\mu/d\varkappa$ is tangent to the corresponding world line and $\varkappa$ is an affine parameter along the path. Let $\tilde{k}^\mu = dx^\mu/d\tilde{\varkappa}$ represent the unperturbed light ray whose trajectory is given by

$$x(t) = a, \qquad y(t) = b, \qquad z(t) = \mathfrak{z} + t - t_0, \tag{7.179}$$

where a, b and $\mathfrak{z}$ are constants. To simplify matters in this case, we can choose $\tilde{\varkappa} = t - t_0$, so that $\tilde{k}^\mu = (1,0,0,1)$.

A comment is in order here regarding the physical significance of $\mathfrak{z}$. In the regime of geometric optics, eqn (7.178) with $k^\mu = dx^\mu/d\varkappa$ represents the equation of motion of the light particle ("photon") along the null ray. At $t = t_0$, $\mathfrak{z}$ indicates the position of the unperturbed photon along the z axis away from the (x, y) plane.

To calculate the deflection of light from eqn (7.178), we consider the net deviation Δk^μ,

$$\Delta k^\mu = k^\mu(+\infty) - k^\mu(-\infty) = -\int_{-\infty}^{\infty} {}^0\Gamma^\mu_{\alpha\beta}\,k^\alpha k^\beta\,d\varkappa, \tag{7.180}$$

where $k^\mu(-\infty) = \tilde{k}^\mu$. The integrand here is computed along the null geodesic. To linear order, however, the calculation can be performed along the unperturbed light ray, namely,

$$\Delta k^\mu = -\int_{-\infty}^{\infty} \mathbb{L}^\mu(t_0 + \tilde{\varkappa}, a, b, \mathfrak{z} + \tilde{\varkappa})\,d\tilde{\varkappa}, \tag{7.181}$$

where $\tilde{\varkappa} = t - t_0$ and

$$\mathbb{L}^\mu(x) := {}^0\Gamma^\mu_{\alpha\beta}(x)\,\tilde{k}^\alpha \tilde{k}^\beta. \tag{7.182}$$

Here, the Christoffel symbols,

$$
{}^{0}\Gamma^{\mu}_{\alpha\beta} = \frac{1}{2}\eta^{\mu\nu}(h_{\nu\alpha,\beta} + h_{\nu\beta,\alpha} - h_{\alpha\beta,\nu}),
\tag{7.183}
$$

are determined from eqns (7.174)–(7.177). It follows from a detailed calculation that $\mathbb{L}^{\mu}(t_0 + \tilde{\varkappa}, a, b, \mathfrak{z} + \tilde{\varkappa})$ can be represented as

$$
\mathbb{L}^{0} = 2\gamma^2 \sum_{j} m_j \frac{1}{u_j} \frac{d\phi(u_j)}{du_j} \left[\gamma^2\tilde{\varkappa} - \beta^3\gamma^2(a - x_j) + (1 + \beta^2)(\mathfrak{z} - z_j)\right],
\tag{7.184}
$$

$$
\mathbb{L}^{1} = 2\gamma^2 \sum_{j} m_j \frac{1}{u_j} \frac{d\phi(u_j)}{du_j} \left[\beta\gamma^2\tilde{\varkappa} + (1 - \beta^2\gamma^2)(a - x_j) + 2\beta(\mathfrak{z} - z_j)\right],
\tag{7.185}
$$

$$
\mathbb{L}^{2} = 2\gamma^2 \sum_{j} m_j \frac{1}{u_j} \frac{d\phi(u_j)}{du_j} (b - y_j),
\tag{7.186}
$$

$$
\mathbb{L}^{3} = 2\beta\gamma^2 \sum_{j} m_j \frac{1}{u_j} \frac{d\phi(u_j)}{du_j} \left[(a - x_j) + \beta(\mathfrak{z} - z_j)\right].
\tag{7.187}
$$

In principle, the integration in eqn (7.181) can now be carried through to determine the net deviation of the ray due to the gravitational attraction of the moving source; however, this calculation would involve

$$
\frac{1}{r}\frac{d\phi}{dr} = \frac{G}{r^3}\left[1 + \alpha_0 - \alpha_0(1 + \frac{1}{2}\mu_0 r)e^{-\mu_0 r}\right].
\tag{7.188}
$$

We address the problem of calculating the relevant integrals in Appendix D. Using the results of Appendix D, we find that for $\beta \neq 0$,

$$
\Delta k^0 = \beta\Delta k^1 = \Delta k^3 = -4\beta\gamma G \sum_{j} \frac{m_j \mathcal{X}_j}{\mathcal{X}_j^2 + \mathcal{Y}_j^2}\left[1 + \alpha_0 - \alpha_0\,\mathfrak{I}\big(\mu_0\sqrt{\mathcal{X}_j^2 + \mathcal{Y}_j^2}\big)\right],
\tag{7.189}
$$

$$
\Delta k^2 = -4\gamma G \sum_{j} \frac{m_j \mathcal{Y}_j}{\mathcal{X}_j^2 + \mathcal{Y}_j^2}\left[1 + \alpha_0 - \alpha_0\,\mathfrak{I}\big(\mu_0\sqrt{\mathcal{X}_j^2 + \mathcal{Y}_j^2}\big)\right],
\tag{7.190}
$$

where

$$
\mathcal{X}_j := (a - x_j) + \beta(\mathfrak{z} - z_j), \qquad \mathcal{Y}_j := b - y_j.
\tag{7.191}
$$

Moreover, $\mathfrak{I}(x) := \mathcal{J}_2(x) + (x/2)\mathcal{J}_1(x)$, where $\mathcal{J}_1$ and $\mathcal{J}_2$ are discussed in Appendix D; indeed,

$$
\mathfrak{I}(x) = \int_{0}^{\infty} \frac{(1 + \frac{1}{2}x\cosh v)\,e^{-x\cosh v}}{\cosh^2 v}\,dv,
\tag{7.192}
$$

so that $\mathfrak{I}(0) = 1$ and $\mathfrak{I}(\infty) = 0$. For $\alpha_0 = 0$, formulas (7.189)–(7.191) extend the results of previous work on light deflection in GR; see Wucknitz and Sperhake (2004) and the references cited therein.

With z as the line-of-sight coordinate, the overall effect of the deflection of the light ray in the plane of the sky can be expressed via the angles $\check{\alpha} = -(\Delta k^1, \Delta k^2)$, where

$$\check{\alpha} = 4\gamma G \sum_j m_j \frac{(\mathcal{X}_j, \mathcal{Y}_j)}{\mathcal{X}_j{}^2 + \mathcal{Y}_j{}^2} \left[1 + \alpha_0 - \alpha_0 \, \Im \left(\mu_0 \sqrt{\mathcal{X}_j{}^2 + \mathcal{Y}_j{}^2} \right) \right]. \tag{7.193}$$

Other than an overall factor of γ, the effect of the motion of the gravitational source appears here in $\beta(\mathfrak{z} - z_j)$ contained in $\mathcal{X}_j$.

The end result for the deflection angle $\check{\alpha}$, and hence $\mathcal{X}_j$ and $\mathcal{Y}_j$, is independent of t_0 and any specific configuration of the lens and the photon. To illustrate this important point, we note that the photon crosses the (x, y) plane at time $\bar{t}_0 = t_0 - \mathfrak{z}$, when the point mass m_j, say, is at $(\bar{x}_j, \bar{y}_j, \bar{z}_j)$; then, repeating our calculation in this case would yield $\mathcal{X}_j = (a - \bar{x}_j) - \beta \bar{z}_j$ and $\mathcal{Y}_j = b - \bar{y}_j$. These are the same quantities as given in eqn (7.191), since the lens has moved during the time interval $\mathfrak{z}$; that is, $x_j = \bar{x}_j + \beta \mathfrak{z}$, $y_j = \bar{y}_j$ and $z_j = \bar{z}_j$.

Let us now suppose that the gravitational lens is thin; that is, the extent of the deflecting mass in the z direction is small (Schneider, Ehlers and Falco 1999). Therefore, we may neglect $\beta z_j = \beta \bar{z}_j$ in $\mathcal{X}_j$, since in practice $|\beta| \ll 1$. Then, at the instant that the unperturbed photon crosses the lens plane, it is possible to express eqn (7.193) for a moving extended lens in a form that can be incorporated into the standard lens equation, namely,

$$\check{\alpha}(\boldsymbol{\theta}) = \frac{4G}{c^2} \int \frac{\boldsymbol{\theta} - \bar{\boldsymbol{\theta}}}{|\boldsymbol{\theta} - \bar{\boldsymbol{\theta}}|^2} \left[1 + \alpha_0 - \alpha_0 \, \Im(\mu_0 |\boldsymbol{\theta} - \bar{\boldsymbol{\theta}}|) \right] \bar{\Sigma}(\bar{\boldsymbol{\theta}}) \, d^2\bar{\theta}, \tag{7.194}$$

where $\bar{\Sigma}(\bar{\boldsymbol{\theta}})$ is the surface mass density of the deflecting source ("thin lens") and the integration is carried over the lens plane, which essentially coincides with the (x, y) plane. Thus, in eqn (7.194),

$$\boldsymbol{\theta} = (a, b), \qquad \bar{\boldsymbol{\theta}} = (\bar{x}, \bar{y}), \tag{7.195}$$

where $\boldsymbol{\theta}$ is the unperturbed position of the photon as it crosses the lens plane and $\bar{\boldsymbol{\theta}}$ indicates the position of a point of the extended lens at that instant. Furthermore, it is possible to write $\check{\alpha} = \nabla \check{\Psi}$, where the lensing potential $\check{\Psi}$ is given by

$$\check{\Psi}(\boldsymbol{\theta}) = \frac{4G}{c^2} \int \left[\ln |\boldsymbol{\theta} - \bar{\boldsymbol{\theta}}| + \alpha_0 \, \mathfrak{N}(\mu_0 |\boldsymbol{\theta} - \bar{\boldsymbol{\theta}}|) \right] \bar{\Sigma}(\bar{\boldsymbol{\theta}}) \, d^2\bar{\theta}. \tag{7.196}$$

Here, the first term in the integrand is the GR result, which follows from $\nabla \ln |\mathbf{x}| = \mathbf{x}/|\mathbf{x}|^2$, while the nonlocal contribution to the lensing potential involves $\mathfrak{N}$, which is related to $\Im$ via $d\mathfrak{N}/dx = [1 - \Im(x)]/x$.

It follows from these results that in the theoretical interpretation of gravitational lensing data in accordance with nonlocal gravity, due account must be taken of the existence of the repulsive "Yukawa" part of the gravitational potential as well. This may lead to the resolution of problems associated with light deflection by colliding clusters of galaxies. However, the confrontation of the nonlocal gravity theory with lensing data would require a separate detailed investigation.

7.8   Gravitation and Nonlocality

We have introduced nonlocality into gravitational theory via a *constitutive ansatz*. The manner in which this has been done can lead to different nonlocal theories. In particular, our ansatz consists of two distinct parts: a *nonlocal* connection between $N_{\mu\nu\rho}$ and $X_{\mu\nu\rho}$ involving a scalar kernel as well as a *local* connection between $X_{\mu\nu\rho}$ and the torsion tensor. As described in detail in Appendix A, the physical interpretation of the theory can depend on the nature of the local connection between $X_{\mu\nu\rho}$ and $\mathfrak{C}_{\mu\nu\rho}$. For instance, choosing $X_{\mu\nu\rho} = \mathfrak{C}_{[\mu\nu\rho]}$ implies that linearized nonlocal gravity is essentially equivalent to linearized GR, so that the effects of nonlocality are expected to appear in the nonlinear regime of the theory; see Appendix A. However, gravitation is the weakest force in nature; therefore, the linear approximation of nonlocal gravity contains many of the possible physical applications of the theory including the Newtonian regime of nonlocal gravity as well as linearized gravitational radiation.

In retrospect, the local connection adopted in this work in eqn (6.109), namely, $X_{\mu\nu\rho} = \mathfrak{C}_{\mu\nu\rho} + \check{p}\,(\check{C}_\mu\,g_{\nu\rho} - \check{C}_\nu\,g_{\mu\rho})$ with $\check{p}$ a non-zero constant, makes it possible to have nonlocality in the linear regime. It follows from Section 7.6 that $\check{p}$ does not affect the gravitational field of a *stationary* source in linearized nonlocal gravity; therefore, $\check{p}$ is expected to be significant for the gravitational field of time-varying sources in their near zones. Furthermore, eqn (6.109) makes it possible to recover the Tohline–Kuhn modified gravity approach to the problem of the "flat" rotation curves of the spiral galaxies and to place this scheme within the fully relativistic framework of nonlocal gravity. In this way the *effective dark matter* of nonlocal gravity has been identified with the hypothetical *dark matter* invoked in astrophysics. *Henceforward, we take the view that what appears as dark matter in astrophysics and cosmology is in fact the nonlocal aspect of the gravitational interaction.* This is indeed a possibility due to the persistent negative result of experiments that have searched for the particles of dark matter.

Dark matter is currently indispensable for explaining: (i) gravitational dynamics of galaxies and clusters of galaxies (Zwicky 1933, 1937; Rubin and Ford 1970; Roberts and Whitehurst 1975; Sofue and Rubin 2001; Seigar 2015; Harvey *et al.* 2015), (ii) gravitational lensing observations in general and the Bullet Cluster (Clowe *et al.* 2006; Clowe, Randall and Markevitch 2007) in particular and (iii) the formation of structure in cosmology and the large scale structure of the universe. The nonlocal character of gravity, however, cannot yet replace dark matter on all physical scales. We emphasize that nonlocal gravity theory is so far in the early stages of development and only some of its implications have been confronted with observation; see Chapter 8. Moreover, a beginning has recently been made in the development of nonlocal Newtonian cosmology; see Chapter 10.

7.9   Appendix A: Constitutive Relation of NLG

This appendix is devoted to a discussion of the constitutive relation of nonlocal gravity. More precisely, we wish to examine the *local* connection between $X_{\mu\nu\rho}$ and the torsion tensor in eqn (6.109) and its implications for linearized NLG. Ultimately, of course, the confrontation of the theory with observation can determine the right relation.

Imagine, for instance, the possibility of choosing $X_{\mu\nu\rho} = \mathfrak{C}_{[\mu\nu\rho]}$. Returning to the general form of the linearized field eqns (7.21)–(7.22), we have in this case

$$X_{(\mu}{}^{\sigma}{}_{\nu)} = 0, \qquad X_{[\mu}{}^{\sigma}{}_{\nu]} = \frac{1}{2}\eta^{\sigma\rho}\,\phi_{[\mu\rho,\nu]}, \qquad (7.197)$$

since in the linear approximation $\mathfrak{C}_{[\mu\rho\nu]} = \frac{1}{2}\phi_{[\mu\rho,\nu]}$. Thus eqn (7.21) is the same here as in the linearized Einstein equation of GR and eqn (7.22) takes the form

$$\eta^{\sigma\rho}\,\partial_{\sigma}\int K(x-y)\,\phi_{[\mu\rho,\nu]}(y)\,d^4y = 0. \qquad (7.198)$$

In this case, we have a *complete* separation of the 10 dynamic metric variables $\bar{h}_{\mu\nu}$ from the 6 tetrad variables $\phi_{\mu\nu}$. The integral constraints (7.198) can be satisfied with

$$\phi_{\mu\nu} = 0. \qquad (7.199)$$

Thus at the linear level, this theory of nonlocal gravity is essentially equivalent to local GR; therefore, the connection between nonlocal gravity and dark matter disappears in this case.

In connection with the separation of the metric variables from the tetrad variables, let us consider the possibility that

$$X_{\mu\nu\rho} = \mathfrak{C}_{\mu\nu\rho} + \frac{1}{2}\mathfrak{C}_{\rho\mu\nu}. \qquad (7.200)$$

It is useful to note that we now have in eqns (7.21)–(7.22),

$$X_{(\mu}{}^{\sigma}{}_{\nu)} = \mathfrak{C}_{(\mu}{}^{\sigma}{}_{\nu)}, \qquad X_{[\mu}{}^{\sigma}{}_{\nu]} = \frac{3}{4}\eta^{\sigma\rho}\,\phi_{[\mu\rho,\nu]}. \qquad (7.201)$$

The constraint equations in this case contain the secondary tetrad variables $\phi_{\mu\nu}$ exclusively. Thus to simplify matters, one can again assume that $\phi_{\mu\nu} = 0$; then, the constraint equations are satisfied and the ten dynamic nonlocal field equations depend solely upon $\bar{h}_{\mu\nu}$. However, we note that in this case $X_{\mu\nu\rho} \neq -X_{\nu\mu\rho}$, so that $\mathcal{N}_{\mu\nu}$ in eqn (6.121) does not in general transform as a tensor under arbitrary coordinate transformations. Thus this case violates the basic geometric structure of nonlocal gravity theory.

Clearly, one can concoct other combinations and study their consequences; however, the rest of this appendix is devoted to a detailed discussion of the difficulty associated with the simplest possibility, namely, $X_{\mu\nu\rho} = \mathfrak{C}_{\mu\nu\rho}$, adopted, along with the possibility that $T_{\mu\nu} \neq T_{\nu\mu}$, in Hehl and Mashhoon (2009a, 2009b) and Mashhoon (2011b). In nonlocal gravity (NLG), $T_{\mu\nu} = T_{\nu\mu}$ as in GR; however, $X_{\mu\nu\rho} = \mathfrak{C}_{\mu\nu\rho}$ then leads, in a manner that is independent of any gauge condition, to a contradiction. In other words, the field equations in this case can be obtained from eqns (7.23)–(7.43) for $\check{p} = 0$, and we recall here that $S_{0\mu} = 0$. Let us take

$$^{0}G_{00} = \kappa\,T_{00} \qquad (7.202)$$

from the set of field equations for the metric variables and write it using eqn (7.8) as

$$\bar{h}_{00,i}{}^{i} - \bar{h}_{ij,}{}^{ij} = -2\kappa \, \mathcal{T}_{00}, \tag{7.203}$$

where $\mathcal{T}_{00}$ is the total energy density of the source defined by eqn (7.61). Next, we take eqn (7.38) from the set of integral constraint equations, namely,

$$\int K(x - y) \, \delta(x^0 - y^0 - |\mathbf{x} - \mathbf{y}|) \mathcal{W}_i(y) \, d^4y = 0, \tag{7.204}$$

where, in agreement with eqn (7.36), $\mathcal{W}_i$ is given by

$$\mathcal{W}_i = -\phi_{ij,}{}^{j} - \left(\bar{h}_{00,i} - \bar{h}_{ij,}{}^{j} \right). \tag{7.205}$$

Integrating over the temporal coordinate in eqn (7.204), we find

$$\int K(|\mathbf{x} - \mathbf{y}|, \mathbf{x} - \mathbf{y}) \, \mathcal{W}_i(x^0 - |\mathbf{x} - \mathbf{y}|, \mathbf{y}) \, d^3y = 0. \tag{7.206}$$

We note that

$$\delta^{ik} \mathcal{W}_{i,k} = -\bar{h}_{00,i}{}^{i} + \bar{h}_{ij,}{}^{ij}, \tag{7.207}$$

since $\phi_{ij} = -\phi_{ji}$. Hence, we find from eqn (7.203) the interesting result that

$$\delta^{ij} \mathcal{W}_{i,j} = 2\kappa \, \mathcal{T}_{00}. \tag{7.208}$$

To demonstrate that eqn (7.208) is in general incompatible with eqn (7.206), we apply the partial derivative operator $\partial/\partial x^j$ to eqn (7.206). To simplify the calculation, let us define auxiliary functions $\bar{\eta}$ and $\bar{F}$ by

$$\bar{\eta} := x^0 - |\mathbf{x} - \mathbf{y}|, \qquad \bar{F}(\mathbf{x} - \mathbf{y}) := K(|\mathbf{x} - \mathbf{y}|, \mathbf{x} - \mathbf{y}). \tag{7.209}$$

Then, we have that

$$\frac{\partial \bar{\eta}}{\partial x^j} = -\frac{\partial \bar{\eta}}{\partial y^j}, \qquad \frac{\partial \bar{F}}{\partial x^j} = -\frac{\partial \bar{F}}{\partial y^j}. \tag{7.210}$$

Hence, taking the derivative of eqn (7.206) results in

$$\partial_j \int \bar{F} \, \mathcal{W}_i \, d^3y = \int \left[-\frac{\partial \bar{F}}{\partial y^j} \mathcal{W}_i(\bar{\eta}, \mathbf{y}) + \bar{F} \frac{\partial \bar{\eta}}{\partial x^j} \mathcal{W}_{i,0}(\bar{\eta}, \mathbf{y}) \right] d^3y = 0. \tag{7.211}$$

Using integration by parts, we find that

$$\int \frac{\partial}{\partial y^j} (\bar{F} \mathcal{W}_i) \, d^3y = \int \bar{F} \left[\frac{\partial}{\partial y^j} \mathcal{W}_i(\bar{\eta}, \mathbf{y}) + \frac{\partial \bar{\eta}}{\partial x^j} \mathcal{W}_{i,0}(\bar{\eta}, \mathbf{y}) \right] d^3y. \tag{7.212}$$

From

$$\frac{\partial}{\partial y^j} \mathcal{W}_i(\bar{\eta}, \mathbf{y}) = \frac{\partial \bar{\eta}}{\partial y^j} \mathcal{W}_{i,0}(\bar{\eta}, \mathbf{y}) + \mathcal{W}_{i,j}(\bar{\eta}, \mathbf{y}) \tag{7.213}$$

and eqn (7.210), we see that in eqn (7.212) terms involving $\mathcal{W}_{i,0}$ cancel; thus, eqn (7.212) can be written as

$$\int \frac{\partial}{\partial y^j} (\bar{F} \mathcal{W}_i) \, d^3y = \int \bar{F} \, \mathcal{W}_{i,j} \, d^3y. \tag{7.214}$$

Taking the trace of this equation and using Gauss's theorem, we finally get from eqn (7.208) that

$$\int \bar{F}\left(\delta^{ij}\,\mathcal{W}_{i,j}\right)d^3y = 2\kappa \int K(|\mathbf{x}-\mathbf{y}|,\mathbf{x}-\mathbf{y})\,\mathcal{T}_{00}(\bar{\eta},\mathbf{y})\,d^3y = 0. \qquad (7.215)$$

This important result can also be expressed as

$$\int W(x-y)\mathcal{T}_{00}(y)\,d^4y = 0, \qquad (7.216)$$

where kernel W is given by eqn (7.67).

The source of the gravitational field has been assumed to be finite and isolated in space, but is otherwise arbitrary. It is conceivable that eqn (7.216) could be satisfied for rather special source configurations. In general, however, eqn (7.216) is not satisfied for an arbitrary source, which indicates that a solution of the field equation does not exist. We have thus shown, without using any gauge condition, that the symmetric metric part of the field equation of NLG is in general incompatible with the antisymmetric tetrad part for $X_{\mu\nu\rho} = \mathfrak{C}_{\mu\nu\rho}$. The incompatibility proof can be directly extended to constitutive relations of the forms $X_{\mu\nu\rho} = \mathfrak{C}_{\mu\nu\rho} + p'\,\mathfrak{C}_{[\mu\nu\rho]}$ and $X_{\mu\nu\rho} = \mathfrak{C}_{\mu\nu\rho} + p''\,E_{\mu\nu\rho\sigma}\,C^\sigma$, where $p' \neq 0$ and $p'' \neq 0$ are constant parameters.

Let us now consider the constitutive relation adopted in the present paper. Then, instead of eqn (7.202), we have

$$^0G_{00} = \kappa\,\mathcal{T}_{00} - \check{p}\,\mathcal{U}_{00}, \qquad (7.217)$$

where

$$U_{00} = \int K(x-y)\check{C}_{0,0}(y)\,d^4y, \qquad \mathcal{U}_{00} = -\int R(x-y)\check{C}_{0,0}(y)\,d^4y \qquad (7.218)$$

and we have used here the reciprocity relation (7.55). It follows from eqns (7.8) and (7.207) that

$$\delta^{ij}\,\mathcal{W}_{i,j} = 2\kappa\,\mathcal{T}_{00} + 2\check{p}\int R(x-y)\check{C}_{0,0}(y)\,d^4y. \qquad (7.219)$$

Next, the relevant integral constraint is in this case $S_{[i\,0]} = \check{p}\,U_{[i\,0]}$, or

$$\int K(|\mathbf{x}-\mathbf{y}|,\mathbf{x}-\mathbf{y})\,\mathcal{W}_i(x^0 - |\mathbf{x}-\mathbf{y}|,\mathbf{y})\,d^3y = 4\check{p}\,U_{[i\,0]}. \qquad (7.220)$$

Hence, using the approach adopted above for the $\check{p} = 0$ case, we have

$$\int K(|\mathbf{x}-\mathbf{y}|,\mathbf{x}-\mathbf{y})\left(\delta^{ij}\,\mathcal{W}_{i,j}\right)(x^0 - |\mathbf{x}-\mathbf{y}|,\mathbf{y})\,d^3y = 4\check{p}\,\delta^{ij}\partial_j\,U_{[i\,0]}. \qquad (7.221)$$

It follows from eqn (7.219) that

$$\kappa\int K_c(x-y)\,\mathcal{T}_{00}(y)\,d^4y + \check{p}\int\int K_c(x-z)R(z-y)\check{C}_{0,0}(y)\,d^4y\,d^4z = 2\check{p}\,\delta^{ij}\partial_j\,U_{[i\,0]}, \qquad (7.222)$$

where K_c is the light-cone kernel defined by eqn (7.37). Calculating $U_{[i\,0]}$ from eqn (7.32) and using $\check{C}^\sigma{}_{,\sigma} = 0$, we find

$$\delta^{ij}\partial_j \int K(|\mathbf{x}-\mathbf{y}|, \mathbf{x}-\mathbf{y})\,\check{C}_i(\bar{\eta},\mathbf{y})\,d^3y = \int K_c(x-y)\check{C}_{0,0}(y)\,d^4y. \tag{7.223}$$

Moreover,

$$\delta^{ij}\partial_j \int K(x-y)\,\check{C}_{[i,0]}(y)\,d^4y = \frac{1}{2}\partial_\sigma \int K(x-y)(\check{C}^\sigma{}_{,0} - \check{C}_{0,}{}^\sigma)(y)\,d^4y, \tag{7.224}$$

which, after using Gauss's theorem and $\check{C}^\sigma{}_{,\sigma} = 0$, results in

$$\delta^{ij}\partial_j \int K(x-y)\,\check{C}_{[i,0]}(y)\,d^4y = -\frac{1}{2}\int K(x-y)(\Box\,\check{C}_0)(y)\,d^4y. \tag{7.225}$$

Putting all these results together and using the definition of kernel W in eqn (7.67), we finally arrive at a nonlocal integral constraint for $\check{C}_0$,

$$\kappa \int W(x-y)\,T_{00}(y)\,d^4y = -\check{p}\int [\,W(x-y)\,\check{C}_{0,0}(y) + K(x-y)\,\Box\,\check{C}_0(y)\,]\,d^4y. \tag{7.226}$$

In principle, this equation for $\check{C}_0$ can be solved—for example, via Fourier analysis—in terms of T_{00}, the energy density of the gravitational source. In this way, for $\check{p} \neq 0$, we avoid the contradiction that we encountered in eqn (7.216).

7.10　Appendix B: Liouville–Neumann Method

Consider a linear integral equation of the second kind given by

$$\phi(x) + \lambda \int_a^b K(x,y)\,\phi(y)\,dy = f(x), \tag{7.227}$$

where a, b and λ are constants. We work in the space of continuous functions on the interval $[a,b]$. If kernel $K(x,y)$ identically vanishes for $y > x$, then the integral equation reduces to a Volterra equation discussed in Section 2.9. In the general case under consideration here, we have a Fredholm equation and we seek a solution of this Fredholm equation by the Liouville–Neumann method of successive substitutions. That is, we first take the integral term to the right-hand side of eqn (7.227) and replace ϕ in the integrand by its value given by this equation. Repeating this process eventually leads to an infinite *Liouville–Neumann series*, namely,

$$\phi(x) = f(x) - \lambda \int_a^b K(x,y)f(y)dy$$
$$+ \lambda^2 \int_a^b \int_a^b K(x,z)\,K(z,y)f(y)dy\,dz + \cdots. \tag{7.228}$$

If this series is uniformly convergent for sufficiently small $|\lambda|$, we have a solution of the integral eqn (7.227). This solution is *unique* in the space of real continuous functions on the interval $[a, b]$; a generalization of this result to the space of square-integrable functions is contained in Tricomi (1957).

We can express the unique solution of eqn (7.227) in the form

$$f(x) + \lambda \int_a^b R(x, y) f(y) \, dy = \phi(x), \tag{7.229}$$

where $R(x, y)$ is the kernel *reciprocal* to $K(x, y)$. To get a formula for R, let us define the iterated kernels K_n, $n = 1, 2, \ldots$, by

$$K_1(x, y) = -K(x, y),$$

$$K_{n+1}(x, y) = \int_a^b K_n(x, z) \, K_1(z, y) dz. \tag{7.230}$$

These functions occur in the infinite series of eqn (7.228). In fact, we can write eqn (7.228) as

$$\phi(x) = f(x) + \int_a^b \left[\sum_{n=1}^\infty \lambda^n K_n(x, y) \right] f(y) \, dy, \tag{7.231}$$

so that the *reciprocal kernel* $R(x, y)$ is given by

$$R(x, y) = \sum_{n=1}^\infty \lambda^{n-1} K_n(x, y). \tag{7.232}$$

This infinite series of the iterated kernels is essentially the *Neumann series*; we have suppressed here the dependence of the reciprocal kernel on λ for the sake of simplicity.

The reciprocity between eqns (7.227) and (7.229) leads to

$$K(x, y) + R(x, y) = -\lambda \int_a^b K(x, z) R(z, y) dz$$

$$= -\lambda \int_a^b R(x, z) K(z, y) dz. \tag{7.233}$$

A more direct demonstration of these equations can be obtained from eqn (7.232) and

$$K_{n+p}(x, y) = \int_a^b K_n(x, z) K_p(z, y) dz, \tag{7.234}$$

where $p = 1, 2, \ldots$. To prove this latter relation, we proceed by induction and note that this equation is true for $p = 1$ by definition of iterated kernels in eqn (7.230). Assuming that eqn (7.234) is valid for p, we must then prove that it also holds for $p + 1$, that is,

$$K_{n+p+1}(x, y) = \int_a^b K_n(x, z) K_{p+1}(z, y) dz. \tag{7.235}$$

To show this, we start from the definition of iterated kernels, namely,

$$K_{n+p+1}(x,y) = \int_a^b K_{n+p}(x,z)K_1(z,y)dz, \tag{7.236}$$

and we express the integrand using eqn (7.234) as

$$K_{n+p+1}(x,y) = \int_a^b \left[\int_a^b K_n(x,w)\,K_p(w,z)\,dw \right] K_1(z,y)\,dz. \tag{7.237}$$

Interchanging the order of integration in the double integral, we find

$$K_{n+p+1}(x,y) = \int_a^b K_n(x,w) \left[\int_a^b K_p(w,z)\,K_1(z,y)\,dz \right] dw. \tag{7.238}$$

By the definition of iterated kernels, we have

$$K_{p+1}(w,y) = \int_a^b K_p(w,z)\,K_1(z,y)\,dz, \tag{7.239}$$

which, when substituted in eqn (7.238), leads to eqn (7.235) and this completes the proof of eqn (7.234).

Finally, let us note that if $K(x,y)$ is a convolution kernel, that is, $K(x,y) = \bar{k}(x-y)$, or symmetric, that is, $K(x,y) = K(y,x)$, then all iterated kernels as well as $R(x,y)$ would be likewise of the convolution type or symmetric, respectively.

7.11   Appendix C: Calculation of $\hat{q}(|\boldsymbol{\xi}|)$

Reciprocal kernels q_1 and q_2, given by eqns (7.98) and (7.99), respectively, are both L^1 and L^2 and their Fourier transforms ($\hat{q}_1$ and $\hat{q}_2$) are dimensionless functions that can be explicitly calculated in terms of the *exponential integral function E_1*, as will be demonstrated in the last part of this section (Mashhoon 2013b). However, to ensure that they also satisfy the requirement that $1 + \hat{q}(|\boldsymbol{\xi}|) \neq 0$ given in eqn (7.88), we first follow a different approach here (Chicone and Mashhoon 2012). To this end, we note that if in the calculation of the spatial Fourier transform $q(\mathbf{x})$ is only a function of the radial variable $r = |\mathbf{x}|$, then we can introduce spherical polar coordinates (r, ϑ, φ) and imagine that the coordinate system is so oriented that $\boldsymbol{\xi}$ points along the polar axis. The angular integrations can now be simply carried out using the fact that

$$\int_0^\pi e^{-i\xi r \cos\vartheta} \sin\vartheta\, d\vartheta = 2\,\frac{\sin(\xi r)}{\xi r}. \tag{7.240}$$

The result is then

$$\hat{q}(\xi) = \frac{4\pi}{\xi} \int_0^\infty rq(r)\sin(\xi r)\,dr, \tag{7.241}$$

where $\xi : 0 \to \infty$ is the magnitude of $\boldsymbol{\xi}$.

In general, the Fourier transform $\hat{q}(\boldsymbol{\xi})$ of a square-integrable function $q(\mathbf{x})$ is square integrable. We wish to use the Fourier transform method of Section 7.4 to calculate kernel $\chi(r)$. In case $1 + \hat{q} \neq 0$ and $\hat{\chi}(|\boldsymbol{\xi}|)$ is an L^2 function, the Fourier transform method becomes applicable here and we have

$$\chi(r) = -\frac{1}{2\pi^2 r} \int_0^\infty \frac{\xi \hat{q}(\xi)}{1 + \hat{q}(\xi)} \sin(\xi r) \, d\xi. \tag{7.242}$$

Let us first consider q_1 and note that eqns (7.98) and (7.241) imply

$$\hat{q}_1(\xi) = \frac{1}{\lambda_0 \, \xi} \int_0^\infty (\mu_0 + \frac{1}{a_0 + r}) \, e^{-\mu_0 \, r} \sin(\xi r) dr. \tag{7.243}$$

It follows from formula 3.893 on page 477 of Gradshteyn and Ryzhik (1980) that

$$\int_0^\infty e^{-\mu_0 \, r} \sin(\xi r) \, dr = \frac{\xi}{\mu_0^2 + \xi^2}; \tag{7.244}$$

hence,

$$\hat{q}_1(\xi) = \frac{1}{\lambda_0} \frac{\mu_0}{\mu_0^2 + \xi^2} + \frac{1}{\lambda_0 \, \xi} \int_0^\infty \frac{e^{-\mu_0 \, r}}{a_0 + r} \sin(\xi r) \, dr. \tag{7.245}$$

At this point, it proves useful to digress briefly here and discuss a lemma regarding the Fourier sine transform. Consider the integral

$$\mathfrak{g}(\xi) = \int_0^\infty \mathfrak{h}(x, \xi) \sin(\xi x) \, dx. \tag{7.246}$$

For each $\xi \in (0, \infty)$, let $\mathfrak{h}(x, \xi)$ be a smooth positive integrable function that monotonically decreases over the interval of integration; then, $\mathfrak{g}(\xi) > 0$. To prove this result for each $\xi > 0$, we divide the integration interval in eqn (7.246) into segments $(2\pi\xi^{-1}n, 2\pi\xi^{-1}n + 2\pi\xi^{-1})$ for $n = 0, 1, 2, \ldots$. In each such segment, the corresponding sine function, $\sin(\xi x)$, goes through a complete cycle and is positive in the first half and negative in the second half. On the other hand, the monotonically decreasing function $\mathfrak{h}(x, \xi) > 0$ is consistently larger in the first half of the cycle than in the second half; therefore, the result of the integration over each full cycle is positive and consequently $\mathfrak{g}(\xi) > 0$. For $\xi \to 0$, $\sin(\xi x) \to 0$ and hence $\mathfrak{g}(0) = 0$, while for $\xi \to \infty$, the integration segments shrink to zero and $\mathfrak{g}$ tends to 0 in the limit as $\xi \to \infty$, if the corresponding limit of $\mathfrak{h}(x, \xi)$ is finite everywhere over the integration domain. This latter conclusion is, of course, a variation on the Riemann–Lebesgue lemma.

Returning now to eqn (7.245), we note that $\exp(-\mu_0 r)/(a_0 + r)$ is a smooth positive integrable function that decreases monotonically for $r : 0 \to \infty$. Thus the above lemma can be used to conclude that for $0 \leq \xi < \infty$, $\hat{q}_1(\xi) > 0$, while $\hat{q}_1(\xi) \to 0$ as $\xi \to \infty$ by the Riemann–Lebesgue lemma. It follows that $|\hat{\chi}_1(\xi)| \leq |\hat{q}_1(\xi)|$, so that $\hat{\chi}_1$ is in L^2 as well and the nonlocal kernel χ_1 can, in principle, be determined via eqn (7.242). To illustrate this point, we consider an example involving a particular choice of parameters, namely, $\mu_0 \lambda_0 = 0.1$ and $a_0/\lambda_0 = 0.001$, so that $\alpha_0 := 2/(\mu_0 \lambda_0) = 20$ in this case. In Fig. 7.3a, we plot $\hat{q}_1(\xi)$ versus $\lambda_0 \xi$ and in Fig. 7.3b, $-\lambda_0^3 \chi(r)$ is plotted versus r/λ_0.

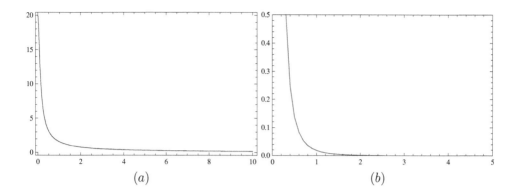

(a) (b)

Fig. 7.3 (a) Plot of $\hat{q}_1$ versus $\lambda_0 \xi$ for the reciprocal kernel q_1 given in eqn (7.98). The parameter values are $\mu_0 \lambda_0 = 0.1$ and $a_0/\lambda_0 = 0.001$. As noted in this appendix, $\hat{q}_1$ is always greater than $\hat{q}_2$, but this is hardly noticeable for the parameter values under consideration in this case. For instance, at $\lambda_0 \xi = 2$, $\hat{q}_1 \approx 0.779$ here, while $\hat{q}_2 \approx 0.773$ in Fig. 7.4a. (b) Plot of the corresponding dimensionless kernel $-\lambda_0^3 \chi_1$ versus r/λ_0. The function $-\lambda_0^3 \chi_1$ starts from ∞ at $r = 0$ and drops off to nearly zero very fast; in fact, for $r/\lambda_0 \geq 2.5$ it is essentially zero at the level of accuracy of this plot. Reproduced from Chicone, C. and Mashhoon, B., 2012, "Nonlocal Gravity: Modified Poisson's Equation", *J. Math. Phys.* **53**, 042501 (17 pages), with the permission of AIP Publishing. DOI: 10.1063/1.3702449

Next, we consider kernel q_2 and note that it can be expressed as

$$q_2(r) = \frac{1}{4\pi\lambda_0} \frac{d}{dr} \left[-\frac{\exp(-\mu_0 r)}{a_0 + r} \right]. \tag{7.247}$$

Reciprocal kernel $q_2(r)$ is smooth and positive everywhere and rapidly decreases to zero at infinity; indeed, $q_2(r)$ is integrable as well as square integrable. *It is possible to show that for any $\xi \geq 0$, $\hat{q}_2(\xi) > -a_0/\lambda_0$.*

Substituting eqn (7.247) in eqn (7.241) and integrating by parts, we find

$$\hat{q}_2(\xi) = \frac{1}{\lambda_0 \xi} \int_0^\infty \frac{e^{-\mu_0 r}}{a_0 + r} \frac{d}{dr} [r \sin(\xi r)] \, dr. \tag{7.248}$$

Next, we differentiate $r \sin(\xi r)$ and note that

$$\sin(\xi r) + \xi r \cos(\xi r) = [\sin(\xi r) - a_0 \xi \cos(\xi r)] + (a_0 + r) \xi \cos(\xi r). \tag{7.249}$$

Therefore, eqn (7.248) can be written as

$$\hat{q}_2(\xi) = \mathfrak{U} + \frac{1}{\lambda_0} \int_0^\infty e^{-\mu_0 r} \cos(\xi r) \, dr. \tag{7.250}$$

Here,

$$\int_0^\infty e^{-\mu_0 r} \cos(\xi r) \, dr = \frac{\mu_0}{\mu_0^2 + \xi^2} \tag{7.251}$$

follows from formula 3.893 on page 477 of Gradshteyn and Ryzhik (1980) and

$$\mathfrak{U} = \frac{1}{\lambda_0\,\xi} \int_0^\infty \frac{e^{-\mu_0 r}}{a_0 + r} \left[\sin(\xi r) - a_0 \xi \cos(\xi r)\right] dr. \tag{7.252}$$

Let us now introduce an angle $\bar{\gamma}$ connected with $a_0\,\xi$ such that

$$a_0\,\xi := \tan\bar{\gamma}, \tag{7.253}$$

and note that as $\xi : 0 \to \infty$, we have $\bar{\gamma} : 0 \to \pi/2$ and $\bar{\gamma}/\xi : a_0 \to 0$. It is useful to introduce a new variable X in eqn (7.252), $r = X + \bar{\gamma}/\xi$, since

$$\sin(\xi r) - a_0\,\xi\cos(\xi r) = \frac{1}{\cos\bar{\gamma}}\sin(\xi r - \bar{\gamma}). \tag{7.254}$$

As $\xi : 0 \to \infty$, $\cos\bar{\gamma}$ decreases from 1 to 0, so that for any finite value of ξ, $\cos\bar{\gamma} > 0$. Then, eqn (7.252) can be written as

$$\mathfrak{U} = \frac{e^{-\mu_0\bar{\gamma}/\xi}}{\lambda_0\,\xi\cos\bar{\gamma}} \int_{-\bar{\gamma}/\xi}^\infty \frac{e^{-\mu_0 X}}{a_0 + X + \bar{\gamma}/\xi} \sin(\xi X)\,dX. \tag{7.255}$$

In this expression, the integration from $X = -\bar{\gamma}/\xi$ to ∞ can be expressed as a sum of two terms, one from 0 to ∞ and the other from $X = -\bar{\gamma}/\xi$ to 0. That is, $\mathfrak{U} = \mathfrak{U}_1 + \mathfrak{U}_2$, where

$$\mathfrak{U}_1 = \frac{e^{-\mu_0\bar{\gamma}/\xi}}{\lambda_0\,\xi\cos\bar{\gamma}} \int_0^\infty \frac{e^{-\mu_0 X}}{a_0 + X + \bar{\gamma}/\xi} \sin(\xi X)\,dX \tag{7.256}$$

is positive by the Fourier sine lemma, since $\exp(-\mu_0 X)/(a_0 + X + \bar{\gamma}/\xi)$ is a smooth positive integrable function that monotonically decreases for $X : 0 \to \infty$. Moreover,

$$\mathfrak{U}_2 = \frac{e^{-\mu_0\bar{\gamma}/\xi}}{\lambda_0\,\xi\cos\bar{\gamma}} \int_{-\bar{\gamma}/\xi}^0 \frac{e^{-\mu_0 X}}{a_0 + X + \bar{\gamma}/\xi} \sin(\xi X)\,dX \tag{7.257}$$

turns out to be negative. To show this, let us introduce a new variable $\xi X = -Y$ into eqn (7.257); then, we have

$$-\lambda_0\,\xi\cos\bar{\gamma}\,\mathfrak{U}_2(\xi) = \int_0^{\bar{\gamma}} \frac{e^{-\frac{\mu_0}{\xi}(\bar{\gamma}-Y)}}{(\bar{\gamma} - Y) + a_0\,\xi} \sin Y\,dY. \tag{7.258}$$

The right-hand side of this equation involves an integrand that increases monotonically from 0 to $\sin\bar{\gamma}/(a_0\,\xi)$ as $Y : 0 \to \bar{\gamma}$. Thus the right-hand side of eqn (7.258) is less than $\bar{\gamma}\sin\bar{\gamma}/(a_0\,\xi)$, which is equal to $\bar{\gamma}\cos\bar{\gamma}$ by eqn (7.253); consequently, $\mathfrak{U}_2(\xi) > -\bar{\gamma}/(\lambda_0\,\xi)$. As $0 \leq \bar{\gamma}/\xi \leq a_0$, we find that $\mathfrak{U}_2(\xi) > -a_0/\lambda_0$. We thus conclude that $\hat{q}_2 > -a_0/\lambda_0$ and

$$1 + \hat{q}_2(\xi) > 1 - a_0/\lambda_0. \tag{7.259}$$

Hence the Fourier transform method is applicable to kernel q_2 if $a_0 < \lambda_0$. It then follows from eqn (7.87) that $|\hat{\chi}_2| < |\hat{q}_2|/(1 - a_0/\lambda_0)$, so that $\hat{\chi}_2(\xi)$ is in L^2 as well, and we can find the nonlocal kernel χ_2 from eqn (7.242). This is illustrated in Fig.7.4 via

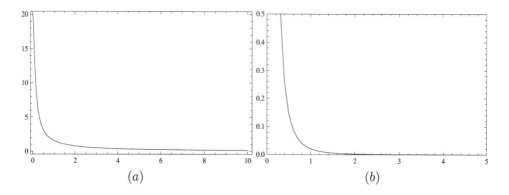

Fig. 7.4 (a) Plot of $\hat{q}_2$ versus $\lambda_0\,\xi$ for the reciprocal kernel q_2 given in eqn (7.99). The parameter values are $\mu_0\,\lambda_0 = 0.1$ and $a_0/\lambda_0 = 0.001$, just as in Fig. 7.3. The function $\hat{q}_2$ starts from $\hat{q}_2(0) \approx 20$ and rapidly falls off initially, but then slowly decreases to zero as $\lambda_0\xi \to \infty$. (b) Plot of the corresponding dimensionless kernel $-\lambda_0^3\chi_2$ versus r/λ_0. The function $-\lambda_0^3\chi_2$ starts from ≈ 80101 at $r = 0$ and drops off to nearly zero very fast; in fact, for $r/\lambda_0 \geq 2.5$ it is essentially zero at the level of accuracy of this plot. Though this figure appears to be indistinguishable from Fig. 7.3b in the plotted range, their numerical values are indeed different. Reproduced from Chicone, C. and Mashhoon, B., 2012, "Nonlocal Gravity: Modified Poisson's Equation", *J. Math. Phys.* **53**, 042501 (17 pages), with the permission of AIP Publishing. DOI: 10.1063/1.3702449

an example with the same choice of parameters as in Fig. 7.3. We plot $\hat{q}_2$ versus $\lambda_0\,\xi$ in Fig. 7.4a and $-\lambda_0^3\,\chi_2(r)$ versus r/λ_0 in Fig. 7.4b. We note that with this choice of parameters, $\hat{q}_2$ is a positive function; in fact, we expect that this would be the case if a_0/λ_0 is sufficiently small compared to unity.

Kernel $q_0(r)$ can be obtained from either q_1 or q_2 by setting $a_0 = 0$. Moreover, $\hat{q}_0$ can be obtained from eqn (7.245) with $a_0 = 0$, namely,

$$\hat{q}_0(\xi) = \frac{1}{\lambda_0}\frac{\mu_0}{\mu_0^2 + \xi^2} + \frac{1}{\lambda_0\,\xi}\int_0^\infty \frac{e^{-\mu_0\,r}}{r}\sin(\xi r)\,dr, \tag{7.260}$$

so that

$$\hat{q}_0 = \frac{\mu_0}{\lambda_0(\mu_0^2 + \xi^2)} + \frac{1}{\lambda_0\,\xi}\arctan\left(\frac{\xi}{\mu_0}\right). \tag{7.261}$$

Here we have used the fact that for real values of the constants p_1 and p_2,

$$\int_0^\infty e^{-p_1 x}\sin(p_2 x)\frac{dx}{x} = \arctan\left(\frac{p_2}{p_1}\right), \quad p_1 \geq 0; \tag{7.262}$$

see formula 3.941 on page 489 of Gradshteyn and Ryzhik (1980). Thus $\hat{q}_0 > 0$ and $\hat{q}_0(0) = 2/(\mu_0\,\lambda_0) := \alpha_0$, while for $\lambda_0\,\xi \gg 1$, $\hat{q}_0 \sim \pi/(2\,\lambda_0\,\xi)$. This asymptotic behavior is similar to the Fourier transform of the Kuhn kernel

$$\hat{q}_K = \frac{\pi}{2\,\lambda_{TK}\,\xi}, \tag{7.263}$$

since q_0 and q_K have much the same behavior for $r \to 0$.

Consider next the smooth positive monotonically decreasing functions given by

$$r\left[q_0(r) - q_1(r)\right] = \frac{a_0}{4\pi\lambda_0} \frac{1}{r\left(a_0 + r\right)} e^{-\mu_0 r} \tag{7.264}$$

and

$$r\left[q_1(r) - q_2(r)\right] = \frac{a_0}{4\pi\lambda_0} \frac{1 + \mu_0(a_0 + r)}{(a_0 + r)^2} e^{-\mu_0 r}. \tag{7.265}$$

The Fourier sine lemma is immediately applicable to these functions and we find from eqn (7.241) that

$$\hat{q}_0(\xi) > \hat{q}_1(\xi) > \hat{q}_2(\xi), \tag{7.266}$$

and they all vanish as $\xi \to \infty$ in accordance with the Riemann–Lebesgue lemma. Furthermore, for $i = 1, 2$, we have

$$\hat{q}_i(0) = \alpha_0 - \mathcal{E}_i(\infty) = \alpha_0\, \epsilon_i(\varsigma). \tag{7.267}$$

Returning now to eqns (7.240) and (7.241), we note that the reciprocal kernel $q(r)$ can be expressed in terms of its Fourier transform $\hat{q}(\xi)$ as

$$q(r) = \frac{1}{2\pi^2 r} \int_0^\infty \xi \hat{q}(\xi) \sin(\xi r)\, d\xi, \tag{7.268}$$

$$q(0) = \frac{1}{2\pi^2} \int_0^\infty \xi^2 \hat{q}(\xi)\, d\xi. \tag{7.269}$$

Moreover, it follows from eqn (7.242) that

$$\chi(0) = -\frac{1}{2\pi^2} \int_0^\infty \frac{\xi^2 \hat{q}(\xi)}{1 + \hat{q}(\xi)}\, d\xi. \tag{7.270}$$

Equations (7.269) and (7.270) then imply

$$\chi(0) + q(0) = \frac{1}{2\pi^2} \int_0^\infty \frac{\xi^2 \hat{q}^2(\xi)}{1 + \hat{q}(\xi)}\, d\xi. \tag{7.271}$$

We recall that $\hat{q}(\xi)$ is square integrable over the whole $\boldsymbol{\xi}$ space; hence, $\xi^2 \hat{q}^2(\xi)$ is integrable over the radial coordinate $\xi : 0 \to \infty$ and the right-hand side of eqn (7.271) is finite. It then follows from eqn (7.271) that $-\chi_1(0) = \infty$, since $q_1(0) = \infty$, while $\chi_2(0)$ is finite due to the finiteness of $q_2(0)$, in agreement with the numerical results of Fig. 7.3b and Fig. 7.4b.

Finally, let us write eqn (7.245) as

$$\hat{q}_1(\xi) = \frac{1}{\lambda_0} \frac{\mu_0}{\mu_0^2 + \xi^2} - \frac{1}{\lambda_0 \xi} \operatorname{Im} \int_0^\infty \frac{e^{-(\mu_0 + i\xi) r}}{a_0 + r}\, dr. \tag{7.272}$$

Introducing here a new variable t, $a_0 + r := a_0\, t$, such that $t : 1 \to \infty$ as $r : 0 \to \infty$, we can write eqn (7.272) as

$$\hat{q}_1(\xi) = \frac{1}{\lambda_0} \frac{\mu_0}{\mu_0^2 + \xi^2} - \frac{1}{\lambda_0 \xi} \operatorname{Im}\left[e^{Z_0} E_1(Z_0)\right], \tag{7.273}$$

where

$$Z_0 = a_0(\mu_0 + i\,\xi) = \varsigma + i\,a_0\,\xi \tag{7.274}$$

and $E_1(z)$ is the exponential integral function defined for a complex number z with positive real part, $\operatorname{Re}(z) > 0$, by (Abramowitz and Stegun 1964)

$$E_1(z) = \int_1^\infty \frac{e^{-z\,t}}{t}\,dt. \tag{7.275}$$

For explicit computations, one may use the expansion (Abramowitz and Stegun 1964)

$$E_1(z) = -C - \ln z - \sum_{n=1}^\infty \frac{(-1)^n z^n}{n\,n!}, \tag{7.276}$$

where $C = 0.577\ldots$ is Euler's constant. Next, to calculate $\hat{q}_2(\xi)$, we write eqn (7.248) as

$$\hat{q}_2(\xi) = \frac{1}{\lambda_0 \xi} \int_0^\infty \frac{e^{-\mu_0 r}}{a_0 + r} \left[\sin(\xi r) + \xi r \cos(\xi r)\right]\,dr. \tag{7.277}$$

Using eqns (7.245) and (7.251), eqn (7.277) can be expressed as

$$\hat{q}_2(\xi) = \hat{q}_1(\xi) - \frac{a_0}{\lambda_0} \int_0^\infty \frac{e^{-\mu_0 r}}{a_0 + r} \cos(\xi r)\,dr. \tag{7.278}$$

Following the same procedure as in the case of eqn (7.245), we find

$$\hat{q}_2(\xi) = \hat{q}_1(\xi) - \frac{a_0}{\lambda_0} \operatorname{Re}\left[e^{Z_0} E_1(Z_0)\right]. \tag{7.279}$$

7.12   Appendix D: Light Deflection Integrals

In eqns (7.184)–(7.187) of Section 7.7, consider

$$\frac{1}{r}\frac{d\phi}{dr} = \frac{G(1+\alpha_0)}{r^3} - \alpha_0 G\left(1 + \frac{1}{2}\mu_0 r\right)\frac{e^{-\mu_0 r}}{r^3}, \tag{7.280}$$

where the first part on the right-hand side is simply due to Newtonian attraction augmented by $1 + \alpha_0$, while the second repulsive "Yukawa" part is due to the fading of spatial memory in nonlocal gravity. To compute the net deflection of light, the integrals due to the first part of eqn (7.280) are simpler and we therefore treat them first.

Let $w(X) > 0$ be given by

$$w(X) = \mathcal{A} + 2\,\mathcal{B}\,X + \mathcal{C}\,X^2, \tag{7.281}$$

where $\tilde{\Delta} := \mathcal{A}\mathcal{C} - \mathcal{B}^2 \neq 0$. It is then straightforward to verify that

$$\int \frac{dX}{w^{3/2}} = \frac{B + CX}{\tilde{\Delta}\, w^{1/2}}, \qquad \int \frac{X\, dX}{w^{3/2}} = -\frac{A + BX}{\tilde{\Delta}\, w^{1/2}}, \tag{7.282}$$

where only positive square roots are considered throughout. Let us now assume that $C > 0$ and $\tilde{\Delta} > 0$, so that

$$C\, w(X) = (C X + B)^2 + \tilde{\Delta}. \tag{7.283}$$

Hence, $w > 0$ for $X : -\infty \to +\infty$. In this case, we have

$$\mathcal{I}_1 = \int_{-\infty}^{\infty} \frac{dX}{w^{3/2}} = \frac{2\, C^{1/2}}{\tilde{\Delta}}, \qquad \mathcal{I}_2 = \int_{-\infty}^{\infty} \frac{X\, dX}{w^{3/2}} = -\frac{2\, B}{\tilde{\Delta}\, C^{1/2}}. \tag{7.284}$$

For the problem of light deflection discussed in Section 7.7, we have $X = t - t_0$ and $w(t - t_0) = u_j^2$, where u_j is given by eqn (7.177); that is, along the unperturbed ray,

$$u_j^2 = \mathcal{A}_j + 2\, \mathcal{B}_j\, (t - t_0) + \mathcal{C}_j\, (t - t_0)^2, \tag{7.285}$$

where

$$\mathcal{A}_j = \gamma^2 (a - x_j)^2 + (b - y_j)^2 + (\zeta - z_j)^2, \qquad \mathcal{B}_j = -\beta\gamma^2 (a - x_j) + (\zeta - z_j) \tag{7.286}$$

and $\mathcal{C}_j = \gamma^2$. Moreover, we find that $\tilde{\Delta}_j = \mathcal{A}_j \mathcal{C}_j - \mathcal{B}_j^2 = \gamma^2 (\mathcal{X}_j^2 + \mathcal{Y}_j^2)$, where $\mathcal{X}_j$ and $\mathcal{Y}_j$ are defined in eqn (7.191) and $\tilde{\Delta}_j$, by assumption, never vanishes. Thus the conditions for the applicability of eqn (7.284) are satisfied. We thus find that the integrals for the first part of eqn (7.280) are given by

$$\mathcal{I}_1 = \int_{-\infty}^{\infty} \frac{dX}{u_j^3} = \frac{2\,\gamma^{-1}}{\mathcal{X}_j^2 + \mathcal{Y}_j^2}, \qquad \mathcal{I}_2 = \int_{-\infty}^{\infty} \frac{X\, dX}{u_j^3} = \frac{2\,\gamma^{-1}[\,\beta\,\mathcal{X}_j - (\zeta - z_j)\,]}{\mathcal{X}_j^2 + \mathcal{Y}_j^2}, \tag{7.287}$$

which, together with the results given below for the second part of eqn (7.280), eventually lead to eqns (7.189)–(7.190) of Section 7.7.

To treat the integration of the second ("Yukawa") part of eqn (7.280), let us first note that

$$u_j = \left(\hat{u}_j^2 + \hat{\Delta}_j^2\right)^{1/2}, \tag{7.288}$$

where

$$\hat{u}_j = \gamma X + \gamma^{-1} \mathcal{B}_j, \qquad \hat{\Delta}_j = (\mathcal{X}_j^2 + \mathcal{Y}_j^2)^{1/2} \tag{7.289}$$

and, as before, $X = t - t_0$. As $X : -\infty \to +\infty$, $\hat{u}_j$ also goes from $-\infty$ to $+\infty$; therefore, it proves useful to introduce a new variable $v : -\infty \to +\infty$ such that

$$\hat{u}_j = \hat{\Delta}_j \sinh v, \qquad u_j = \hat{\Delta}_j \cosh v. \tag{7.290}$$

The calculation of the integrals for the second part then ultimately reduces to the determination of $\mathcal{J}_1(\vartheta_j)$ and $\mathcal{J}_2(\vartheta_j)$, where

$$\vartheta_j := \mu_0\, \hat{\Delta}_j > 0 \tag{7.291}$$

and

$$\mathcal{J}_n(\vartheta) := \int_0^{\infty} \frac{e^{-\vartheta \cosh v}}{\cosh^n v}\, dv \tag{7.292}$$

for $n = 1, 2, 3, \dots$. It is interesting to observe that $\mathcal{J}_n(0) = (\sqrt{\pi}/2)\Gamma(\frac{n}{2})/\Gamma(\frac{n+1}{2})$ and $\mathcal{J}_n(\infty) = 0$.

It is useful to express $\mathcal{J}_n(x)$ for $x \geq 0$ as

$$\mathcal{J}_n(x) = \int_0^{\frac{\pi}{2}} \cos^{n-1}\varphi\, e^{-x\sec\varphi}\, d\varphi, \tag{7.293}$$

where, in eqn (7.292), we have employed $\cosh\upsilon = \sec\varphi$, $\sinh\upsilon = \tan\varphi$, $d\upsilon = d\varphi/\cos\varphi$ and $\varphi : 0 \to \pi/2$ as $\upsilon : 0 \to \infty$.

We note that $\mathcal{J}_1(x)$, which is a special case of Sievert's integral (Abramowitz and Stegun 1964), is *not* analytic about $x = 0$. Moreover

$$\mathcal{J}_1'' - \mathcal{J}_1 = \int_0^{\frac{\pi}{2}} \tan^2\varphi\, e^{-x\sec\varphi}\, d\varphi, \tag{7.294}$$

where $\mathcal{J}_1'(x) = d\mathcal{J}_1/dx$, etc. The right-hand side of eqn (7.294) can be evaluated using integration by parts; that is,

$$\int_0^{\frac{\pi}{2}} \sin\varphi\, e^{-x\sec\varphi}\, \frac{d}{d\varphi}\left(\frac{1}{\cos\varphi}\right) d\varphi = -\mathcal{J}_1 + x\,(\mathcal{J}_1' - \mathcal{J}_1'''). \tag{7.295}$$

It follows that $\mathcal{J}_1'(x)$ satisfies the modified Bessel differential equation of order zero, namely,

$$x\,\mathfrak{b}''(x) + \mathfrak{b}'(x) - x\,\mathfrak{b}(x) = 0. \tag{7.296}$$

The solutions of this equation are $I_0(x)$ and $K_0(x)$, which are the modified Bessel functions of order zero; see Abramowitz and Stegun (1964). In fact, $I_0(x)$,

$$I_0(x) = \sum_{k=0}^{\infty} \frac{x^{2k}}{(2^k\,k!)^2}, \tag{7.297}$$

is regular at $x = 0$ and valid everywhere. Furthermore,

$$K_0(x) = -(\ln\frac{x}{2} + C)\,I_0(x) + \sum_{k=1}^{\infty} \beta_k \frac{x^{2k}}{(2^k\,k!)^2}, \tag{7.298}$$

where $C = 0.577\ldots$ is Euler's constant and

$$\beta_k = \sum_{n=1}^{k} \frac{1}{n}. \tag{7.299}$$

For $x : 0 \to \infty$, $K_0(x)$ behaves as $-\ln x$ for $x \to 0$, but then rapidly decreases monotonically with increasing x and vanishes exponentially as $x \to \infty$. In fact,

$$K_0(x) \sim \sqrt{\frac{\pi}{2x}}\, e^{-x} \tag{7.300}$$

for $x \to \infty$ (Abramowitz and Stegun 1964).

To determine $\mathcal{J}_1$ and $\mathcal{J}_2$ in terms of Bessel functions, let us first note that $\mathcal{J}_1(0) = \pi/2$ and $\mathcal{J}_2(0) = 1$. Moreover, for $0 < |\epsilon| \ll 1$, we find from eqn (7.292) that for $\vartheta > 0$,

$$\mathcal{J}_1(\vartheta + \epsilon) = \mathcal{J}_1(\vartheta) - \epsilon K_0(\vartheta) + \cdots, \tag{7.301}$$

$$\mathcal{J}_2(\vartheta + \epsilon) = \mathcal{J}_2(\vartheta) - \epsilon \mathcal{J}_1(\vartheta) + \frac{1}{2}\epsilon^2 K_0(\vartheta) + \cdots, \tag{7.302}$$

where $K_0(\vartheta)$ is given by (Abramowitz and Stegun 1964)

$$K_0(\vartheta) = \int_0^\infty e^{-\vartheta \cosh v} \, dv. \tag{7.303}$$

It follows from eqns (7.301)–(7.302) that

$$\frac{d\mathcal{J}_1}{d\vartheta} = -K_0(\vartheta), \qquad \frac{d\mathcal{J}_2}{d\vartheta} = -\mathcal{J}_1(\vartheta). \tag{7.304}$$

Therefore, we have

$$\mathcal{J}_1(\vartheta) = \frac{\pi}{2} - \int_0^\vartheta K_0(x) \, dx \tag{7.305}$$

and

$$\mathcal{J}_2(\vartheta) = 1 - \int_0^\vartheta \mathcal{J}_1(x) \, dx. \tag{7.306}$$

Let us now define a new function $\mathcal{B}(x)$,

$$\mathcal{B}(x) := -x + \int_0^x I_0(t) \, dt, \tag{7.307}$$

so that we have

$$\mathcal{B}(x) = \sum_{k=1}^\infty \frac{x^{2k+1}}{(2k+1)\,(2^k\,k!)^2}. \tag{7.308}$$

It is then possible to use $I_0(x) = 1 + d\mathcal{B}/dx$ in eqn (7.298) and subsequently express eqn (7.305) as

$$\mathcal{J}_1(x) = \frac{\pi}{2} - x + x\left(\ln\frac{x}{2} + C\right) + \int_0^x \left(\ln\frac{t}{2} + C\right)\frac{d\mathcal{B}}{dt}\,dt - \sum_{k=1}^\infty \beta_k \frac{x^{2k+1}}{(2k+1)\,(2^k\,k!)^2}. \tag{7.309}$$

Finally, we find via integration by parts that

$$\mathcal{J}_1(x) = \frac{\pi}{2} - x + \left(\ln\frac{x}{2} + C\right)[x + \mathcal{B}(x)]$$
$$- \sum_{k=1}^\infty \left(\beta_k + \frac{1}{2k+1}\right)\frac{x^{2k+1}}{(2k+1)\,(2^k\,k!)^2} \tag{7.310}$$

and $\mathcal{J}_2(x)$ can be explicitly determined from eqn (7.306).

8
Nonlocal Gravity and Dark Matter

The main purpose of this chapter is to compare nonlocal gravity theory with observation. In this way, we try to determine the parameters λ_0, a_0 and μ_0 of the reciprocal kernel q in the Newtonian regime of nonlocal gravity. Our starting point is the modified inverse square force law. We recall from Chapter 7 that in the Newtonian regime of nonlocal gravity, the force of gravity on point mass m due to point mass m' is given by

$$\mathbf{F}(\mathbf{r}) = -Gmm' \frac{\hat{\mathbf{r}}}{r^2} \left\{ [1 - \mathcal{E}(r) + \alpha_0] - \alpha_0 \left(1 + \frac{1}{2}\mu_0\, r\right) e^{-\mu_0\, r} \right\}, \qquad (8.1)$$

where $\mathbf{r} = \mathbf{x}_m - \mathbf{x}_{m'}$, $r = |\mathbf{r}|$ and $\hat{\mathbf{r}} = \mathbf{r}/r$. This force is always *attractive and conservative*. It consists of two parts: an enhanced Newtonian part and a repulsive "Yukawa" part. The latter is due to the fact that spatial gravitational memory fades exponentially with distance.

As described in detail in Chapter 7, to recover the Tohline–Kuhn phenomenological scheme and thereby explain the "flat" rotation curves of spiral galaxies in a manner that would be consistent with nonlocal gravity theory, we find two simple possible functional forms for the reciprocal kernel of the nonlocal Poisson equation in the Newtonian regime, namely, q_1 and q_2 given in eqns (7.98) and (7.99), respectively. If we employ q_1, then $\mathcal{E}$ in eqn (8.1) is given by

$$\mathcal{E}_1(r) = \frac{a_0}{\lambda_0}\, e^\varsigma \left[E_1(\varsigma) - E_1(\varsigma + \mu_0 r) \right]. \qquad (8.2)$$

On the other hand, if we employ q_2, then $\mathcal{E}$ in the force law is given by

$$\mathcal{E}_2(r) = \frac{a_0}{\lambda_0} \left\{ -\frac{r}{r + a_0} e^{-\mu_0 r} + 2\, e^\varsigma \left[E_1(\varsigma) - E_1(\varsigma + \mu_0 r) \right] \right\}. \qquad (8.3)$$

Here, $\varsigma := \mu_0\, a_0$, $\lambda_0 = 2/(\alpha_0\, \mu_0)$ and $E_1(x)$ is the *exponential integral function* (Abramowitz and Stegun 1964).

We note that $\mathcal{E}_1(r)$ and $\mathcal{E}_2(r)$ start from zero at $r = 0$ and monotonically increase as $r \to \infty$; furthermore, they asymptotically approach $\mathcal{E}_1(\infty) = \mathcal{E}_\infty$ and $\mathcal{E}_2(\infty) = 2\,\mathcal{E}_\infty$, respectively, where $2\,\mathcal{E}_\infty = \alpha_0\, \varsigma\, E_1(\varsigma)\, \exp\varsigma$. It follows that as $\mu_0\, r \to \infty$ such that the Yukawa part of the gravitational force can be neglected, we simply have the Newtonian inverse square force law augmented by the constant factor $1 + \alpha_0\, \epsilon_i(\varsigma)$, where $i = 1, 2$,

$$\alpha_0\, \epsilon_i(\varsigma) = \alpha_0 - \mathcal{E}_i(\infty) > 0 \qquad (8.4)$$

and $\epsilon_1(\varsigma)$ and $\epsilon_2(\varsigma)$ are plotted in Fig. 7.2.

Nonlocal Gravity. Bahram Mashhoon. © Bahram Mashhoon 2017. Published 2017 by Oxford University Press.

Let us briefly digress here and mention that if $a_0 = 0$, $q_1 = q_2 = q_0$, where q_0 is given by eqn (7.100), and $\mathcal{E}(r)$ in eqn (8.1) vanishes. Then, the gravitational force can be expressed as

$$\mathbf{F}(\mathbf{r}) = -Gmm'\,\hat{\mathbf{r}}\left[\frac{1}{r^2} + \frac{1}{\lambda_0\,r}\bar{U}(\mu_0 r)\right], \tag{8.5}$$

where $\bar{U}(x)$ is given by

$$\bar{U}(x) = \frac{2e^{-x}}{x}\left(e^x - 1 - \frac{1}{2}x\right). \tag{8.6}$$

For $x : 0 \to \infty$, $\bar{U}(x)$ is a positive function that starts from $\bar{U} = 1$ at $x = 0$ and then decreases monotonically as x increases and tends to zero as $x \to \infty$. For $0 < x \ll 1$, we can write

$$\bar{U}(x) = 1 - \frac{1}{6}x^2 + \frac{1}{12}x^3 - \frac{1}{40}x^4 + O(x^5), \tag{8.7}$$

so that neglecting terms of order $(\mu_0 r)^2$ and higher in the expansion of $\bar{U}(\mu_0 r)$, we find that eqn (8.5) essentially reduces to the Tohline–Kuhn force (7.91). The implications of the Tohline–Kuhn force for gravitational two-body systems have been the subject of a number of recent investigations (Fabris and Pereira Campos 2009; Blome et al. 2010; Rahvar and Mashhoon 2014; Lu et al. 2014). However, it appears that the extension of the $1/r$ part of the Tohline–Kuhn force (7.91) within the Solar System can likely be ruled out by current observational data regarding the perihelion precession of planetary orbits (Iorio 2015; Deng and Xie 2015). On the other hand, nonlocal gravity is characterized by the short-distance parameter $a_0 \neq 0$ and the following section is devoted to a discussion of the short-distance behavior of nonlocal gravity in the Solar System.

It follows from the connection between the Newtonian regime of nonlocal gravity and the Tohline–Kuhn approach that the basic nonlocality parameter λ_0 is expected to be a galactic length of order 1 kpc. Moreover, the quantities $\mathcal{E}_1(r)$ and $\mathcal{E}_2(r)$ are proportional to a_0/λ_0, where the short-distance parameter a_0 is such that $a_0 < \lambda_0 < \mu_0^{-1}$. In this chapter, we wish to explore the consequences of the new force law for gravitational systems and attempt to determine the reciprocal kernel q.

8.1 Solar System

It is interesting to work out the implications of nonlocal gravity for the gravitational physics of the Solar System. To this end, it appears that $\mathcal{E}_1(r)$ and $\mathcal{E}_2(r)$ are crucial for the discussion of the short-distance behavior of the gravitational force. To investigate this point, let us first find their Taylor expansions about $r = 0$. From eqns (8.2) and (8.3), it is straightforward to show by repeated differentiation that

$$\mathcal{E}_1(r) = \frac{r}{\lambda_0}\left[1 - \frac{1}{2}W_1(\varsigma)\left(\frac{r}{a_0}\right) + \frac{1}{3}W_2(\varsigma)\left(\frac{r}{a_0}\right)^2 - \cdots\right] \tag{8.8}$$

and

$$\mathcal{E}_2(r) = \frac{r}{\lambda_0}\left[1 - \frac{1}{3}W_2(\varsigma)\left(\frac{r}{a_0}\right)^2 + \cdots\right], \tag{8.9}$$

where
$$W_1(\varsigma) = 1 + \varsigma, \qquad W_2(\varsigma) = 1 + \varsigma + \frac{1}{2}\varsigma^2. \tag{8.10}$$

Thus, we find from eqn (8.1) that for $r < a_0$, we have the following expansions in powers of r/a_0,

$$\mathbf{F}_1(\mathbf{r}) = -Gmm'\frac{\hat{\mathbf{r}}}{r^2}\left[1 + \frac{1}{2}(1+\varsigma)\frac{r^2}{\lambda_0 a_0} - \frac{1}{3}(1+\varsigma+\varsigma^2)\frac{r^3}{\lambda_0 a_0^2} + \cdots\right] \tag{8.11}$$

and

$$\mathbf{F}_2(\mathbf{r}) = -Gmm'\frac{\hat{\mathbf{r}}}{r^2}\left[1 + \frac{1}{3}(1+\varsigma)\frac{r^3}{\lambda_0 a_0^2} + \cdots\right]. \tag{8.12}$$

It is remarkable that in the square brackets in eqns (8.11) and (8.12), the linear r/λ_0 term, which is the leading term in both $\mathcal{E}_1(r)$ and $\mathcal{E}_2(r)$, is absent; in fact, this term is simply canceled by the corresponding Tohline–Kuhn term. In other words, the presence of the short-range parameter $a_0 \neq 0$ in effect shields the near-field region from the influence of the $1/r$ part of the Tohline–Kuhn force.

Let us now apply these results to the relative motion of a gravitational binary system with a major axis that is small in comparison with a_0. In this case, the *main* deviation from the Newtonian inverse square force law in the two-body system, $\delta\mathbf{F}$, can be expressed as an expansion in powers of r/a_0. The result could be either of the form

$$\delta\mathbf{F}_1(\mathbf{r}) = -\frac{1}{2}\frac{Gmm'}{\lambda_0 a_0}(1+\varsigma)\hat{\mathbf{r}} + \frac{1}{3}\frac{Gmm'}{\lambda_0 a_0}(1+\varsigma+\varsigma^2)\frac{r}{a_0}\hat{\mathbf{r}} + \cdots \tag{8.13}$$

if kernel q_1 is employed, or

$$\delta\mathbf{F}_2(\mathbf{r}) = -\frac{1}{3}\frac{Gmm'}{\lambda_0 a_0}(1+\varsigma)\frac{r}{a_0}\hat{\mathbf{r}} + \cdots \tag{8.14}$$

if kernel q_2 is employed. Here, $a_0 < \lambda_0 < \mu_0^{-1}$; indeed, let us note that with $\lambda_0 \approx 3$ kpc and $\mu_0^{-1} \approx 17$ kpc, we expect that $\varsigma = \mu_0 a_0$, $0 < \varsigma < 1/5$, would be rather small in comparison with unity.

8.1.1   Kepler system

Imagine a Keplerian two-body system of point particles with a *radial* perturbing acceleration $\mathcal{A}_g = \delta\mathbf{F}/\bar{m}$,

$$\frac{d^2\mathbf{r}}{dt^2} + \frac{GM\mathbf{r}}{r^3} = \mathcal{A}_g, \tag{8.15}$$

where $M = m + m'$ is the total mass and $\bar{m}$ is the reduced mass of the system. The orbital angular momentum of the system is then conserved and the orbit remains planar. Consider first the case where the radial acceleration is of the form $\mathcal{A}_g = \eta_g \, \mathbf{r}$, where η_g is a constant. It can be shown using the Lagrange planetary equations, when averaged over the fast Keplerian motion with orbital frequency ω_K, $\omega_K^2 = GM/\bar{A}^3$, that the orbit keeps its shape but slowly precesses; that is, the semimajor axis of the orbit $\bar{A}$ and the orbital eccentricity e remain constant on the average, but there is a slow pericenter precession whose frequency is given by $\Omega_g \,\hat{\ell}$, where (Kerr, Hauck and Mashhoon 2003)

$$\Omega_g = \frac{3}{2}\frac{\eta_g}{\omega_K}\sqrt{1-e^2} \tag{8.16}$$

and $\hat{\ell}$ is the unit orbital angular momentum vector.

This case is reminiscent of the orbital perturbation due to the presence of a cosmological constant. Moreover, eqn (8.16) can also be obtained from the study of the average precession of the Runge–Lenz vector due to the presence of the perturbing acceleration (Kerr, Hauck and Mashhoon 2003).

Similarly, if the perturbing acceleration is radial and *constant*, namely, $\boldsymbol{A}_g = \eta'_g\,\hat{\mathbf{r}}$, then, as before, the shape of the orbit remains constant on the average, but there is a slow pericenter precession of frequency $\Omega'_g\,\hat{\ell}$, where

$$\Omega'_g = \frac{\eta'_g}{\omega_K\,\bar{A}}\sqrt{1-e^2}. \tag{8.17}$$

This result has been noted before in connection with the studies of the Pioneer anomaly (Iorio and Giudice 2006; Sanders 2006; Sereno and Jetzer 2006).

It follows from the results of the previous section that in nonlocal gravity the orbit on average remains planar and keeps its shape, but slowly precesses. If the reciprocal kernel of nonlocal gravity in the Newtonian regime is q_1, then $\delta\mathbf{F}_1(\mathbf{r}) = \bar{m}\,(\eta' + \eta_1\,r)\,\hat{\mathbf{r}}$, where

$$\eta' = -\frac{1}{2}\frac{GM}{\lambda_0\,a_0}(1+\varsigma),\quad \eta_1 = \frac{1}{3}\frac{GM}{\lambda_0\,a_0^2}(1+\varsigma+\varsigma^2). \tag{8.18}$$

Thus, superposing small perturbations, we get for the pericenter advance in this case that

$$\Omega_1 = -\frac{1}{2}\,\omega_K\,\frac{\bar{A}^2}{\lambda_0\,a_0}\left[1+\varsigma-\frac{\bar{A}}{a_0}(1+\varsigma+\varsigma^2)\right]\sqrt{1-e^2}. \tag{8.19}$$

On the other hand, if the reciprocal kernel turns out to be q_2, then $\delta\mathbf{F}_2(\mathbf{r}) = \bar{m}\,\eta_2\,\mathbf{r}$, where

$$\eta_2 = -\frac{1}{3}\frac{GM}{\lambda_0\,a_0^2}(1+\varsigma) \tag{8.20}$$

and hence the pericenter advance is given by

$$\Omega_2 = -\frac{1}{2}\,\omega_K\,\frac{\bar{A}^3}{\lambda_0\,a_0^2}(1+\varsigma)\sqrt{1-e^2}. \tag{8.21}$$

Thus, in either case, the rate of advance of the pericenter turns out to be *negative*.

It is interesting to explore the implications of these results for the Solar System.

8.1.2   Perihelion precession

Thus far we have dealt with the force between point particles. To apply our results to realistic systems, such as the core of galaxies, binary pulsars or the Solar System, we need to investigate the influence of the finite size of an astronomical body on the attractive gravitational force that it can generate. To simplify matters, imagine a point mass m outside a spherically symmetric body of radius $\bar{R}_0$ that has uniform mass density. Let $\bar{R}$ be the distance between m and the center of the sphere so that $\bar{R} > \bar{R}_0$. If the force of gravity is radial, we expect by symmetry that the net force on m would be along the line joining the center of the sphere to m. Under what conditions would the spherical body act on m as though its mass were concentrated at its center? It turns out that, in addition to Newton's law of gravity, any radial force that is proportional to distance would work just as well, so that in general the desired two-body force can be any linear superposition of these forces such as in the case of kernel q_2 and eqn (8.12). On the other hand, in connection with kernel q_1 and eqn (8.11), we find, after a detailed but straightforward calculation, that for a *constant* radial force the same is true, except that the strength of the constant force is thereby reduced by a factor of

$$1 - \frac{1}{5}\left(\frac{\bar{R}_0}{\bar{R}}\right)^2. \tag{8.22}$$

This factor is nearly unity in most applications of interest here and we therefore assume that we can treat uniform spherical bodies like point particles for the sake of simplicity. This means that we can approximately apply the results of the previous section to the influence of the Sun on the motion of a planet in the Solar System.

The recent advances in the study of precession of perihelia of planetary orbits have been reviewed by Iorio (2015). In absolute magnitude, for instance, the extra perihelion shift of Mercury and Saturn due to nonlocal gravity would be expected to be less than about 10 and 2 milliarcseconds per century, respectively; otherwise, the effect of nonlocality would have already shown up in high-precision ephemerides (Fienga *et al.* 2015; Pitjeva and Pitjev 2013), barring certain exceptional circumstances. Thus if the kernel of nonlocal gravity is q_1, the nonlocal contribution to the perihelion precession Ω_1 is expected to be such that its absolute magnitude for Mercury and Saturn would be less than about 10^{-2} and 2×10^{-3} seconds of arc per century, respectively. In general, the inequality involving $|\Omega_i|$ under consideration here for q_i, $i = 1, 2$, gives a lower limit on a_0 that increases with $\bar{A}$ as $\bar{A}^{1/2}$ or $\bar{A}^{3/4}$ depending on whether we choose q_1 or q_2, respectively. Thus the lower limit on a_0 can become more significant the farther the planetary orbit is from the Sun.

For the orbit of Mercury, $\bar{A} \approx 6 \times 10^{12}$ cm and $e \approx 0.2$; moreover, the orbital period is ≈ 0.24 yr. If the reciprocal kernel is q_1, it follows from eqn (8.19) and $\lambda_0 \approx 3$ kpc that in this case, $a_0 \gtrsim 7 \times 10^{13}$ cm. Similarly, if the kernel is q_2, we find from eqn (8.21) that in this case, $a_0 \gtrsim 2 \times 10^{13}$ cm.

For the orbit of Saturn, the orbital period is about 29.5 yr, $\bar{A} \approx 1.4 \times 10^{14}$ cm and $e \approx 0.056$. In a similar way, we find that if the reciprocal kernel is q_1, $a_0 \gtrsim 2 \times 10^{15}$ cm. However, if the reciprocal kernel is q_2, then $a_0 \gtrsim 5.5 \times 10^{14}$ cm. These preliminary lower limits can be significantly strengthened if, in the analysis of planetary data, Newton's law of gravity is replaced by either $\mathbf{F}_1$, given in eqn (8.11), or $\mathbf{F}_2$, given in eqn (8.12), depending upon whether the reciprocal kernel of nonlocal gravity is chosen to be q_1 or q_2, respectively.

8.1.3 Search for modifications of gravity

We note that the radius of a star or a planet is generally much smaller than the length scales a_0, λ_0 and μ_0^{-1} that appear in the nonlocal contribution to the gravitational force. It is therefore reasonable to approximate the exterior gravitational force due to a star or a planet by assuming that its mass is concentrated at its center. Hence, one can employ eqn (8.1) in astronomical systems such as binary pulsars and the Solar System, where possible nonlocal deviations from general relativity may become measurable in the future. In fact, nonlocal gravity in the Solar System could be tested experimentally via ESA's Gaia mission, launched in 2013, or other possible missions dedicated to measuring deviations from general relativity in the Solar System (Hees et al. 2015; Buscaino et al. 2015). In this connection, we note that in eqns (8.13) and (8.14),

$$\frac{1}{2}\frac{G M_\odot}{\lambda_0\, a_0}\,(1+\varsigma) \approx \left(\frac{10^{18}\,\text{cm}}{a_0}\right) 10^{-14}\,\text{cm s}^{-2}, \tag{8.23}$$

which, combined with the lower limits on a_0 mentioned above, is at least three orders of magnitude smaller than the acceleration involved in the Pioneer anomaly ($\sim 10^{-7}$ cm s^{-2}). It follows that even with dedicated missions it would still be very difficult to measure directly the nonlocal modification to the gravitational acceleration in the Solar System.

8.1.4 Gravitational deflection of light

Light rays follow null geodesics in nonlocal gravity. Consider the propagation of a light ray with impact parameter $\bar{\varrho}$ in the gravitational field generated by a point mass $\bar{M}$ that is essentially fixed at $r = 0$. It is well known that in the linear post-Newtonian approximation, the total deflection angle of the light ray is twice the Newtonian expectation. Therefore, if $\bar{\Delta}$ is the net deflection angle, we have for $i = 1, 2$,

$$\bar{\Delta}_i = \frac{4G\bar{M}\,\bar{\varrho}}{c^2}\int_0^{\frac{\pi}{2}} f_i\!\left(\frac{\bar{\varrho}}{\sin\vartheta}\right)\frac{d\vartheta}{\sin\vartheta}, \tag{8.24}$$

where $f_i(r)$ is given by

$$f_i(r) = \frac{1}{r^2}\left[1 - \mathcal{E}_i(r) + \alpha_0\right] - \frac{\alpha_0}{r^2}\left(1 + \frac{1}{2}\mu_0\, r\right) e^{-\mu_0\, r}, \tag{8.25}$$

see eqn (7.118). Here, $\bar{\varrho} = r\sin\vartheta$ is the fixed impact parameter and $\vartheta : 0 \to \pi$ is the corresponding scattering angle.

For $a_0 = 0$, the reciprocal kernel is then q_0 and the net deflection angle $\bar{\Delta}_0$ has been studied in some detail (Blome et al. 2010; Rahvar and Mashhoon 2014). For our present purposes, $\bar{\Delta}_0$ can be expressed as

$$\bar{\Delta}_0 = \frac{4G\bar{M}}{c^2\,\bar{\varrho}}\left[1 + \alpha_0 - \alpha_0\,\mathfrak{J}(\varrho)\right], \tag{8.26}$$

where ϱ is the *dimensionless impact parameter*,

$$\varrho := \mu_0\,\bar{\varrho}. \tag{8.27}$$

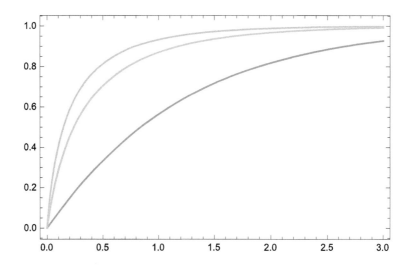

Fig. 8.1 The figure depicts the graphs of the functions $\mathfrak{I}_0(\varrho)$, $\mathfrak{I}_1(\varrho, \varsigma)$ and $\mathfrak{I}_2(\varrho, \varsigma)$ for $\varsigma = 0.1$ and $\varrho : 0 \to \infty$. The graph of $\mathfrak{I}_1(\varrho, \varsigma)$ lies above the graph of $\mathfrak{I}_2(\varrho, \varsigma)$, which in turn lies above the graph of $\mathfrak{I}_0(\varrho)$. Reproduced from Chicone, C. and Mashhoon, B., 2016, "Nonlocal Gravity in the Solar System", *Class. Quantum Grav.* **33**, 075005 (21 pages), with the permission of IOP Publishing. DOI: 10.1088/0264-9381/33/7/075005

In eqn (8.26), $\mathfrak{I}$ is the integral defined in eqn (7.192) and can be expressed as

$$\mathfrak{I}(x) = \int_0^{\frac{\pi}{2}} \left(\cos\varphi + \frac{1}{2}x\right) e^{-x \sec\varphi}\, d\varphi, \tag{8.28}$$

where φ here is related to ϑ in eqn (8.24) via $\vartheta + \varphi = \pi/2$.

It proves useful to define $\mathfrak{I}_0$ and the Einstein deflection angle $\bar{\Delta}_E$ by

$$\mathfrak{I}_0 := 1 - \mathfrak{I}, \qquad \bar{\Delta}_E := \frac{4G\bar{M}}{c^2\,\bar{\varrho}}. \tag{8.29}$$

For dimensionless impact parameter $\varrho : 0 \to \infty$, we note that $\mathfrak{I}_0(\varrho) : 0 \to 1$; that is, $\mathfrak{I}_0(\varrho)$ monotonically increases from zero and asymptotically approaches unity as $\varrho \to \infty$. For $0 < \varrho \ll 1$, $\mathfrak{I}_0(\varrho) \approx \pi\varrho/4$ and hence $\bar{\Delta}_0$ differs from the Einstein deflection angle $\bar{\Delta}_E$ by a constant angle that is proportional to the mass of the source and coincides with the result derived from the Tohline–Kuhn force law (Blome *et al.* 2010; Rahvar and Mashhoon 2014). It is indeed smaller than the Einstein deflection angle by a factor of $\sim 10^{-11}$ for light rays passing near the rim of the Sun.

In nonlocal gravity, $a_0 > 0$ and we find from eqns (8.24)–(8.26) that $\bar{\Delta}_i$ is given by

$$\bar{\Delta}_i = \bar{\Delta}_0 - \bar{\Delta}_E \int_0^{\frac{\pi}{2}} \cos\varphi\, \mathcal{E}_i\left(\frac{\bar{\varrho}}{\cos\varphi}\right) d\varphi. \tag{8.30}$$

We can therefore write

$$\bar{\Delta}_i = \bar{\Delta}_E \left[1 + \alpha_0\, \mathfrak{I}_0(\varrho) - \mathcal{E}_i(\infty)\, \mathfrak{I}_i(\varrho, \varsigma)\right], \tag{8.31}$$

where

$$\Im_i(\varrho,\varsigma) := \frac{1}{\mathcal{E}_i(\infty)} \int_0^{\frac{\pi}{2}} \cos\varphi \, \mathcal{E}_i\Big(\frac{\bar{\varrho}}{\cos\varphi}\Big) \, d\varphi. \tag{8.32}$$

The functions $\mathcal{E}_1(r)$ and $\mathcal{E}_2(r)$ are given in eqns (8.2) and (8.3), respectively. It turns out that for $r/a_0 \ll 1$, $\mathcal{E}_i(r) \approx r/\lambda_0$ by eqns (8.8) and (8.9); hence, for $0 < \varrho \ll 1$, $\mathcal{E}_i(\infty)\,\Im_i(\varrho,\varsigma) \approx a_0(\pi\varrho/4)$. As expected, this term cancels the other (Tohline–Kuhn) term, $\alpha_0\,\Im_0(\varrho) \approx a_0(\pi\varrho/4)$, in eqn (8.31). Moreover, for $\varrho : 0 \to \infty$, we note that $\Im_i(\varrho,\varsigma) : 0 \to 1$; that is, $\Im_i(\varrho,\varsigma)$ monotonically increases from zero at $\varrho = 0$ and asymptotically approaches unity as $\varrho \to \infty$. Thus for $\varrho \gg 1$, i.e. large impact parameters $\bar{\varrho} \gg \mu_0^{-1}$, $\bar{\Delta}_i \to \bar{\Delta}_E[1 + \alpha_0 - \mathcal{E}_i(\infty)]$, where $\alpha_0 - \mathcal{E}_i(\infty) = \alpha_0\,\epsilon_i(\varsigma)$ in accordance with eqn (8.4); that is, the extra deflection angle takes due account of the net effective dark matter associated with $\bar{M}$.

The new integral, $\Im_i(\varrho,\varsigma)$, can be expressed in terms of $d\mathcal{E}_i/dr$. Using integration by parts, eqn (8.32) can be written as

$$\Im_i(\varrho,\varsigma) = 1 - \frac{\bar{\varrho}}{\mathcal{E}_i(\infty)} \int_0^{\frac{\pi}{2}} \tan^2\varphi \, \frac{d\mathcal{E}_i}{dr}\Big(\frac{\bar{\varrho}}{\cos\varphi}\Big) \, d\varphi, \tag{8.33}$$

where $d\mathcal{E}_1/dr$ and $d\mathcal{E}_2/dr$ are given by eqns (7.126) and (7.127), respectively. More explicitly, we have

$$\Im_1(\varrho,\varsigma) = 1 - \frac{\varrho}{e^\varsigma\,E_1(\varsigma)} \int_0^{\frac{\pi}{2}} \frac{\sin^2\varphi}{(\varrho + \varsigma\cos\varphi)\cos\varphi}\, e^{-\varrho\sec\varphi}\, d\varphi \tag{8.34}$$

and

$$\Im_2(\varrho,\varsigma) = \Im_1(\varrho,\varsigma) - \frac{1}{2}\,\frac{\varrho}{e^\varsigma\,E_1(\varsigma)} \int_0^{\frac{\pi}{2}} \frac{\cos\varphi}{\varrho + \varsigma\cos\varphi}\, e^{-\varrho\sec\varphi}\, d\varphi. \tag{8.35}$$

We plot $\Im_0(\varrho)$, $\Im_1(\varrho,\varsigma)$ and $\Im_2(\varrho,\varsigma)$ for $\varsigma = 0.1$ and $\varrho : 0 \to \infty$ in Fig. 8.1.

It is possible to express the net deflection angle as

$$\bar{\Delta}_i = \bar{\Delta}_E\,[1 + \alpha_0\,\mathfrak{D}_i(\varrho,\varsigma)], \tag{8.36}$$

where $\mathfrak{D}_1$ and $\mathfrak{D}_2$ are given by

$$\mathfrak{D}_1(\varrho,\varsigma) = \Im_0(\varrho) - \frac{1}{2}\,\varsigma\,e^\varsigma\,E_1(\varsigma)\,\Im_1(\varrho,\varsigma) \tag{8.37}$$

and

$$\mathfrak{D}_2(\varrho,\varsigma) = \Im_0(\varrho) - \varsigma\,e^\varsigma\,E_1(\varsigma)\,\Im_2(\varrho,\varsigma), \tag{8.38}$$

respectively.

It now remains to discuss the influence of $a_0 > 0$ on the gravitational deflection of starlight by the Sun ($\bar{M} = M_\odot$). If the reciprocal kernel is $q_i, i = 1, 2$, then the net deflection angle due to nonlocality is $\alpha_0\,\mathfrak{D}_i(\varrho,\varsigma)$ times the Einstein deflection angle $\bar{\Delta}_E$, in accordance with eqn (8.36). For light rays passing near the rim of the Sun, the dimensionless impact parameter is very small ($\varrho \sim 10^{-12}$). Moreover, using the lower limits placed on a_0 in the previous section, we note that $\varsigma :\sim 4 \times 10^{-8} \to 0.2$ in

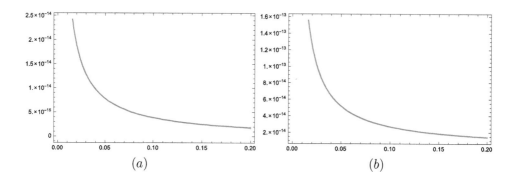

Fig. 8.2 (a) The figure depicts the graph of the function $\varsigma \mapsto \mathfrak{D}_1(10^{-8},\varsigma)$. We note that for $\varsigma \approx 4 \times 10^{-8}$, $\mathfrak{D}_1 \approx 1.6 \times 10^{-9}$ in this case. (b) The figure depicts the graph of the function $\varsigma \mapsto \mathfrak{D}_2(10^{-7},\varsigma)$. We note that for $\varsigma \approx 10^{-8}$, $\mathfrak{D}_2 \approx 5.6 \times 10^{-8}$ in this case. Reproduced from Chicone, C. and Mashhoon, B., 2016, "Nonlocal Gravity in the Solar System", *Class. Quantum Grav.* **33**, 075005 (21 pages), with the permission of IOP Publishing. DOI: 10.1088/0264-9381/33/7/075005

$\mathfrak{D}_1$, while $\varsigma : \sim 10^{-8} \to 0.2$ in $\mathfrak{D}_2$. Our numerical results indicate that $|\mathfrak{D}_1|$ and $|\mathfrak{D}_2|$ are negligibly small compared to unity. For instance, for $\varrho = 10^{-12}$ we find both $\mathfrak{D}_1$ and $\mathfrak{D}_2$ to be $\approx -10^{-15}$. To illustrate the situation, we plot $\mathfrak{D}_1$ and $\mathfrak{D}_2$ in Fig. 8.2. It appears from these results that the nonlocal contribution to the gravitational bending of light in the Solar System is utterly negligible at the present time.

It is important to point out that $\mathfrak{I}_0(\varrho)$ and $\mathfrak{I}_i(\varrho,\varsigma)$ are not analytic about $\varrho = 0$, so that they cannot be expanded in a Taylor series about $\varrho = 0$. The behavior of these functions for $\varrho \to 0$ can in principle be determined using asymptotic approximation methods (Bleistein and Handelsman 1986). A simple case is illustrated in Section 7.12.

8.1.5   Gravitational time delay

The general expressions for the gravitational potentials due to a point mass $\bar{M}$ at $r = 0$ corresponding to the reciprocal kernels q_1 and q_2 are given in eqns (7.134) and (7.135), respectively. Within the Solar System, $\mu_0 r \ll 1$ and we can therefore use expansions in powers of this small quantity as in eqn (7.136). Neglecting terms of order $(r/a_0)^2$ and higher, we find

$$\Phi_1(r) \approx -\frac{G\bar{M}}{r} - \frac{G\bar{M}}{\lambda_0}\left[1 + e^{\varsigma} E_1(\varsigma)\right] + \frac{1}{2}\frac{G\bar{M}}{\lambda_0}(1+\varsigma)\frac{r}{a_0} \tag{8.39}$$

and

$$\Phi_2(r) \approx -\frac{G\bar{M}}{r} - \frac{G\bar{M}}{\lambda_0} e^{\varsigma} E_1(\varsigma). \tag{8.40}$$

The nonlocal contribution to the gravitational potential is extremely small within the Solar System. To illustrate this point consider, for instance, the gravitational shift of the frequency of light, which involves the difference in the potential at two spatially separated events. In the approximation scheme under consideration here, the contribution to the shift in the potential due to nonlocality is non-zero only in the case of Φ_1 and is given by

$$\frac{1}{2} \frac{G\bar{M}}{\lambda_0} (1 + \varsigma) \frac{r_2 - r_1}{a_0}, \tag{8.41}$$

where r_1 and r_2 are the radial positions of the events under consideration. This is rather small in absolute magnitude when compared with the corresponding shift of the Newtonian potential; that is, at a distance of ten astronomical units, say, we have $(10 \text{ AU})^2/(\lambda_0 a_0) \lesssim 10^{-9}$ based on the lower limit on a_0 already established using the perihelion precession of Saturn. Therefore, we conclude that the relative contribution of nonlocality to the gravitational shift of the frequency of light is very small within the Solar System.

Consider next the gravitational time delay $\bar{D}$ of a light signal that travels from event $P_1 : (ct_1, \mathbf{r}_1)$ to event $P_2 : (ct_2, \mathbf{r}_2)$. Then, $\bar{D} = t_2 - t_1 - |\mathbf{r}_2 - \mathbf{r}_1|/c$ is given by

$$\bar{D}_i = -\frac{2}{c^3} \int_{P_1}^{P_2} \Phi_i \, d\sigma, \tag{8.42}$$

where $\sigma : 0 \rightarrow |\mathbf{r}_2 - \mathbf{r}_1|$ is the distance along a straight line from P_1 to P_2. It is in general straightforward to compute $\bar{D}_i$ for nonlocal gravity in the Solar System. However, to simplify matters, we consider only the time delay due to Φ_2, which is

$$\bar{D}_2 = \frac{2G\bar{M}}{c^3} \left[\ln \frac{r_2 + \hat{\mathbf{n}} \cdot \mathbf{r}_2}{r_1 + \hat{\mathbf{n}} \cdot \mathbf{r}_1} + e^\varsigma E_1(\varsigma) \frac{|\mathbf{r}_2 - \mathbf{r}_1|}{\lambda_0} \right], \tag{8.43}$$

where $\hat{\mathbf{n}} = (\mathbf{r}_2 - \mathbf{r}_1)/|\mathbf{r}_2 - \mathbf{r}_1|$. The result is simply the sum of the Shapiro time delay and the nonlocal contribution to signal retardation. We recall that $\varsigma : \sim 10^{-8} \rightarrow 0.2$ in this case; moreover, it follows from eqn (7.121) that for $0 < \varsigma \ll 1$, $E_1(\varsigma) \approx -C - \ln \varsigma$. If $|\mathbf{r}_2 - \mathbf{r}_1|$ is about an astronomical unit, then $|\mathbf{r}_2 - \mathbf{r}_1|/\lambda_0 \sim 10^{-9}$; therefore, the nonlocal effect is rather small and probably difficult to measure, since there are uncertainties due to clock stability as well as the existence of the interplanetary medium (Shapiro 1980).

Before leaving our discussion of the gravitational physics of the Solar System on the basis of the linear regime of nonlocal gravity, we need to emphasize that *nonlocal gravity is consistent with the observational data regarding the Solar System*. Indeed, we have placed preliminary lower limits on the short-range parameter a_0 on the basis of current data regarding the perihelion precession of planetary orbits in the Solar System. For instance, for Saturn, a preliminary lower limit of $a_0 \gtrsim 2 \times 10^{15}$ cm can be established for reciprocal kernel q_1, while $a_0 \gtrsim 5.5 \times 10^{14}$ cm for reciprocal kernel q_2. Each such kernel contains three parameters that all have dimensions of length: λ_0, μ_0^{-1} and a_0. Furthermore, we have $a_0 < \lambda_0 < \mu_0^{-1}$. We now proceed to the determination of long-range parameters λ_0 and μ_0^{-1}. The aim of this approach is to determine a unique *universal* reciprocal kernel with three parameters that would be valid for the present epoch.

8.2 Spiral Galaxies

To model the rotation curves of nearby spiral galaxies in accordance with nonlocal gravity theory, we can use eqn (8.1) instead of the gravitational inverse square force law and assume that there is no actual dark matter. In other words, a spiral galaxy essentially consists of baryonic matter in the form of stars and interstellar gas. We ignore dust in our analysis, as the mass of the dust is at most a few percent of the mass of the interstellar matter. To simplify matters further, we neglect the short-range parameter a_0 in our model (Rahvar and Mashhoon 2014).

What is actually measured via astronomical observation is the surface density of the luminosity of a galaxy, which decreases from the center towards the edge of the galaxy. From the galactic luminosity, we can obtain the total mass of stars in the galaxy using the stellar mass-to-light ratio $\Upsilon_\star$. Moreover, the mass of the gaseous component of the galaxy can be obtained from the total mass of the hydrogen gas, M_H, via $M_{gas} = \frac{4}{3} M_H$, due to the hydrogen to helium abundance in Big Bang nucleosynthesis, while M_H can be determined from the 21 cm radiation of the atomic hydrogen.

We choose a sample of twelve spiral galaxies from The HI Nearby Galaxy Survey (THINGS) catalog. The available observational data for these galaxies include their rotation curves, which have been measured by the Doppler effect using 21 cm radiation of neutral hydrogen gas. Moreover, for our sample of galaxies we use the stellar luminosity distributions given in the Spitzer Infrared Nearby Galaxies Survey (SINGS) catalog. We then employ the gravitational force law (8.1) to fit the observational data for the twelve spiral galaxies of our sample of THINGS catalog. In this way, the best-fitting values for the parameters α_0 and μ_0 of the reciprocal kernel q_0 of nonlocal gravity theory are determined from the combination of observational data for the twelve spiral galaxies of our sample and the end results are (Rahvar and Mashhoon 2014)

$$\alpha_0 = 10.94 \pm 2.56, \qquad \mu_0 = 0.059 \pm 0.028 \text{ kpc}^{-1}, \qquad \lambda_0 \approx 3 \pm 2 \text{ kpc}, \qquad (8.44)$$

where λ_0 has been evaluated using $\lambda_0 = 2/(\alpha_0 \mu_0)$.

To ensure that the nonlocal model is consistent with the astrophysics of star formation as well as the Tully–Fisher relation (Tully and Fisher 1977), we first fix the force law (8.1) (with $a_0 = 0$) by adopting the best values of α_0 and μ_0 given in eqn (8.44), namely, $\alpha_0 \approx 11$ and $\mu_0^{-1} \approx 17$ kpc. We use the resulting nonlocal gravity theory to fit the observed rotation curves of twenty-seven spiral galaxies from the Ursa Major cluster. In this study, we let the stellar mass-to-light ratio $\Upsilon_\star$ be the only free parameter. The Ursa Major Cluster of Galaxies is a spiral-rich member of the Virgo Supercluster and is located at a distance of about 18.6 Mpc. The best-fitting stellar mass-to-light ratio $\Upsilon_\star$ turns out to be in good agreement with astrophysical models; moreover, our results are consistent with the Tully–Fisher relation, which is essentially an empirically established linear proportionality between the intrinsic infrared luminosity of a spiral galaxy and v_c^4, where v_c is the corresponding asymptotic rotation speed. The results of this analysis are presented in detail in Rahvar and Mashhoon (2014).

The kernel of nonlocal gravity χ as well as the corresponding reciprocal kernel q are *universal* functions that must be applicable to all gravitational systems at the present epoch. To test the approximate validity of the tentative values of the parameters α_0

and μ_0 given in eqn (8.44), we next turn to the internal dynamics of rich clusters of galaxies.

8.3 Chandra X-Ray Clusters of Galaxies

Clusters of galaxies are filled with hot ionized gas that is luminous in X rays. The low-density gas contains $\sim 10^{-3}$ atoms/cm^3 and has a temperature of order 10^8 K. Most of the electrons originate from hydrogen and helium atoms that are fully ionized. We assume, as usual, that in the plasma the mass density of protons is nearly $\frac{3}{4}\rho_g$ and the mass density of helium ions is nearly $\frac{1}{4}\rho_g$, where ρ_g is the mass density of the cluster gas. The gas contains most of the baryonic mass of rich clusters and is roughly at the *virial temperature* T_v that is related to the radial (i.e. line-of-sight) velocity dispersion σ_r of the galaxies in the cluster, namely, $k_B T_v \approx \mu_p m_p \sigma_r^2$, where μ_p is the *mean* atomic mass of the plasma (electrons and ions) in amu, $\mu_p \approx 0.6$, and m_p is the proton mass. To simplify matters, we can imagine the cluster gas to be spherically symmetric and in equilibrium. It is then possible to employ the gas density profile $\rho_g(r)$ and the corresponding temperature profile $T_v(r)$, obtained from the observational data provided by the Chandra X-ray telescope, to find the magnitude of the Newtonian gravitational acceleration $\mathfrak{g}_c(r)$ inside the cluster. In equilibrium, the gas pressure gradient is balanced by the gravitational attraction, so that $dp_g/dr = -\rho_g\,\mathfrak{g}_c(r)$, where the gas pressure p_g is given by $p_g/(k_B T_v) = \rho_g/(\mu_p m_p)$ in accordance with the ideal gas law (Sarazin 1988). It follows from these equations that for a spherically symmetric system in hydrostatic equilibrium, the magnitude of Newtonian gravitational acceleration is related to the observable parameters by (Sarazin 1988)

$$\mathfrak{g}_c(r) = -\frac{k_B T_v(r)}{\mu_p m_p\, r}\left(\frac{d\ln\rho_g(r)}{d\ln r} + \frac{d\ln T_v(r)}{d\ln r}\right). \tag{8.45}$$

In order to have correct dynamics with only baryonic matter and no actual dark matter, the right-hand side of this equation should be proportional to the right-hand side of eqn (8.1) of our nonlocal gravity model. To simplify matters, we neglect, as in the previous section, the short-range parameter a_0 in eqn (8.1). Then, for $\mu_0 r \gg 1$, the force between two point particles reduces to the Newtonian inverse square force law augmented by the constant factor of $1 + \alpha_0$. Thus in this case, a homogeneous spherical shell of matter exerts no force on a test particle in its interior but attracts an exterior test particle as though the mass of the shell were concentrated at its center. The radius of a cluster is of the order of 1000 kpc; therefore, in the outer parts of the cluster and away from the central region that has the highest mass density, consider the determination of the Newtonian acceleration of gravity on a point mass at fixed radius r. Except for the total mass interior to a sphere of radius $\sim \mu_0^{-1} \approx 17$ kpc about our fixed position, we can safely neglect the repulsive Yukawa force in eqn (8.1) in comparison with the attractive Newtonian force. As the mass within the sphere of radius $\sim \mu_0^{-1} \approx 17$ kpc is very small compared to the mass of the cluster, we can simply estimate $\mathfrak{g}_c(r)$ using the Newtonian inverse square force law augmented by $1 + \alpha_0$. It follows that at a radial distance r well beyond the central regions of a cluster, we have the approximate expression

$$\mathfrak{g}_c(r) \approx G\left(1 + \alpha_0\right) \frac{M_c(r)}{r^2},$$ (8.46)

where $\alpha_0 \approx 11$ and $M_c(r)$ is the actual mass of the cluster within a sphere of radius r. From eqns (8.45) and (8.46), it is possible to determine $M_c(r)$ once the gas density and temperature profiles are known. The mass of the cluster is dominated by its baryonic content; therefore, let $M_b(r)$ be the total mass of the gas and stars within a sphere of radius r. The mass of the gas is obtained by integrating ρ_g over the volume of a sphere of radius r. We would expect that $M_c(r) \approx M_b(r)$ if the gravitational force that balances gas pressure within the cluster were correctly represented by nonlocal gravity (Rahvar and Mashhoon 2014).

To compare $M_c(r)$ of a cluster with its baryonic counterpart $M_b(r)$, we choose a sample of ten rich nearby clusters of galaxies from the Chandra catalog. We compute $M_c(r)$ for each cluster from eqns (8.45) and (8.46) using its gas density and temperature profiles and then compare the resulting mass to the corresponding baryonic mass of the cluster. The details of the calculations are contained in Rahvar and Mashhoon (2014). For each cluster, we find that the actual mass of the cluster obtained from nonlocal gravity is consistent with the measured baryonic mass.

If the (reciprocal) kernel of nonlocal gravity in the Newtonian regime is truly universal, we should be able to apply the theory to all *nearby* gravitational many-body systems and make predictions about their structures. In this endeavor, gravitational systems at cosmological distances are excluded, since kernels χ and q are purely spatial and do not reflect the fading of gravitational memory with time that should be an important characteristic feature of nonlocal gravity during the expansion of the universe. To make predictions about nearby systems, we first need to extend the virial theorem to nonlocal gravity.

In general, the virial theorem of Newtonian physics establishes a simple linear proportionality between the time averages of the kinetic and potential energies of an isolated material system for which the potential energy is a homogeneous function of spatial coordinates. For an isolated *gravitational* N-body system, the significance of the virial theorem has to do with the circumstance that the kinetic energy is a sum of terms each proportional to the mass of a body in the system, while the potential energy is a sum of terms each proportional to the product of two masses in the system. Thus, under favorable conditions, the virial theorem can be used to connect the total *dynamic* mass of an isolated relaxed gravitational system with its average internal motion.

8.4   Virial Theorem

Consider an idealized isolated system of N Newtonian point particles with fixed masses m_i, $i = 1, 2, \ldots, N$. We assume that the particles occupy a finite region of space and interact with each other only gravitationally such that the center of mass of the isolated system is at rest in a global inertial frame and the isolated system permanently occupies a compact region of space. The equation of motion of the particle with mass m_i and state $(\mathbf{x}_i, \mathbf{v}_i)$ is then

$$m_i \frac{d\mathbf{v}_i}{dt} = -\sum_j{}' \frac{G\, m_i\, m_j\, (\mathbf{x}_i - \mathbf{x}_j)}{|\mathbf{x}_i - \mathbf{x}_j|^3} [1 + \mathbb{N}(|\mathbf{x}_i - \mathbf{x}_j|)] \tag{8.47}$$

for $j = 1, 2, \ldots, N$, but the case $j = i$ is excluded in the sum by convention. In fact, a prime over the summation sign indicates that in the sum $j \neq i$. Here $1 + \mathbb{N}(r)$ is a *universal* function that is inside the curly brackets in eqn (8.1) and the contribution of nonlocality, $\mathbb{N}(r)$, is given by

$$\mathbb{N}(r) := \alpha_0 \left[1 - \left(1 + \frac{1}{2}\mu_0\, r\right) e^{-\mu_0\, r} \right] - \mathcal{E}(r). \tag{8.48}$$

Consider next the quantities

$$\mathbb{I} = \frac{1}{2}\sum_i m_i\, x_i^2, \qquad \frac{d\mathbb{I}}{dt} = \sum_i m_i\, \mathbf{x}_i \cdot \mathbf{v}_i, \tag{8.49}$$

where $x_i = |\mathbf{x}_i|$ and

$$\frac{d^2\mathbb{I}}{dt^2} = \sum_i m_i\, v_i^2 + \sum_i m_i\, \mathbf{x}_i \cdot \frac{d\mathbf{v}_i}{dt}. \tag{8.50}$$

It follows from eqn (8.47) that

$$\sum_i m_i\, \mathbf{x}_i \cdot \frac{d\mathbf{v}_i}{dt} = -\sum_{i,j}{}' \frac{G\, m_i\, m_j\, (\mathbf{x}_i - \mathbf{x}_j) \cdot \mathbf{x}_i}{|\mathbf{x}_i - \mathbf{x}_j|^3} [1 + \mathbb{N}(|\mathbf{x}_i - \mathbf{x}_j|)]. \tag{8.51}$$

Exchanging i and j in the expression on the right-hand side of eqn (8.51), we get

$$\sum_i m_i\, \mathbf{x}_i \cdot \frac{d\mathbf{v}_i}{dt} = \sum_{i,j}{}' \frac{G\, m_i\, m_j\, (\mathbf{x}_i - \mathbf{x}_j) \cdot \mathbf{x}_j}{|\mathbf{x}_i - \mathbf{x}_j|^3} [1 + \mathbb{N}(|\mathbf{x}_i - \mathbf{x}_j|)]. \tag{8.52}$$

Adding eqns (8.51) and (8.52) results in

$$\sum_i m_i\, \mathbf{x}_i \cdot \frac{d\mathbf{v}_i}{dt} = -\frac{1}{2}\sum_{i,j}{}' \frac{G\, m_i\, m_j}{|\mathbf{x}_i - \mathbf{x}_j|} [1 + \mathbb{N}(|\mathbf{x}_i - \mathbf{x}_j|)]. \tag{8.53}$$

Using this result, eqn (8.50) takes the form

$$\frac{d^2\mathbb{I}}{dt^2} = \sum_i m_i\, v_i^2 - \frac{1}{2}\sum_{i,j}{}' \frac{G\, m_i\, m_j}{|\mathbf{x}_i - \mathbf{x}_j|} [1 + \mathbb{N}(|\mathbf{x}_i - \mathbf{x}_j|)]. \tag{8.54}$$

Let us recall here that $\mathbb{N}$ is defined by eqn (8.48); moreover, the net kinetic energy and the Newtonian gravitational potential energy of the system are given by

$$\mathbb{T} = \frac{1}{2}\sum_i m_i\, v_i^2, \qquad \mathbb{W}_N = -\frac{1}{2}\sum_{i,j}{}' \frac{G\, m_i\, m_j}{|\mathbf{x}_i - \mathbf{x}_j|}. \tag{8.55}$$

Hence,

$$\frac{d^2\,\mathbb{I}}{dt^2} = 2\,\mathbb{T} + \mathbb{W}_N + \mathbb{D}, \tag{8.56}$$

where

$$\mathbb{D} = -\frac{1}{2}\sum_{i,j}{}' \frac{G\,m_i\,m_j}{|\mathbf{x}_i - \mathbf{x}_j|}\,\mathbb{N}(|\mathbf{x}_i - \mathbf{x}_j|). \tag{8.57}$$

Finally, we are interested in the average of eqn (8.56) over time. Let $< f >$ denote the time average of f, where

$$< f > = \lim_{\tau\to\infty}\frac{1}{\tau}\int_0^\tau f(t)\,dt. \tag{8.58}$$

Then, it follows from averaging eqn (8.56) over time that

$$2 < \mathbb{T} > = - < \mathbb{W}_N > - < \mathbb{D} >, \tag{8.59}$$

since $d\,\mathbb{I}/dt$, which is the sum of $m\,\mathbf{x}\cdot\mathbf{v}$ over all particles in the system, is a bounded function of time and hence the time average of $d^2\,\mathbb{I}/dt^2$ vanishes. This is clearly based on the assumption that the spatial coordinates and velocities of all particles indeed remain finite for all time. Equation (8.59) expresses the *virial theorem* in nonlocal Newtonian gravity.

It is important to digress here and re-examine some of the assumptions involved in our derivation of the virial theorem. In general, any consequence of the gravitational interaction involves the whole mass–energy content of the observable universe due to the universality of the gravitational interaction; therefore, an astronomical system may be considered isolated only to the extent that the tidal influence of the rest of the observable universe on the internal dynamics of the system can be neglected. Moreover, the parameters of the force law (8.1) refer to the present epoch and hence the virial theorem (8.59) ignores cosmological evolution. Thus the temporal average over an infinite period of time in eqn (8.59) must be reinterpreted here to mean that the relatively isolated system under consideration has evolved under its own gravity such that it is at the present epoch in a steady equilibrium state. In other words, the system is currently in virial equilibrium. Finally, we recall that a point particle of mass m in eqn (8.59) could reasonably represent a star of mass m as well, where the mass of the star is assumed to be concentrated at its center.

The deviation of the virial theorem (8.59) from the Newtonian result is contained in $< \mathbb{D} >$, where $\mathbb{D}$ is given by eqn (8.57). More explicitly, we have

$$\mathbb{D} = -\frac{1}{2}\sum_{i,j}{}' \frac{G\,m_i\,m_j}{|\mathbf{x}_i - \mathbf{x}_j|}\left[\alpha_0 - \alpha_0\left(1 + \frac{1}{2}\,\mu_0\,|\mathbf{x}_i - \mathbf{x}_j|\right)e^{-\mu_0\,|\mathbf{x}_i - \mathbf{x}_j|} - \mathcal{E}(|\mathbf{x}_i - \mathbf{x}_j|)\right]. \tag{8.60}$$

It proves useful at this point to study some of the properties of the function $\mathbb{N}$, which is the contribution of nonlocality that is inside the square brackets in eqn (8.60). The argument of this function is $|\mathbf{x}_i - \mathbf{x}_j| > 0$ for $i \neq j$; therefore, $|\mathbf{x}_i - \mathbf{x}_j|$ varies over the interval $(0, \mathcal{D}_0]$, where $\mathcal{D}_0$ is the largest possible distance between any two baryonic point masses in the system. Thus $\mathbb{N}(r)$, in the context of the virial theorem, is defined for the interval $0 < r \leq \mathcal{D}_0$, where $\mathcal{D}_0$ is the diameter of the smallest sphere that completely encloses the *baryonic* system for all time. In general, however, $\mathbb{N}(0) = 0$ and $\mathbb{N}(\infty) = \alpha_0 - \mathcal{E}(\infty) > 0$, where $\mathcal{E}(\infty) = \mathcal{E}_\infty$ or $2\mathcal{E}_\infty$, depending on whether we use reciprocal kernel q_1 or q_2, respectively. Moreover, $d\,\mathbb{N}(r)/dr$ is given by

$$\frac{d}{dr}\mathbb{N}_1(r) = \frac{1}{2}\alpha_0\,\mu_0\,\frac{r\,[1 + \mu_0\,(a_0 + r)]}{a_0 + r}\,e^{-\mu_0\,r}, \qquad (8.61)$$

if we use q_1 or

$$\frac{d}{dr}\mathbb{N}_2(r) = \frac{1}{2}\alpha_0\,\mu_0\,\frac{r^2\,[1 + \mu_0\,(a_0 + r)]}{(a_0 + r)^2}\,e^{-\mu_0\,r}, \qquad (8.62)$$

if we use q_2. Writing $\exp(\mu_0\,r) = 1 + \mu_0\,r + \mathcal{R}$, where $\mathcal{R} > 0$ represents the remainder of the power series, it is straightforward to see that for $r \geq 0$ and $n = 1, 2, \ldots,$

$$e^{\mu_0\,r}\,(a_0 + r)^n > r^n\,[1 + \mu_0\,(a_0 + r)]. \qquad (8.63)$$

This result, for $n = 1$ and $n = 2$, implies that the right-hand sides of eqns (8.61) and (8.62), respectively, are less than $\alpha_0\,\mu_0/2$. Therefore, it follows that in general

$$\frac{d}{dr}\mathbb{N}(r) < \frac{1}{2}\alpha_0\,\mu_0. \qquad (8.64)$$

Moreover, for $r > 0$, eqn (8.64) implies

$$\mathbb{N}(r) = \int_0^r \frac{d\,\mathbb{N}(x)}{dx}\,dx < \frac{1}{2}\alpha_0\,\mu_0\,r. \qquad (8.65)$$

We conclude that $\mathbb{N}$ is a monotonically increasing function of r that is zero at $r = 0$ with a slope that vanishes at $r = 0$. For $r \gg \mu_0^{-1}$, $\mathbb{N}(r)$ asymptotically approaches a constant $\alpha_0\,\epsilon(\varsigma) := \alpha_0 - \mathcal{E}(\infty)$. Here $\epsilon(\varsigma)$ is either $\epsilon_1(\varsigma)$ or $\epsilon_2(\varsigma)$ depending on whether we use q_1 or q_2, respectively, see eqn (8.4). The functions $\epsilon_1(\varsigma)$ and $\epsilon_2(\varsigma)$ are plotted in Fig. 7.2.

8.5 Dark Matter

Most of the matter in the universe is currently thought to be in the form of certain elusive particles that have not been directly detected (Aprile *et al.* 2012; Akerib *et al.* 2014; Agnese *et al.* 2014; Baudis 2015). The existence and properties of this *dark matter* have thus far been deduced only through its gravity. We are interested here in dark matter only as it pertains to nearby stellar systems such as galaxies and clusters of galaxies (Zwicky 1933, 1937; Rubin and Ford 1970; Roberts and Whitehurst 1975; Sofue and Rubin 2001; Seigar 2015; Harvey *et al.* 2015). We mention that dark matter is also essential in the explanation of gravitational lensing observations (Clowe *et al.* 2006; Clowe, Randall and Markevitch 2007), see Section 7.7, and in the solution of the problem of structure formation in cosmology, see Chapter 10.

Actual (mainly baryonic) mass is observationally estimated for astronomical systems using the mass-to-light ratio. However, it turns out that the dynamic mass of the system is usually larger and this observational fact is normally attributed to the possible existence of non-baryonic dark matter. Let M_b be the baryonic mass and M_{DM} be the mass of the non-baryonic dark matter needed to explain the gravitational dynamics of the system. Then,

$$f_{DM} := \frac{M_{DM}}{M_b} \tag{8.66}$$

is the dark matter fraction and $M_b + M_{DM} = M_b\,(1 + f_{DM})$ is the dynamic mass of the system.

In observational astrophysics, the virial theorem of Newtonian gravity is interpreted to be a relationship between the dynamic (virial) mass of the entire system and its average internal motion deduced from the rotation curve or velocity dispersion of the bound collection of masses in virial equilibrium. Therefore, regardless of how the net amount of dark matter in galaxies and clusters of galaxies is operationally estimated and the corresponding f_{DM} is thereby determined, for sufficiently isolated self-gravitating astronomical systems in virial equilibrium, we must have

$$2 < \mathbb{T} > = -(1 + f_{DM}) < \mathbb{W}_N > . \tag{8.67}$$

In other words, virial theorem (8.67) is employed in astronomy to infer in some way the total dynamic mass of the system. Indeed, Zwicky first noted the need for dark matter in his application of the standard virial theorem of Newtonian gravity to the Coma Cluster of Galaxies (Zwicky 1933, 1937).

8.6   Effective Dark Matter

A significant physical consequence of nonlocal gravity theory is that it appears to simulate dark matter. In particular, in the Newtonian regime of nonlocal gravity, the Poisson equation is modified such that the density of ordinary matter ρ is accompanied by a term ρ_D that is obtained from the folding (convolution) of ρ with the reciprocal kernel of nonlocal gravity. Thus ρ_D has the interpretation of the density of *effective dark matter* and $\rho + \rho_D$ is the density of the *effective dynamic mass*; see Section 7.4.

The virial theorem makes it possible to elucidate in a simple way the manner in which nonlocality can simulate dark matter. It follows from a comparison of eqns (8.59) and (8.67) that nonlocal gravity can account for this "excess mass" if

$$< \mathbb{D} > = f_{DM} < \mathbb{W}_N > . \tag{8.68}$$

It is interesting to apply the virial theorem of nonlocal gravity to sufficiently isolated astronomical N-body systems. The configurations that we briefly consider below consist of clusters of galaxies with diameters $\mathcal{D}_0 \gg \mu_0^{-1} \approx 17$ kpc, galaxies with $\mathcal{D}_0 \sim \mu_0^{-1}$ and globular star clusters with $\mathcal{D}_0 \ll \mu_0^{-1}$. The results presented in this section follow from certain general properties of the function $\mathbb{N}(r)$ and are completely independent of how the baryonic matter is distributed within the astronomical system under consideration.

8.6.1 Clusters of galaxies: $f_{DM} \approx \alpha_0 \, \epsilon(\varsigma)$

Consider, for example, a cluster of galaxies, where nearly all of the relevant distances are much larger than $\mu_0^{-1} \approx 17$ kpc. In this case, $\mu_0 \, r \gg 1$ and hence $\mathbb{N}$ approaches its asymptotic value, namely,

$$\mathbb{N} \approx \alpha_0 \, \epsilon(\varsigma), \tag{8.69}$$

where $\epsilon(\varsigma) = \epsilon_1(\varsigma)$ or $\epsilon_2(\varsigma)$, defined in eqn (8.4), depending on whether we use q_1 or q_2, respectively. Hence, eqn (8.57) can be written as

$$< \mathbb{D} > \approx \alpha_0 \, \epsilon(\varsigma) \; < \mathbb{W}_N > . \tag{8.70}$$

It then follows from eqn (8.68) that for galaxy clusters

$$f_{DM} \approx \alpha_0 \, \epsilon(\varsigma) \tag{8.71}$$

in nonlocal gravity. We recall that $\epsilon(\varsigma)$ is only weakly sensitive to the magnitude of a_0. It follows from $\alpha_0 \approx 11$ and Fig. 7.2 that f_{DM} for galaxy clusters is about 10, in general agreement with observational data (Rahvar and Mashhoon 2014). This theoretical result is essentially equivalent to the work on galaxy clusters contained in (Rahvar and Mashhoon 2014), except that eqn (8.71) takes into account the existence of the short-range parameter a_0. In connection with the effective dark matter fraction f_{DM}, *nonlocal gravity thus predicts a universal result given by eqn (8.71) for all nearby clusters of galaxies that are sufficiently isolated and in virial equilibrium.*

8.6.2 Galaxies: $\frac{\mathcal{D}_0}{f_{DM}} > \lambda_0$

Consider next a sufficiently isolated galaxy of diameter $\mathcal{D}_0$ in virial equilibrium. In this case, we recall that $\mathbb{N}(r)$ is a monotonically increasing function of r, so that for $0 < r \leq \mathcal{D}_0$, eqn (8.65) implies

$$\mathbb{N}(r) \leq \mathbb{N}(\mathcal{D}_0) < \frac{1}{2} \, \alpha_0 \, \mu_0 \, \mathcal{D}_0. \tag{8.72}$$

Therefore, it follows from eqn (8.57) that in this case

$$\mathbb{D} > (\frac{1}{2} \, \alpha_0 \, \mu_0 \, \mathcal{D}_0) \, \mathbb{W}_N. \tag{8.73}$$

The virial theorem for nonlocal gravity in the case of an isolated galaxy is then

$$2 < \mathbb{T} > + < \mathbb{W}_N > < -(\frac{1}{2} \, \alpha_0 \, \mu_0 \, \mathcal{D}_0) \, < \mathbb{W}_N >, \tag{8.74}$$

which means, when compared with eqn (8.67), that

$$f_{DM} < \frac{1}{2} \, \alpha_0 \, \mu_0 \, \mathcal{D}_0. \tag{8.75}$$

Let us note that

$$\frac{1}{2} \, \alpha_0 \, \mu_0 = \frac{1}{\lambda_0}, \tag{8.76}$$

where λ_0 is the basic nonlocality length scale. Its exact value is not known; however, from the results of Rahvar and Mashhoon (2014), we have, tentatively, $\lambda_0 \approx 3 \pm 2$ kpc.

If we formally let $\lambda_0 \to \infty$, then eqn (8.75), namely, $f_{DM} < \mathcal{D}_0/\lambda_0$, implies that in this case nonlocality and the effective dark matter both disappear, as expected. Therefore, for a sufficiently isolated galaxy in virial equilibrium, the ratio of its baryonic diameter to dark matter fraction f_{DM} must always be above a fixed length λ_0 of about 3 ± 2 kpc; that is,

$$\frac{\mathcal{D}_0}{f_{DM}} > \lambda_0. \tag{8.77}$$

To illustrate relation (8.77), consider, for instance, the Andromeda Galaxy (M31) with a diameter $\mathcal{D}_0$ of about 67 kpc. In this case, we have $f_{DM} \approx 12.7$ (Barmby *et al.* 2006, 2007), so that for this spiral galaxy

$$\frac{\mathcal{D}_0}{f_{DM}} \text{ (Andromeda Galaxy)} \approx 5.3 \text{ kpc}. \tag{8.78}$$

More recently, the distribution of dark matter in M31 has been the subject of further study (Tamm *et al.* 2012). Similarly, for the Triangulum Galaxy (M33), we have $\mathcal{D}_0 \approx 34$ kpc and $f_{DM} \approx 5$ (Corbelli 2003), so that

$$\frac{\mathcal{D}_0}{f_{DM}} \text{ (Triangulum Galaxy)} \approx 6.8 \text{ kpc}. \tag{8.79}$$

Turning next to an elliptical galaxy, namely, the massive E0 galaxy NGC 1407, we have $\mathcal{D}_0 \approx 160$ kpc and $f_{DM} \approx 31$ (Pota *et al.* 2015), so that

$$\frac{\mathcal{D}_0}{f_{DM}} \text{ (NGC 1407)} \approx 5.2 \text{ kpc}. \tag{8.80}$$

Moreover, for the intermediate-luminosity elliptical galaxy NGC 4494, which has a half-light radius of $R_e \approx 3.77$ kpc, the dark matter fraction has been found to be $f_{DM} = 0.6 \pm 0.1$ (Morganti *et al.* 2013). Assuming that the baryonic system has a radius of $2\,R_e$, we have $\mathcal{D}_0 = 4\,R_e \approx 15$ kpc and $f_{DM} \approx 0.6$; hence,

$$\frac{\mathcal{D}_0}{f_{DM}} \text{ (NGC 4494)} \approx 25 \text{ kpc}. \tag{8.81}$$

We emphasize that the results presented here are valid for the present epoch in the expansion of the universe. Observations indicate, however, that the diameters of massive galaxies can increase with decreasing redshift z (Peralta de Arriba *et al.* 2014). For a discussion of such *massive compact early-type galaxies*, see Section 10.6.

Finally, it is interesting to consider f_{DM} at the other extreme, namely, for the case of globular star clusters. The diameter of a globular star cluster is about 40 pc. We can therefore conclude from eqn (8.77) with $\lambda_0 \approx 3$ kpc that for globular star clusters

$$f_{DM} \text{ (globular star cluster)} \lesssim 10^{-2}. \tag{8.82}$$

Thus according to the virial theorem of nonlocal gravity, less than about one per cent of the mass of a globular star cluster must appear as effective dark matter if the system is sufficiently isolated and is in virial equilibrium. It is not clear to what extent such systems can be considered isolated. It is usually assumed that observational data are consistent with the existence of almost no dark matter in globular star clusters. However, a recent investigation of six Galactic globular clusters has led to the conclusion that $f_{DM} \approx 0.4$ (Sollima, Bellazzini and Lee 2012). The resolution of this discrepancy is beyond the scope of the present book.

Isolated dwarf galaxies with diameters $\mathcal{D}_0 \ll \mu_0^{-1}$ would similarly be expected to contain a relatively small percentage of effective dark matter. One encounters a significant discrepancy here as well, see Oh *et al.* (2015). In connection with *dwarf spheroidal galaxies*, we note that such systems are seldom isolated and the tidal influence of a much larger neighboring galaxy on the dynamics of the dwarf spheroidal galaxy cannot be ignored (Kuhn and Miller 1989; Fleck and Kuhn 2003; Muñoz *et al.* 2005).

To ascertain whether nonlocal gravity is capable of resolving the difficulties associated with globular star clusters and isolated dwarf galaxies, it is necessary to employ eqn (8.1) with *nonzero short-range parameter* a_0 in order to analyze the internal gravitational dynamics of these systems. Perhaps in this way the magnitude of a_0 and a unique reciprocal kernel q can be determined.

8.7 Galaxies and Nonlocal Gravity

Nonlocal gravity theory predicts that the amount of effective dark matter in a sufficiently isolated nearby galaxy in virial equilibrium is such that f_{DM} has an upper bound, $\mathcal{D}_0/\lambda_0$, that is completely independent of the distribution of baryonic matter in the galaxy. However, it is possible to derive an *improved* upper bound for f_{DM}, which does depend on how baryons are distributed within the galaxy. To this end, we note that eqn (8.57) for $\mathbb{D}$ and $\mathbb{N}(r) < r/\lambda_0$ imply

$$\mathbb{D} > -\frac{1}{2} \sum_{i,j}' \frac{G\, m_i\, m_j}{\lambda_0}. \tag{8.83}$$

This inequality holds for $<\mathbb{D}>$ as well; therefore, if follows from this result together with eqn (8.68) that

$$<\mathbb{W}_N> f_{DM} > -\frac{1}{2} \sum_{i,j}' \frac{G\, m_i\, m_j}{\lambda_0}. \tag{8.84}$$

Let us define a characteristic length, R_{av}, for the average extent of the distribution of baryons in the galaxy via

$$R_{av} <\mathbb{W}_N> := -\frac{1}{2} \sum_{i,j}' G\, m_i\, m_j. \tag{8.85}$$

Then, it follows from eqns (8.84) and (8.85) that

$$f_{DM} < \frac{R_{av}}{\lambda_0}. \tag{8.86}$$

Clearly, R_{av} depends upon the density of baryons in the galaxy. In the Newtonian gravitational potential energy in eqn (8.85), $0 < |\mathbf{x}_i - \mathbf{x}_j| \leq \mathcal{D}_0$; therefore, in general,

$R_{av} \leq \mathcal{D}_0$ and hence we recover from the new inequality, namely, $f_{DM} < R_{av}/\lambda_0$, our previous less tight but more general result $f_{DM} < \mathcal{D}_0/\lambda_0$.

Dark matter is currently required in astrophysics for explaining the gravitational dynamics of galaxies as well as clusters of galaxies, gravitational lensing observations and structure formation in cosmology. It s important to emphasize here that only some of the implications of nonlocal gravity theory have thus far been confronted with observation (Rahvar and Mashhoon 2014; Chicone and Mashhoon 2016a).

9
Linearized Gravitational Waves in Nonlocal Gravity

The main field equation of nonlocal gravity in its linearized form has been presented in Chapter 7. In a similar way as in general relativity (GR), the general linear approximation of nonlocal gravity can be used to study nonlocal post-Newtonian gravity as well as gravitational radiation. The nonlocal modification of Poisson's equation of Newtonian gravitation theory as well as certain post-Newtonian effects have been studied in detail in the last two chapters. Therefore, the present chapter is devoted to the treatment of linearized gravitational waves in nonlocal general relativity. As the treatment of linearized gravitational waves in GR is well known, we concentrate here on an examination of the nonlocal deviations of the theory from the standard general relativistic analysis.

Loss of orbital energy due to the emission of gravitational radiation can explain the steady orbital decay rate of the Hulse–Taylor binary pulsar as well as similar relativistic binary systems (Blanchet 2014); indeed, this circumstance provides *indirect* evidence for radiative reaction and the existence of gravitational waves within the context of GR. The same result can be obtained within the framework of $GR_{||}$, the teleparallel equivalent of GR (Schweizer and Straumann 1979; Schweizer, Straumann and Wipf 1980; Muench, Gronwald and Hehl 1998). Moreover, nonlocal gravity involves a basic galactic length scale $\lambda_0 \approx 3$ kpc that is very much larger than the orbital size of a relativistic binary pulsar, so that nonlocal effects are likely to be negligibly small. Indeed, as explained in detail in Chapter 8, the relative influence of nonlocality on a binary star system is expected to be of order $2\bar{A}/\lambda_0$, where $2\bar{A}$ is the major axis of the binary system. For the Hulse–Taylor relativistic binary pulsar PSR B1913+16, $\bar{A} \approx 2 \times 10^{11}$ cm and $2\bar{A}/\lambda_0 \approx 3 \times 10^{-11}$, which is extremely small compared to unity. Therefore, we expect that the compatibility of the gravitational radiation damping with $GR_{||}$ (Schweizer and Straumann 1979; Schweizer, Straumann and Wipf 1980; Muench, Gronwald and Hehl 1998) would simply extend to the nonlocal generalization of $GR_{||}$. Hence, nonlocal gravity is consistent with observational data regarding the orbital decay of relativistic binary systems.

Current approaches for the *direct* detection of gravitational waves have been reviewed by Riles (2013). The galactic scale associated with nonlocality, $\lambda_0 \approx 3$ kpc, corresponds to a characteristic frequency of $c/\lambda_0 \approx 3 \times 10^{-12}$ Hz. However, present laboratory efforts involving interferometers are directed at detecting gravitational waves of dominant frequency $\gtrsim 1$ Hz and corresponding dominant wavelength λ, where $\lambda/\lambda_0 \lesssim 3 \times 10^{-12}$. Moreover, future space-based interferometers may be able to detect low-frequency ($\sim 10^{-4}$ Hz) gravitational waves from astrophysical sources. On the other hand, pulsar timing residuals from an ensemble of highly stable pulsars can be used to search for a stochastic background of very low frequency ($\sim$ several nHz) gravitational waves (Riles 2013). Thus current observational possibilities all involve gravitational radiation of wavelengths that are rather short in comparison with λ_0; indeed, in all cases of interest,

$$\frac{\lambda}{\lambda_0} \lesssim 10^{-3}. \tag{9.1}$$

The first direct detection of gravitational radiation by the Earth-based LIGO detectors has been recently reported (Abbott *et al.* 2016a). The signal, GW150914, detected on 14 September 2015, had an amplitude of about 10^{-21} and frequencies that ranged from about 35 Hz to about 250 Hz. The analysis of observational data indicates that the radiation originated from a binary black hole merger (Abbott *et al.* 2016a). Moreover, the two detectors of LIGO simultaneously observed a second event on 26 December 2015 (Abbott *et al.* 2016b). The second signal, GW151226, had a peak amplitude of about 3.4×10^{-22} and frequencies that ranged from about 35 Hz to about 450 Hz. The observational data indicate that the second gravitational wave signal also originated from the merger of two black holes (Abbott *et al.* 2016b).

The main purpose of this chapter is to discuss linearized gravitational waves— namely, their generation, propagation and detection—within the framework of nonlocal gravity. We expect that in nonlocal gravity, the treatment of extremely low-frequency ($\sim 10^{-12}$ Hz) gravitational waves with wavelengths of order λ_0 would be quite different than in general relativity. In fact, for radiation of frequency $\gtrsim 10^{-8}$ Hz, which is the frequency range that is the focus of current observational searches, the corresponding wavelengths are very small compared to λ_0. We find that in this frequency regime the nonlocal deviations from GR essentially average out and can be safely neglected in practice. On the other hand, nonlocality is expected to play a significant role in the treatment of gravitational waves of frequency $\lesssim c/\lambda_0 \sim 10^{-12}$ Hz.

9.1 Nonlocal Wave Equation

In GR, the source-free field equation in the linear approximation reduces to the wave equation for the trace-reversed potentials $\bar{h}_{\mu\nu}$, $\Box \bar{h}_{\mu\nu} = 0$, once the transverse gauge condition, $\bar{h}^{\mu\nu}{}_{,\nu} = 0$, has been imposed. The linearized field equation of nonlocal gravity with $\bar{h}^{\mu\nu}{}_{,\nu} = 0$ has been treated in detail in Section 7.3. To describe the propagation of *free* gravitational waves in linearized nonlocal gravity, we begin with the linearized field equation.

The gravitational potentials in linearized nonlocal gravity in the transverse gauge consist of ten symmetric perturbations $h_{\mu\nu}$ of the Minkowski metric tensor, or, equivalently, their trace-reversed counterparts $\bar{h}_{\mu\nu}$, and six tetrad variables given by the antisymmetric tensor $\phi_{\mu\nu}$. The field equation then reduces to ten *dynamic* equations involving the metric variables, namely,

$$\Box \bar{h}_{\mu\nu} + 2\,\mathcal{S}_{\mu\nu} = -2\,\kappa\,\mathcal{T}_{\mu\nu} + 2\,\breve{p}\,\mathcal{U}_{\mu\nu} \tag{9.2}$$

and six source-free integral *constraint* equations given by

$$\int K_c(x-y)\left(-\bar{h}_{00,i}+\bar{h}_{ij,}{}^j-\phi_{ij,}{}^j\right)(y)\,d^4y = 4\,\check{p}\,U_{[i\,0]}(x),\qquad(9.3)$$

$$\int K_c(x-y)\left(\bar{h}_{0i,j}+\phi_{0i,j}-\bar{h}_{0j,i}-\phi_{0j,i}\right)(y)\,d^4y = 4\,\check{p}\,U_{[i\,j]}(x).\qquad(9.4)$$

In these equations, which correspond, respectively, to eqns (7.70), (7.38) and (7.39) of Chapter 7, the source of the gravitational field is $\mathcal{T}_{\mu\nu} = T_{\mu\nu} + T^D_{\mu\nu}$, where $T_{\mu\nu}$ is the *conserved* energy–momentum tensor of matter, $\partial_\nu T^{\mu\nu} = 0$, and $T^D_{\mu\nu}$ is its dark counterpart given by the convolution of $T_{\mu\nu}$ and the reciprocal kernel R. Similarly, $\mathcal{U}_{\mu\nu}$ is the sum of $U_{\mu\nu}$ and its convolution with R; moreover, $U_{\mu\nu}$ and $\mathcal{U}_{\mu\nu}$ are integral expressions involving torsion pseudovector $\check{C}_\alpha$ and its derivatives. The torsion pseudovector is the dual of $C_{[\mu\nu\rho]} = -\phi_{[\mu\nu,\rho]}$. Thus $U_{\mu\nu}$ and $\mathcal{U}_{\mu\nu}$ are nonlocal expressions involving only the derivatives of $\phi_{\mu\nu}$. Finally, $\mathcal{S}_{\mu\nu}$ is the sum of $S_{\mu\nu}$, given by eqn (7.28), and its convolution with R. In fact, $\mathcal{S}_{0\mu} = 0$ and the dynamic eqn (9.2) can be written in components as

$$\Box\bar{h}_{0\mu} = -2\,\kappa\,\mathcal{T}_{0\mu} - 2\,\check{p}\int R(x-y)\,\check{C}_{0,\mu}(y)\,d^4y,\qquad(9.5)$$

$$\Box\bar{h}_{ij}+\int W(x-y)\left[\bar{h}_{ij,0}-\bar{h}_{0(i,j)}+\phi_{0(i,j)}-\delta_{ij}\,\phi_{0k,}{}^k\right](y)\,d^4y = -2\,\kappa\,\mathcal{T}_{ij}+2\,\check{p}\,\mathcal{U}_{(ij)}.\qquad(9.6)$$

We recall that $K_c(x-y)$ is the restriction of the kernel of nonlocal gravity $K(x-y)$ to the light cone; that is, the light-cone kernel K_c is given by

$$K_c(x-y) = K(x-y)\,\delta(x^0-y^0-|\mathbf{x}-\mathbf{y}|).\qquad(9.7)$$

Moreover, we have

$$W(x-y) = K_c(x-y) + \int R(x-z)K_c(z-y)\,d^4z.\qquad(9.8)$$

9.1.1 Source-free field equation

In the absence of any gravitational source, $T_{\mu\nu} = 0$ and hence $\mathcal{T}_{\mu\nu} = 0$. Then, inspection of the field eqns (9.3)–(9.6) reveals that when $\mathcal{T}_{\mu\nu} = 0$, these equations are all satisfied provided

$$\bar{h}_{0\mu} = 0,\qquad \phi_{\mu\nu} = 0\qquad(9.9)$$

and the *nonlocal gravitational wave equation*

$$\Box\bar{h}_{ij} + \int W(x-y)\,\bar{h}_{ij,0}(y)\,d^4y = 0\qquad(9.10)$$

holds.

To satisfy the field equation of nonlocal gravity in this way, the field quantities in display (9.9) only need to be constants. Let us recall in this connection that in Chapter 7 we assumed that, in the absence of the gravitational interaction, the tetrad frame field of the fundamental observers would coincide with the preferred global inertial frame associated with the source. In the complete absence of any sources, this restriction is no longer applicable. It is then possible to redefine the background coordinate system and tetrad frame such that in effect the constant field quantities all vanish as in display (9.9).

To understand the physical import of eqn (9.10), a mechanical analogy turns out to be quite useful: imagine expressing the wave amplitude in eqn (9.10) as a Fourier sum in *space*; then the wave equation is reminiscent of the equation of motion of a linear harmonic oscillator with a dissipation term that is proportional to the velocity of the oscillator. Here, $\partial \bar{h}_{ij}/\partial t$ is suggestive of the "velocity" of the oscillator. With the appropriate sign for the coefficient of the dissipation term, one has a *damped oscillator*. In a similar way, with the proper functional forms for the nonlocal kernels, eqn (9.10) indicates free propagation of gravitational waves with *damping*. The nonlocality of the theory originates from a certain average over past events in spacetime; this memory of the past thus appears to act as a drag that dampens the free propagation of linearized gravitational waves. In nonlocal gravity, memory fades exponentially for events that are distant in space and time. This exponential decay is reflected in the kernels R and K of linearized nonlocal gravity; see Section 7.5. Similarly, the amplitude of gravitational radiation decays exponentially with time as $\exp(-t/t_g)$, where the damping time t_g of the wave amplitude is related to the nonlocal kernel.

In general, as the gravitational waves propagate freely through Minkowski vacuum, the wave amplitude may grow or decay in time due to nonlocality. We expect that with the correct nonlocal kernel, the solutions of our linear nonlocal homogeneous wave equation are well behaved and decay in time leading to the stability of Minkowski spacetime under small perturbations. This issue has been treated in detail in Chicone and Mashhoon (2013) and Mashhoon (2013b). As discussed later, the main result can be demonstrated when a certain approximation is employed; that is, there is no instability and the waves indeed decay in time *when retardation is neglected in the kernel*. However, the general mathematical problem involving retardation remains unsolved.

The nonlocal wave eqn (9.10) is *linear* and $(\bar{h}_{ij})$ are *real* amplitudes of the free gravitational radiation field. For the purposes of the present discussion, we can therefore treat $(\bar{h}_{ij})$ as *complex* amplitudes with the proviso that only their real parts have physical significance. Thus let $\Psi_g(x)$ be an element of the matrix $(\bar{h}_{ij})$ and assume that

$$\Psi_g(x) = e^{-i\omega x^0} \psi_g(\mathbf{x}), \tag{9.11}$$

where $\omega = \omega_R - i\,\omega_I$ is the wave frequency that is in general complex. The nonlocal wave equation can now be expressed as

$$(\nabla^2 + \omega^2)\psi_g(\mathbf{x}) - i\,\omega \int W(x-y)\, e^{i\,\omega\,(x^0-y^0)}\psi_g(\mathbf{y})\, d^4y = 0. \tag{9.12}$$

Writing eqn (9.11) in terms of real and imaginary parts of ω, we find

$$\Psi_g(x) = e^{-i\,\omega_R x^0}\, \psi_g(\mathbf{x})\, e^{-\omega_I\, x^0}. \tag{9.13}$$

We expect that $\omega_I > 0$ and the wave amplitude will exponentially *decay* in time; otherwise, the perturbation will blow up as $x^0 \to \infty$, which is physically unacceptable, as it would indicate an intrinsic instability of Minkowski spacetime within the framework of nonlocal gravity. Using eqns (9.7)–(9.8) and the general expression for the reciprocal kernel R given by eqn (7.143), namely,

$$R(x) = \nu \, e^{-\nu(x^0 - |\mathbf{x}|)} \, u(x^0 - |\mathbf{x}|) \, q(\mathbf{x}), \tag{9.14}$$

it is in principle possible to determine kernel W. One can show that

$$(\nabla^2 + \omega^2)\psi_g(\mathbf{x}) + i\omega \int W_\omega(\mathbf{x} - \mathbf{y}) \, \psi_g(\mathbf{y}) \, d^3y = 0, \tag{9.15}$$

where W_ω is given by

$$W_\omega(\mathbf{z}) = -e^{i\omega |\mathbf{z}|} K(|\mathbf{z}|, \mathbf{z})$$
$$- \frac{\nu}{\nu - i\omega} \int e^{i\omega\,(|\mathbf{u}| + |\mathbf{z} - \mathbf{u}|)} \, q(\mathbf{z} - \mathbf{u}) \, K(|\mathbf{u}|, \mathbf{u}) \, d^3u. \tag{9.16}$$

Let

$$\hat{\psi}_g(\mathbf{k}) = \int \psi_g(\mathbf{x}) e^{-i\,\mathbf{k}\cdot\mathbf{x}} \, d^3x \tag{9.17}$$

be the spatial Fourier transform of the gravitational wave amplitude ψ_g, where $\mathbf{k}$ is the wave vector. Then,

$$\psi_g(\mathbf{x}) = \frac{1}{(2\pi)^3} \int \hat{\psi}_g(\mathbf{k}) e^{i\,\mathbf{k}\cdot\mathbf{x}} \, d^3k. \tag{9.18}$$

Using the convolution theorem for Fourier transforms, eqn (9.15) can be written as

$$\omega^2 - |\mathbf{k}|^2 + i\omega \, \hat{W}_\omega(\mathbf{k}) = 0 \tag{9.19}$$

provided $\hat{\psi}_g(\mathbf{k}) \neq 0$. The solution of the nonlocal wave equation thus reduces to the solution of eqn (9.19). In the absence of nonlocal kernel K, $W = 0$ and ω is real and is given by $\omega^2 = |\mathbf{k}|^2$, which represents free propagation of gravitational waves at the speed of light. A complete analysis of the solutions of the dispersion relation (9.19) is beyond the scope of this treatment. We turn instead to an approximation procedure, introduced at the end of Section 7.5, in order to show that free gravitational waves are indeed *damped* (i.e. $\omega_I > 0$) as they propagate in vacuum.

9.1.2 Damping

Let us recall from the discussion in Section 7.5 that with our general result for the reciprocal kernel (9.14), it is not possible to obtain an explicit formula for kernel K. To proceed, an approximation scheme was employed in Section 7.5 based on neglecting certain retardation effects. This means, for instance, that in eqns (9.7)–(9.8) we approximate $K_c(x - y)$ by $K(x - y)\delta(x^0 - y^0)$, since $|\mathbf{x} - \mathbf{y}|$ is due to retardation. Similarly, we approximate R by

$$R(x) \approx \nu \, e^{-\nu x^0} \, u(x^0) \, q(\mathbf{x}), \tag{9.20}$$

as in eqn (7.153). Using these expressions in W and computing the integral in eqn (9.12), we find an approximate expression for W_ω, namely,

$$W_\omega(\mathbf{z}) \approx -K(0^+, \mathbf{z}) - \frac{\nu}{\nu - i\omega} \int q(\mathbf{z} - \mathbf{u}) \, K(0^+, \mathbf{u}) \, d^3u. \tag{9.21}$$

Next, in this approximation scheme $K(x)$ can be computed from the reciprocity relation using eqn (9.20) for R and the result is

$$K(x) \approx -\frac{\nu}{(2\pi)^3} \, e^{-\nu x^0} u(x^0) \int \hat{q}(\mathbf{k}) e^{i\,\mathbf{k}\cdot\mathbf{x}} \, e^{-\nu \hat{q}\, x^0} \, d^3k; \tag{9.22}$$

see eqn (7.157). Hence,

$$-K(0^+, \mathbf{z}) \approx \nu \, q(\mathbf{z}). \tag{9.23}$$

Substituting eqn (9.23) in eqn (9.21), we find in this way that

$$W_\omega(\mathbf{z}) \approx \nu \, q(\mathbf{z}) + \frac{\nu^2}{\nu - i\omega} \int q(\mathbf{z} - \mathbf{u}) \, q(\mathbf{u}) \, d^3u \tag{9.24}$$

and

$$\hat{W}_\omega(\mathbf{k}) \approx \nu \, \hat{q}(\mathbf{k}) \left[1 + \frac{\nu \, \hat{q}(\mathbf{k})}{\nu - i\omega} \right] \tag{9.25}$$

by the convolution theorem for Fourier transforms.

The Fourier transforms of q_1 and q_2 were considered in detail in Section 7.11. It was shown there that $\hat{q}_1(\mathbf{k}) > 0$ and $\hat{q}_2(\mathbf{k}) > -a_0/\lambda_0$. Indeed, $\hat{q}_2(\mathbf{k})$ is positive when a_0/λ_0 is sufficiently small compared to unity. Though, as discussed in detail in Chapter 8, we have only lower limits on a_0 at the present time, we will henceforth assume that a_0/λ_0 is such that $\hat{q}_2(\mathbf{k}) > 0$. Thus in eqn (9.25), $\hat{q}(\mathbf{k}) > 0$ by assumption and ν^{-1}, $\nu^{-1} > \mu_0^{-1}$, is a constant galactic length that is expected to be comparable with $\mu_0^{-1} \approx 17$ kpc.

It is in principle possible to determine ω_R and ω_I from eqns (9.19) and (9.25), namely,

$$\omega^2 - |\mathbf{k}|^2 + i\,\omega\,\nu\,\hat{q}(\mathbf{k}) \left[1 + \frac{\nu \, \hat{q}(\mathbf{k})}{\nu - i\omega} \right] \approx 0, \tag{9.26}$$

though in practice the algebra is prohibitive. Therefore, let us first consider the possibility that $\omega = \omega_R - i\,\omega_I$ is such that $\omega_R = 0$; then, we find that eqn (9.26) reduces to the following cubic equation for ω_I,

$$\omega_I^3 - \nu\,(1 + \hat{q})\,\omega_I^2 + \nu^2 \, \hat{q}\,(1 + \hat{q})\,\omega_I - \nu\,|\mathbf{k}|^2 \approx 0. \tag{9.27}$$

We know that such a cubic equation must have at least one real root. Moreover, according to Descartes' rule of signs, there should be either three positive roots or one positive root. It follows that in any case $\omega_I > 0$. Next, we assume that $\omega_R \neq 0$ in eqn (9.26); then, we find after some algebra that the imaginary part of eqn (9.26) implies

$$\omega_I \approx \frac{1}{2}\,\nu\,\hat{q}(\mathbf{k}) \left[1 + \frac{\nu^2 \, \hat{q}(\mathbf{k})}{\omega_R^2 + (\nu - \omega_I)^2} \right], \tag{9.28}$$

so that $\omega_I > 0$. *Thus, within the framework of our approximation scheme, the amplitude of free gravitational radiation decays in time as a consequence of memory drag.* The extent to which this result may depend upon neglecting retardation has been studied in Chicone and Mashhoon (2013) and Mashhoon (2013b). Equation (9.28) can be used to estimate the damping time $t_g := 1/\omega_I$.

9.1.3 Damping time $t_g = 1/\omega_I$

The basic scale associated with nonlocality, $\lambda_0 \approx 3$ kpc, corresponds to a characteristic frequency of about $c/\lambda_0 \approx 3 \times 10^{-12}$ Hz. Moreover, ν^{-1} is such that $\nu^{-1} > \mu_0^{-1}$. We expect that ν^{-1} and $\mu_0^{-1} \approx 17$ kpc are comparable in magnitude and thus correspond to a frequency of $c\,\mu_0 \approx 5\times 10^{-13}$ Hz. It is intuitively clear that in eqn (9.19), significant deviations from the standard local dispersion relation for gravitational waves can be expected only for such low-frequency waves. According to eqn (9.1), for the current observational possibilities, the wavelengths of interest are very short compared to λ_0 and it turns out that for such wavelengths the contribution of nonlocality to the dispersion relation (9.26) is negligibly small, so that in practice $\omega^2 \approx |\mathbf{k}|^2$.

Using the explicit expressions for $q_1(r)$ and $q_2(r)$, it is possible to express $\hat{q}(\mathbf{k})$ as a function of $|\mathbf{k}|$ in terms of the exponential integral function E_1; see Section 7.11. In fact, $\hat{q}_1$ and $\hat{q}_2$ have rather similar functional forms: they both start from finite positive values at $|\mathbf{k}| = 0$ and monotonically decrease to zero as $|\mathbf{k}| \to \infty$. The damping time can be computed straightforwardly from eqn (9.28); however, the result is rather complicated and depends upon $a_0 |\mathbf{k}|$. Nevertheless, it is possible to provide a simple expression for our main result regarding t_g. To this end, we note that for $a_0 = 0$, $q_1(r)$ and $q_2(r)$ both simplify to $q_0(r)$; in this case, as shown in Section 7.11, $\hat{q}_0 > \hat{q}_1 > \hat{q}_2$ and the explicit functional form for $\hat{q}_0$ has been worked out in Section 7.11. In particular, for $\lambda_0 |\mathbf{k}| \gg 1$, we find that asymptotically, $\hat{q}_0 \sim \pi/(2\,\lambda_0\,|\mathbf{k}|)$. Therefore, $\hat{q} < \hat{q}_0$ and eqn (9.28) implies that for $\lambda \ll \lambda_0$,

$$t_g \approx \frac{2}{\nu\,\hat{q}(\mathbf{k})} \gtrsim \frac{8}{\nu}\left(\frac{\lambda_0}{\lambda}\right). \tag{9.29}$$

Using this simple result, we find that for *current observational possibilities*

$$t_g \gtrsim 4 \times 10^5 \left(\frac{\lambda_0}{\lambda}\right) \text{yr.} \tag{9.30}$$

It is interesting that t_g is proportional to the frequency of gravitational waves. In fact, nonlocality-induced damping could become significant for cosmological gravitational waves with very low frequencies; however, the damping effect is negligible for current experimental efforts involving interferometers. In other words, inspection of eqn (9.30) reveals that t_g is much longer than the age of the universe for gravitational waves that might be detectable with laser interferometers in the foreseeable future. We recall that for such devices, the waves should have dominant frequency $\gtrsim 1$ Hz for Earth-based and $\gtrsim 10^{-4}$ Hz for space-based antennas. However, detection of gravitational waves with dominant frequency of several nHz may be possible with pulsar timing arrays (Riles 2013). In connection with waves of low frequency $\lesssim 10^{-7}$ Hz, a detailed investigation (Mashhoon 2013b) reveals that t_g can be shorter than, or comparable with, the current estimate for the age of the universe. Linearized gravitational waves with very low frequencies $\lesssim 10^{-12}$ Hz would be highly damped in nonlocal gravity. For a more detailed treatment of the damping time see Mashhoon (2013b).

9.2 Propagation of Gravitational Waves with $\lambda \ll \lambda_0$

In the presence of time-varying sources, the dynamic variables $\bar{h}_{\mu\nu}$ and the tetrad variables $\phi_{\mu\nu}$ cannot be disentangled. This is clearly seen in eqn (7.226), which is a nonlocal relation relating T_{00} and $\check{C}_0$. In fact, eqn (7.226) implies that if $T_{00} \neq 0$, then $\check{C}_0$, which is related to the spatial derivatives of ϕ_{ij}, must in general vary in spacetime. The general time-dependent problem involving field eqns (9.3)–(9.6) is intractable. To proceed, we turn to configurations that involve gravitational waves with *short* wavelengths $\lambda \ll \lambda_0$. This regime includes current observational possibilities; see eqn (9.1). It proves interesting to estimate in this case the significance of nonlocal terms in the field equations. To this end, we turn to the familiar case of eqn (9.10), where the relative importance of the nonlocal term for waves of wavelength $\lambda = 2\pi/|\mathbf{k}|$ can be simply estimated by dividing the nonlocal term in the dispersion relation (9.19) by $|\mathbf{k}|^2$. In this way, we get $|\omega\, \hat{W}_\omega(\mathbf{k})|/|\mathbf{k}|^2$, which for wavelengths $\lambda \ll \lambda_0$ can be estimated using eqn (9.25) and the fact that $\hat{q} < \hat{q}_0 \sim \pi/(2\,\lambda_0\,|\mathbf{k}|)$. The result turns out to be $\lesssim \nu\,\lambda^2/(8\pi\,\lambda_0)$, which implies that the relative significance of the nonlocal terms in the field equations is $\lesssim 10^{-2}\,(\lambda/\lambda_0)^2$. For example, for waves of frequency 35 Hz, as in GW150914, $10^{-2}\,(\lambda/\lambda_0)^2 \approx 10^{-28}$, which is negligibly small in comparison with unity. *This means that for such short waves the nonlocal terms in the field equations can be simply neglected.* Hence, in this approximation, the nonlocal constraints disappear and the dynamic field eqn (9.2) basically reduces to

$$\Box\bar{h}_{\mu\nu} \approx -2\,\kappa\,\mathcal{T}_{\mu\nu}. \tag{9.31}$$

Assuming that the source $T_{\mu\nu}$ is isolated, the corresponding dark source is also then expected to be in effect isolated due to the rapid spatial decay of the reciprocal kernel. Far from the source and its dark counterpart, the gravitational potentials in the wave zone satisfy the wave equation $\Box\bar{h}_{\mu\nu} \approx 0$. Far in the wave zone, a fixed detector perceives the emitted gravitational radiation potentials $\bar{h}_{\mu\nu}$ to be essentially *plane* gravitational waves. Therefore, we can follow here essentially the same analysis as in standard GR and choose the remaining gauge degrees of freedom to set $\bar{h}_{0\nu} = 0$ and $\bar{h} = 0$. Thus with a suitable choice of gauge—namely, the TT gauge—$h_{\mu\nu}$ is purely spatial and traceless with $h^{ij}_{,j} = 0$. To see how this comes about, let us first recall that in nonlocal gravity a gauge transformation involves both the dynamic as well as the tetrad variables; see eqn (7.14). In fact, under an infinitesimal coordinate transformation $x'^\mu = x^\mu - \epsilon^\mu(x)$, we have

$$\bar{h}'_{\mu\nu} = \bar{h}_{\mu\nu} + \epsilon_{\mu,\nu} + \epsilon_{\nu,\mu} - \eta_{\mu\nu}\epsilon^\alpha{}_{,\alpha}, \tag{9.32}$$

$$\phi'_{\mu\nu} = \phi_{\mu\nu} + \epsilon_{\mu,\nu} - \epsilon_{\nu,\mu} \tag{9.33}$$

and

$$\bar{h}' = \bar{h} - 2\epsilon^\alpha{}_{,\alpha}, \qquad \bar{h}'^{\mu\nu}{}_{,\nu} = \bar{h}^{\mu\nu}{}_{,\nu} + \Box\epsilon^\mu. \tag{9.34}$$

The tetrad variables enter the linearized field equations of nonlocal gravity only in the integrands of the nonlocal terms that we neglect in our treatment of waves with $\lambda \ll \lambda_0$; therefore, only the gauge freedom of dynamic variables has significance in this analysis.

When we first impose the transverse gauge condition, we must find gauge functions ϵ^μ for which $\bar{h}'^{\mu\nu}{}_{,\nu} = 0$. It follows from eqn (9.34) that we must solve

$$\Box \epsilon^\mu = -\bar{h}^{\mu\nu}{}_{,\nu}. \tag{9.35}$$

Appropriate solutions of this standard inhomogeneous wave equation with suitable boundary conditions can be found to ensure that the transverse gauge condition is indeed satisfied. Next, we start from such trace-reversed potentials $\bar{h}_{\mu\nu}$ that satisfy $\bar{h}^{\mu\nu}{}_{,\nu} = 0$, but then note from eqn (9.34) that such potentials are not unique. A further gauge transformation maintains the transverse gauge condition provided the four gauge functions $f^\mu(x)$ satisfy the wave equation

$$\Box f^\mu(x) = 0. \tag{9.36}$$

We wish to show that the remaining four gauge functions f_μ can now be so chosen as to set $\bar{h}'_{0\mu} = 0$ and $\bar{h}' = 0$ as well. To this end, let us choose the spatial inertial coordinate system (x^1, x^2, x^3) such that the direction of propagation of the waves from the source to the detector coincides, for instance, with the positive x^3 direction. This simplifies the analysis without any loss in generality. Therefore, near the detector, the spherical gravitational wave front can be locally approximated by a plane wave such that $\bar{h}_{0\mu} = \bar{h}_{0\mu}(\zeta)$, where $\zeta := x^3 - x^0$. Setting $\bar{h}'_{0\mu} = 0$ in eqn (9.32), $\bar{h}' = 0$ in eqn (9.34) and replacing ϵ_μ in these equations by $f_\mu(\zeta)$, we find

$$\frac{df_1}{d\zeta} = \bar{h}_{01}, \qquad \frac{df_2}{d\zeta} = \bar{h}_{02}, \tag{9.37}$$

$$\frac{d(f_0 - f_3)}{d\zeta} = -\bar{h}_{03} = \bar{h}_{00}, \qquad \frac{d(f_0 + f_3)}{d\zeta} = \frac{1}{2}\bar{h}. \tag{9.38}$$

Let us note that $\bar{h}_{00} + \bar{h}_{03} = 0$ in eqn (9.38) is consistent with the transverse gauge condition $\bar{h}^{0\nu}{}_{,\nu} = 0$, which implies that $d(\bar{h}_{00} + \bar{h}_{03})/d\zeta = 0$. This relation can be integrated and the integration constant set to zero, as the presence of a non-zero constant here would be inconsistent with the fact that these potentials originate from the distant source of gravitational waves. It is thus evident that f_μ can be so chosen as to render $h'_{\mu\nu}$ purely spatial and traceless as well. Moreover, we note that in this procedure $\bar{h}_{11} - \bar{h}_{22}$ and $\bar{h}_{12}$ remain invariant; that is, $\bar{h}'_{11} - \bar{h}'_{22} = \bar{h}_{11} - \bar{h}_{22}$ and $\bar{h}'_{12} = \bar{h}_{12}$. The last step in the establishment of the TT gauge involves the spatial components of the dynamic field variables h_{ij}. Again, near the receiver in the wave zone, the spherical gravitational waves associated with these potentials locally behave as plane waves and, as before, we can assume that $h_{ij} = h_{ij}(\zeta)$. Then, the transverse gauge condition, $h^{ij}{}_{,j} = 0$, implies that $dh_{i3}/d\zeta = 0$ and hence $h_{i3}(\zeta) = 0$. In this way, we recover the TT gauge of GR, where the two independent states of gravitational radiation are given by $h_{11} = -h_{22}$ and $h_{12} = h_{21}$.

We now turn to the influence of the dark matter source upon the amplitude of emitted gravitational radiation.

9.3 Dark Source

It follows from eqn (9.2) that the dark source contributes to the emission of gravitational radiation. To estimate this contribution for $\lambda \ll \lambda_0$, we turn to eqn (9.31), where $T^{\mu\nu} = T^{\mu\nu} + T_D^{\mu\nu}$ with $\partial_\nu T^{\mu\nu} = 0$ and $\partial_\nu T_D^{\mu\nu} \approx 0$. In other words, the symmetric matter and dark matter energy–momentum tensors are independently conserved in the approximation scheme under consideration here. We will treat the source and its dark counterpart as isolated astronomical systems.

The special retarded solution of the linearized field equation is given by

$$\bar{h}^{\mu\nu}(x^0, \mathbf{x}) \approx \frac{\kappa}{2\pi} \int \frac{T^{\mu\nu}(x^0 - |\mathbf{x} - \mathbf{y}|, \mathbf{y})}{|\mathbf{x} - \mathbf{y}|} d^3y. \tag{9.39}$$

Far away from the source, we can introduce in eqn (9.39) the approximation that $|\mathbf{x} - \mathbf{y}| \approx |\mathbf{x}|$; that is,

$$\bar{h}^{\mu\nu}(x^0, \mathbf{x}) \approx \frac{\kappa}{2\pi|\mathbf{x}|} \int T^{\mu\nu}(x^0 - |\mathbf{x}|, \mathbf{y}) d^3y, \tag{9.40}$$

so that the solution takes the form of a spherical wave approaching the detector. Furthermore, let $\mathbf{n}$ denote the unit vector that represents the direction of propagation from the source to the receiver. Far in the wave zone, the spherical wave front can be *locally* approximated by a plane wave front with wave vector $\mathbf{k} = \omega\mathbf{n}$. The wave amplitude in the TT gauge can be extracted from eqn (9.40) by means of the projection operator $P^i{}_j = \delta^i_j - n^i n_j$, namely,

$$h_{TT}^{ij} = (P^i{}_\ell P^j{}_m - \frac{1}{2} P^{ij} P_{\ell m}) \bar{h}^{\ell m}. \tag{9.41}$$

For instance, if the spatial frame is oriented such that $\mathbf{n}$ points in the positive x^3 direction, the only non-zero components of h_{TT}^{ij} are $h_{TT}^{11} = -h_{TT}^{22} = (\bar{h}^{11} - \bar{h}^{22})/2$ and $h_{TT}^{12} = h_{TT}^{21} = \bar{h}^{12}$.

Next, we recall that the total conserved symmetric energy–momentum tensor $T^{\mu\nu}$ of an isolated system satisfies Laue's theorem, namely,

$$\int T^{ij}(ct, \mathbf{x}) d^3x = \frac{1}{2c^2} \frac{d^2}{dt^2} \int T^{00}(ct, \mathbf{x}) x^i x^j d^3x. \tag{9.42}$$

It follows from eqns (9.40)–(9.42) that, among other things, h_{TT}^{ij} will depend upon the second temporal derivative of the total quadrupole moment of the system. We note here for the sake of completeness that the temporal coordinate of the quadrupole moment is in fact the "retarded" time $ct = x^0 - |\mathbf{x}|$, where $|\mathbf{x}|$ can be treated as a constant for the purposes of the present discussion. Moreover, the quadrupole moment could just as well be replaced by the reduced (i.e. traceless) quadrupole moment in the expression for h_{TT}^{ij}.

We are particularly interested here in the contribution of the dark quadrupole moment

$$Q_D^{ij}(x^0) := \int T_D^{00}(x^0, \mathbf{x}) x^i x^j d^3x \tag{9.43}$$

to the total quadrupole moment of the system. It follows from the definition of dark energy density, namely,

$$T_D^{00}(x) = \int R(x - y) T^{00}(y) \, d^4y, \tag{9.44}$$

that

$$Q_D^{ij}(x^0) = \int d^4y \, T^{00}(y) \int R(x^0 - y^0, \mathbf{x} - \mathbf{y}) x^i x^j \, d^3x. \tag{9.45}$$

To get a manageable expression for the dark quadrupole moment, we use in eqn (9.45) the approximate formula for R given by eqn (9.20), in which retardation has been neglected. Introducing a new variable $\mathbf{z} = \mathbf{x} - \mathbf{y}$ in eqn (9.45) and recalling that q is spherically symmetric and of the general form of either q_1 given by eqn (7.98) or q_2 given by eqn (7.99), we find

$$Q_D^{ij}(x^0) \approx \Sigma_D^{ij}(x^0) \int q(z) \, d^3z + M \int q(z) z^i z^j \, d^3z, \tag{9.46}$$

where

$$\Sigma_D^{ij}(x^0) = \nu \int u(x^0 - y^0) \, e^{-\nu(x^0 - y^0)} \, Q^{ij}(y^0) \, dy^0 \tag{9.47}$$

and

$$M = \int T^{00}(x) \, d^3x, \qquad Q^{ij}(x^0) = \int T^{00}(x^0, \mathbf{x}) x^i x^j \, d^3x. \tag{9.48}$$

Here, M is the mass–energy of the isolated radiating system and is conserved at the linear order, since $\partial_\nu T^{\mu\nu} = 0$. Therefore, the temporal variation of the dark quadrupole moment in eqn (9.46) is given by the term that is proportional to $\hat{q}(0)$,

$$\hat{q}(0) = \int q(z) \, d^3z. \tag{9.49}$$

This quantity has been evaluated in Chapter 7 and has a magnitude of about 10; see eqn (7.267).

Suppose we are interested in the gravitational radiation emitted by an astronomical system whose quadrupole moment Q_S^{ij} varies with time with a dominant frequency of Ω_S that can be detected on Earth via a gravitational wave detector within a reasonable span of time; therefore, we expect that $\Omega_S \gg c\nu$, where $c\nu < c\mu_0 \approx 5 \times 10^{-13}$ Hz. To get an order of magnitude estimate of the significance of the dark quadrupole moment relative to the quadrupole moment of the system in eqn (9.42), we must evaluate Σ_D^{ij}. Expressing a Fourier component of $Q_S^{ij}(y^0)$ as a constant amplitude times $\cos(\Omega_S y^0 + \varphi_S)$, where φ_S is a constant phase, it follows from eqn (9.47), namely,

$$\nu \int u(x^0 - y^0) \, e^{-\nu(x^0 - y^0)} \cos(\Omega_S y^0 + \varphi_S) \, dy^0 = \frac{\nu}{\nu^2 + \Omega_S^2} [\nu \cos(\Omega_S x^0 + \varphi_S)$$
$$+ \Omega_S \sin(\Omega_S x^0 + \varphi_S)] \tag{9.50}$$

that the relative contribution of the time-dependent part of the dark quadrupole moment will be reduced at least by a factor of $c\nu/\Omega_S \ll 1$. For instance, in the case of incident gravitational radiation of frequency about 35 Hz, as in GW150914, $c\nu/\Omega_S \sim 10^{-14}$, which is very small in comparison with unity. We therefore conclude that the contribution of the dark source to h_{TT}^{ij} is essentially negligible for all systems that are currently under consideration as possible sources of gravitational radiation that could be detectable in the near future.

9.4 Gravitational Radiation Flux

Finally, we must compute the flux of gravitational radiation energy at the detector. The energy–stress content of the gravitational field in nonlocal gravity is described by the traceless tensor $\mathcal{E}_{\mu\nu}$,

$$\mathcal{E}_{\mu\nu} = \kappa^{-1}\left[C_{\mu\rho\sigma}(\mathfrak{C}_{\nu}{}^{\rho\sigma} + N_{\nu}{}^{\rho\sigma}) - \frac{1}{4}g_{\mu\nu}\,C_{\alpha\beta\gamma}(\mathfrak{C}^{\alpha\beta\gamma} + N^{\alpha\beta\gamma})\right], \qquad (9.51)$$

which is indeed the *energy–momentum tensor of the gravitational field*; see Section 6.5. In eqn (9.51), $C_{\mu\nu\rho}$ is the torsion tensor, $\mathfrak{C}_{\mu\nu\rho}$ is the auxiliary torsion tensor and $N_{\mu\nu\rho} = -N_{\nu\mu\rho}$ is a nonlocal tensor field that involves the past history of the gravitational field; see eqn (6.107). The purpose of this section is to calculate $\mathcal{E}_{\mu\nu}$ within the context of linearized nonlocal gravity for gravitational waves with $\lambda \ll \lambda_0$.

In *linearized* nonlocal gravity,

$$N_{\mu\nu\rho}(x) = \int K(x - y)\,X_{\mu\nu\rho}(y)\,d^4y \qquad (9.52)$$

and

$$X_{\mu\nu\rho} = \mathfrak{C}_{\mu\nu\rho} + \breve{p}\,(\check{C}_{\mu}\,\eta_{\nu\rho} - \check{C}_{\nu}\,\eta_{\mu\rho}), \qquad (9.53)$$

where $\breve{p} \neq 0$ is a constant dimensionless parameter; see eqn (6.109). Next, we assume as before that $\phi_{\mu\nu} = 0$; moreover, of the other gauge conditions leading to the TT gauge, at this point only $\bar{h}^{\mu\nu}{}_{,\nu} = 0$ and $\bar{h} = 0$ are explicitly imposed here for the sake of simplicity. Hence, we find that

$$C_{\mu\nu\sigma} = \mathfrak{C}_{\mu\nu\sigma} = \frac{1}{2}\,(h_{\sigma\nu,\mu} - h_{\sigma\mu,\nu}) \qquad (9.54)$$

and in linearized nonlocal gravity $\mathcal{E}_{\mu\nu}$ is given by

$$\kappa\,\mathcal{E}_{\mu\nu} = -\frac{1}{4}\,\eta_{\mu\nu}\,C^{\alpha\beta\gamma}(x)\left[C_{\alpha\beta\gamma}(x) + \int K(x-y)\,C_{\alpha\beta\gamma}(y)\,d^4y\right]$$
$$+ C_{\mu}{}^{\alpha\beta}(x)\left[C_{\nu\alpha\beta}(x) + \int K(x-y)\,C_{\nu\alpha\beta}(y)\,d^4y\right]. \qquad (9.55)$$

Let us note that $\mathcal{E}_{\mu\nu}$ is in general traceless and that in the absence of nonlocality (i.e. $K = 0$), it becomes symmetric as well, and its form is then reminiscent of the electromagnetic energy–momentum tensor, as would be expected from the linearized GR$_{\parallel}$ theory; see Section 6.2.

To estimate the nonlocal contribution to $\mathcal{E}_{\mu\nu}$ for gravitational waves with $\lambda \ll \lambda_0$, we work in the Fourier domain and use the convolution theorem for Fourier transforms as well as eqns (7.154) and (7.156) for $\widehat{R}$ and $\widehat{K}$, respectively. As before, we find that the absolute magnitude of the ratio of the nonlocal term to the corresponding local term in $\mathcal{E}_{\mu\nu}$ is $\lesssim \nu \lambda^2/(8\pi \lambda_0)$. Therefore,

$$|\widehat{K}| \sim |\widehat{R}| \lesssim 10^{-2} \left(\frac{\lambda_0}{\lambda}\right)^2, \tag{9.56}$$

which in view of eqn (9.1) is completely negligible in comparison to unity for radiation of dominant frequency $\gtrsim 10^{-8}$ Hz. We therefore conclude that the nonlocal contribution to the energy–momentum tensor of gravitational radiation can be ignored for gravitational waves that may be detectable in the foreseeable future.

For gravitational waves with $\lambda \ll \lambda_0$, we can therefore ignore nonlocal effects and determine the energy flux by comparing the local ($K = 0$) result $\mathcal{E}_{\mu\nu}^{(0)}$ to the corresponding Landau–Lifshitz energy–momentum pseudotensor (Landau and Lifshitz 1971) of gravitational waves in the linear approximation of GR. Under the same explicit gauge conditions for the deviation of the metric tensor from the Minkowski metric tensor, namely, $h = 0$ and $h^{\mu\nu}{}_{,\nu} = 0$, the corresponding Landau–Lifshitz tensor $t_{\mu\nu}^{LL}$, which is in general symmetric, but not traceless, is given by

$$\kappa\, t_{\mu\nu}^{LL} = \frac{1}{2} h_{\mu\alpha,\beta} h_\nu{}^{\alpha,\beta} + \frac{1}{4}\left(h_{\alpha\beta,\mu} h^{\alpha\beta}{}_{,\nu} - \frac{1}{2}\eta_{\mu\nu}\, h_{\alpha\beta,\gamma} h^{\alpha\beta,\gamma}\right)$$
$$- \frac{1}{2}\left(h_{\mu\alpha,\beta} h^{\alpha\beta}{}_{,\nu} + h_{\nu\alpha,\beta} h^{\alpha\beta}{}_{,\mu} - \frac{1}{2}\eta_{\mu\nu}\, h_{\alpha\beta,\gamma} h^{\gamma\alpha,\beta}\right); \tag{9.57}$$

see the Appendix of Mashhoon (1978). It is straightforward to show from eqn (9.55) that $\mathcal{E}_{\mu\nu}^{(0)}$, which is defined to be $\mathcal{E}_{\mu\nu}$ for $K = 0$, is symmetric and traceless, and can be expressed as

$$\kappa\, \mathcal{E}_{\mu\nu}^{(0)} = \frac{1}{4} h_{\mu\alpha,\beta} h_\nu{}^{\alpha,\beta} + \frac{1}{4}\left(h_{\alpha\beta,\mu} h^{\alpha\beta}{}_{,\nu} - \frac{1}{2}\eta_{\mu\nu}\, h_{\alpha\beta,\gamma} h^{\alpha\beta,\gamma}\right)$$
$$- \frac{1}{4}\left(h_{\mu\alpha,\beta} h^{\alpha\beta}{}_{,\nu} + h_{\nu\alpha,\beta} h^{\alpha\beta}{}_{,\mu} - \frac{1}{2}\eta_{\mu\nu}\, h_{\alpha\beta,\gamma} h^{\gamma\alpha,\beta}\right). \tag{9.58}$$

Each of the expressions in eqns (9.57) and (9.58) consists of the same three parts, but they differ in the overall numerical factors in front of the first and last parts: these are both $\frac{1}{2}$ in eqn (9.57), but $\frac{1}{4}$ in eqn (9.58).

For the calculation of the energy flux, we impose the additional gauge condition that $h_{0\mu} = 0$. It then follows from eqns (9.57) and (9.58) that

$$\mathcal{E}_{0k}^{(0)} - t_{0k}^{LL} = \frac{1}{4\kappa} h_{ki,j} h^{ij}{}_{,0}, \tag{9.59}$$

which vanishes in the TT gauge due to the transverse nature of the radiation. For instance, if the spatial axes are so oriented that locally plane waves propagate to the receiver along the x^3 axis, then $k = 3$ in eqn (9.59) and $h_{3i} = 0$ in the TT gauge. We therefore conclude that in the absence of nonlocality, the flux of gravitational radiation energy will be the same as in standard GR, in agreement with previous results (Schweizer, Straumann and Wipf 1980).

10
Nonlocal Newtonian Cosmology

It is interesting to explore the cosmological implications of nonlocal gravity. Cosmology deals with the structure and history of the universe as a whole. To produce possible models of the universe, exact cosmological solutions of the field equation of nonlocal gravity are indispensable. However, nonlocal gravity is in the early stages of development at present and besides the Minkowski spacetime, no exact solution of the theory is known. On the other hand, the general linear approximation of nonlocal gravity beyond Minkowski spacetime has been investigated in detail in Chapters 7, 8 and 9. In particular, preliminary investigations have shown that the implications of the theory in the Newtonian regime appear to be consistent with the gravitational dynamics of the Solar System, spiral galaxies and clusters of galaxies; see Chapter 8. In these studies, a significant feature of nonlocal gravity is that the nonlocal aspect of gravity appears to simulate astrophysical dark matter. On the other hand, dark matter is considered absolutely necessary as well for structure formation in standard models of cosmology. It is therefore of basic importance to investigate whether the nonlocal character of gravity can effectively replace dark matter in cosmological structure formation. To make a beginning in this direction is the main motivation for this chapter.

The remarkable isotropy of the cosmic microwave background radiation indicates a small amplitude ($\delta \approx 10^{-5}$) for the inhomogeneities that must have existed at the epoch of decoupling ($z \approx 10^3$). The tremendous growth of such inhomogeneities from the recombination era to the present time is due to the intrinsic gravitational instability of a nearly homogeneous distribution of matter. The exact manner in which galaxies, clusters of galaxies and eventually the cosmic web have come about is not known; however, it is generally believed that *dark matter* has played a crucial role in this development (Peebles 1993; Zel'dovich and Novikov 1983; Mukhanov 2005; Gurbatov, Saichev and Shandarin 2012).

As is well known, under the assumptions of spatial homogeneity and isotropy, Newtonian cosmology is an excellent approximation to the standard Friedmann–Lemaître–Robertson–Walker (FLRW) cosmological models of general relativity so long as the net pressure, as a source of gravity, can be neglected in comparison with the energy density of the matter content of the universe. More generally, it turns out that after recombination, Newtonian gravitation can be applied in the study of nonrelativistic motion of matter on subhorizon scales (Peebles 1993; Zel'dovich and Novikov 1983; Mukhanov 2005; Gurbatov, Saichev and Shandarin 2012).

Can nonlocal gravity solve the problem of structure formation in cosmology without recourse to dark matter? This issue can be properly addressed within the framework of an exact cosmological model of nonlocal gravity theory. No exact cosmological solution of nonlocal gravity theory is known; therefore, an adequate treatment of this subject remains a task for the future. Nevertheless, it is important to investigate whether nonlocal gravity is even capable of solving the problem of large-scale cosmological structure formation. In this chapter, we take the very first step in the study of this difficult problem by extending the weak-field Newtonian regime of nonlocal gravity to the cosmological domain along the same lines as in general relativity. That is, following the familiar approach of Newtonian cosmology, we consider the dynamics of a large gas cloud in accordance with the Newtonian regime of nonlocal gravity. We study the resulting homogeneous cosmological background and the nonlinear growth of inhomogeneities in such a background as a useful toy model for the present large-scale structure of the universe (Chicone and Mashhoon 2016b).

We now turn to our toy model of nonlocal Newtonian cosmology, which will turn out to be rather similar to that of the standard Milne–McCrea Newtonian cosmology.

10.1 Nonlocal Newtonian Cosmological Model

Nonlocal gravity, which is the nonlocal extension of general relativity, has only been studied in the linear weak-field approximation; see Chapters 7–9. Nothing is known about strong-field situations, such as black holes; in particular, no cosmological solution of nonlocal gravity is known. On the other hand, in certain situations, Newtonian cosmology provides a good approximation to the homogeneous and isotropic FLRW models of relativistic cosmology. Indeed, Milne and McCrea showed in 1934 that the dynamics of the universe given by Newtonian cosmology is the same as in the standard general relativistic FLRW cosmological models so long as pressure can be neglected (Milne 1934; McCrea and Milne 1934).

To proceed, we may tentatively assume that an extension of Newtonian regime of nonlocal gravity to the cosmological domain could be useful. To this end, we imagine the Newtonian dynamics of a large expanding baryonic gas cloud. Let ρ be the (baryonic) gas density; then, it follows from the continuity equation that

$$\partial_t \rho + \nabla \cdot (\rho \, \mathbf{v}) = 0, \tag{10.1}$$

where $\rho \, \mathbf{v}$ is the gas current. Euler's equation of motion for the gas particles can be written as

$$\partial_t \mathbf{v} + (\mathbf{v} \cdot \nabla) \, \mathbf{v} = -\nabla \Phi - \frac{1}{\rho} \nabla p, \tag{10.2}$$

where Φ is the gravitational potential and p is the gas pressure. In the Newtonian regime of nonlocal gravity, Φ satisfies the nonlocal Poisson equation

$$\nabla^2 \Phi(t, \mathbf{x}) + \int \chi(t, \mathbf{x} - \mathbf{y}) \, \nabla^2 \Phi(t, \mathbf{y}) \, d^3 y = 4\pi G \, \rho(t, \mathbf{x}), \tag{10.3}$$

where χ is the *convolution kernel* of nonlocal gravity in the Newtonian regime; see Section 7.4. Poisson's equation of Newtonian gravity has been modified here by the addition of a certain spatial average over the gravitational field involving kernel χ. This kernel represents the spatial *memory* of the gravitational field. In general, memory dies out in space and time, features that must be reflected in the functional form of the kernel.

As discussed in detail in Chapter 7, under reasonable mathematical conditions, it is possible to write eqn (10.3) as

$$\nabla^2 \Phi = 4\pi G \left(\rho + \rho_D \right), \tag{10.4}$$

where the nonlocal aspect of the gravitational interaction simulates dark matter with an effective density ρ_D given by

$$\rho_D(t, \mathbf{x}) = \int q(t, \mathbf{x} - \mathbf{y}) \, \rho(t, \mathbf{y}) \, d^3 y. \tag{10.5}$$

In other words, the density of the *effective* dark matter in this theory is given by the convolution of the matter density with the *reciprocal kernel* q of nonlocal gravity in the Newtonian regime. It remains to determine the reciprocal kernel q.

10.1.1   Reciprocal kernel q at the present epoch

At the present epoch in cosmic evolution, $t = t_0$ and the functional form of $q(t_0, \mathbf{x})$ has been discussed in detail in Chapter 7. In fact, q is spherically symmetric in space and is of the general form of either $q_1(r)$ given by eqn (7.98) or $q_2(r)$ given by eqn (7.99), where $r = |\mathbf{x}|$. Each of these kernels contains three length scales a_0, λ_0 and μ_0^{-1} such that $a_0 < \lambda_0 < \mu_0^{-1}$. The basic scale of nonlocality at the present epoch is a galactic length λ_0 of order 1 kpc, while a_0 is a short-range parameter that controls the behavior of q as $r \to 0$. At the other extreme, $r \to \infty$, q decays exponentially as $\exp(-\mu_0 r)$, indicating the fading of spatial memory with distance. The short-range parameter a_0 is necessary in dealing with the gravitational physics of the Solar System, globular star clusters and isolated dwarf galaxies; however, it may be neglected in dealing with larger systems such as clusters of galaxies. Cosmology deals with the large-scale structure of the universe; therefore, the short-range parameter a_0 may be deemed irrelevant in the cosmological context. With $a_0 = 0$, q_1 and q_2 both reduce to q_0,

$$q_0(t_0, r) = \frac{1}{4\pi\lambda_0} \frac{(1 + \mu_0 r)}{r^2} e^{-\mu_0 r}. \tag{10.6}$$

As in Section 7.4, it proves useful to introduce the dimensionless parameter α_0,

$$\alpha_0 = \int q_0(t_0, r) \, d^3 x = \frac{2}{\lambda_0 \mu_0}, \tag{10.7}$$

so that instead of λ_0 and μ_0, kernel q_0 can be characterized by the dimensionless parameter α_0 and the *spatial memory* ("Yukawa") parameter μ_0. In terms of α_0 and μ_0, we have

$$q_0(t_0, r) = \frac{\alpha_0 \, \mu_0}{8\pi} \frac{(1 + \mu_0 r)}{r^2} e^{-\mu_0 r}. \tag{10.8}$$

Thus nonlocality disappears if $\lambda_0 \to \infty$ or, equivalently, $\alpha_0 \to 0$. As explained in detail in Chapter 8, astrophysical data tentatively indicate that

$$\lambda_0 \approx 3 \, \text{kpc}, \qquad \frac{1}{\mu_0} \approx 17 \, \text{kpc}, \qquad \alpha_0 \approx 11. \tag{10.9}$$

We must now discuss the reciprocal kernel in the context of the expansion of the universe.

10.1.2   Reciprocal kernel q in cosmology

An appropriate reciprocal kernel should properly incorporate the expansion of the universe. That is, the kernel must reflect the gravitational memory of the past history of the universe, satisfy the mathematical requirements of nonlocal gravity discussed in Chapter 7 and reduce to the currently accepted kernel at the present epoch $t = t_0$. Memory dies out in time and space. The fading of memory over time implies that in nonlocal gravity the strength of the gravitational interaction must decrease with cosmic time. We have encountered a similar feature of memory in the context of linearized nonlocal gravity in Section 7.5; in fact, in eqn (7.143), the reciprocal kernel R decays exponentially in time. Equation (7.143) is invariant under translation in time, in accordance with the corresponding symmetry of the background Minkowski spacetime. The situation is different, however, when we deal with Big Bang cosmology. These considerations, within the framework of Newtonian cosmology, lead to the simple supposition that we should only change the overall temporal dependence of kernel (10.8) in the cosmological context and leave its spatial dependence unchanged. Therefore, we will henceforth assume that kernel $q(t, \mathbf{x} - \mathbf{y})$ of *nonlocal Newtonian cosmology* in eqn (10.5) is given by

$$q(t, \mathbf{x} - \mathbf{y}) = \frac{\alpha(t)}{\alpha_0} q_0(t_0, |\mathbf{x} - \mathbf{y}|), \qquad (10.10)$$

where $\alpha(t)$ is a monotonically decreasing function of cosmic time such that $\alpha(0) = \infty$ at the Big Bang singularity $(t = 0)$ and, at the present epoch, $\alpha(t_0) = \alpha_0$.

It is interesting to consider the ratio of the density of effective dark matter to baryonic matter, ρ_D/ρ, for a spatially homogeneous model of matter density $\rho(t)$, which is naturally assumed throughout to be mainly baryonic. We find from eqn (10.5) that

$$\frac{\rho_D}{\rho} = \int_{\mathbb{R}^3} q(t, \mathbf{x}) \, d^3x = \alpha(t), \qquad (10.11)$$

which means that in such a uniform density model the effective dark matter fraction, $f_{DM} = \alpha(t)$, *decreases* with cosmic time. We recall that in an astrophysical system, the dark matter fraction f_{DM} denotes the ratio of the total mass of dark matter to the total mass of the baryonic matter in the system. Indeed, it is natural to assume that relative to baryonic matter, the effective amount of dark matter was more plentiful in the past, since nonlocality is connected with the memory of the past state of the gravitational field and memory fades with time. It remains to specify the exact temporal dependence of $\alpha(t)$.

It is important to compare the memory dependence of our toy model with the currently accepted model of standard cosmology, where f_{DM} for the universe is about 5 and independent of cosmic time. Does f_{DM} actually evolve with cosmic time, or equivalently, with the cosmological redshift z? It appears that this empirical question has not yet been tackled by observational cosmologists as such an issue does not even arise in the current paradigm of dark matter. On the other hand, the assumption that f_{DM} is a decreasing function of cosmic time would imply that the strength of the universal gravitational attraction decreases as the universe expands. This feature of nonlocal gravity may provide a natural explanation for the observed swelling of massive quiescent early-type galaxies discussed in Section 10.6.

10.2   Cosmological Background

To obtain the nonlocal analog of the standard models of Newtonian cosmology, we assume an infinite spatially homogeneous and isotropic perfect fluid medium with $\rho = \bar{\rho}(t)$ and $p = p(\bar{\rho})$ that is expanding uniformly. This expansion of the universe can be expressed via $\mathbf{x} = A(t)\,\boldsymbol{\xi}$, where $A(t)$ is the scale factor, $\mathbf{x}$ denotes the spatial position of the fluid particle at time t and $\boldsymbol{\xi}$ denotes the spatial position of the particle at some fiducial time t_0 such that $A(t_0) = 1$. We take t_0 to be the present epoch. Thus, $\bar{\mathbf{v}} = d\mathbf{x}/dt = H(t)\,\mathbf{x}$, where $H(t) = \dot{A}/A$ is the Hubble parameter and an overdot denotes differentiation with respect to time. Moreover, it follows from eqn (10.11) that for an infinite homogeneous matter distribution $\rho_D = \alpha(t)\rho$. The explicit solution of the system (10.1)–(10.5) is thus given by

$$\bar{\rho} = A^{-3}\rho_0, \qquad \bar{\mathbf{v}} = \dot{A}A^{-1}\mathbf{x}, \qquad \bar{\Phi} = -\frac{1}{2}\ddot{A}A^{-1}r^2, \qquad (10.12)$$

where ρ_0 is the density of (baryonic) matter at the present epoch, $\bar{\Phi}$ is determined up to an integration constant and the scale factor A is a solution of the differential equation

$$A^{-1}\ddot{A} = -\frac{4\pi G\rho_0}{3}A^{-3}(1+\alpha). \qquad (10.13)$$

For the initial data

$$A(t_0) = 1, \qquad \dot{A}(t_0) = H_0 \qquad (10.14)$$

prescribed at the present epoch t_0 using the Hubble constant H_0, we obtain the Hubble flow for this model provided $\alpha(t)$ is known. A remark is in order here regarding the necessity of a spatially infinite cosmological solution. For a spherically symmetric matter distribution of finite radius, a uniformly expanding homogeneous distribution is possible only if the standard Poisson equation of Newtonian gravity is valid, that is, Newton's inverse square law of gravitation is maintained.

To go forward, we must assume a functional form for $\alpha(t)$; for instance, $\alpha(t)/\alpha_0$ could be $1/A(t)$ raised to any positive power. For an expanding universe where the amount of effective dark matter decreases relative to baryonic matter as the universe expands, perhaps the simplest model—the one that we will discuss—is derived from the assumption

$$\alpha(t) = \frac{\alpha_0}{A}, \qquad (10.15)$$

where $\alpha_0 \approx 11$ is the proportionality constant at the present age (t_0) of the universe. With α as in eqn (10.15), eqn (10.13) then takes the form

$$A^{-1}\ddot{A} = -\frac{4\pi G\rho_0}{3}A^{-3}\left(1+\frac{\alpha_0}{A}\right). \qquad (10.16)$$

Equation (10.16) can be integrated once to obtain

$$\frac{1}{2}\dot{A}^2 = \frac{4\pi G\rho_0}{3}\left(\frac{1}{A}+\frac{\alpha_0}{2A^2}\right) + \bar{E}, \qquad (10.17)$$

where the total energy parameter $\bar{E}$ is the constant of integration. For the sake of simplicity, we set $\bar{E} = 0$, just as in the critical case in standard Newtonian cosmology, which corresponds to the spatially flat FLRW universe.

For this analog of the flat FLRW universe, a second integration yields the equation

$$\frac{1}{3}(2A + \alpha_0)^{3/2} - \alpha_0 (2A + \alpha_0)^{1/2} = 2 \left(\frac{4\pi G \rho_0}{3}\right)^{1/2} t + \bar{C} \qquad (10.18)$$

for a new integration constant $\bar{C}$. Only positive square roots are considered throughout. Assuming that the scale factor vanishes at the Big Bang, $t = 0$, $\bar{C}$ can be evaluated from $A(0) = 0$ and the formula for A thus reduces to

$$\frac{1}{3}(2A + \alpha_0)^{3/2} - \alpha_0 (2A + \alpha_0)^{1/2} = 2 \left(\frac{4\pi G \rho_0}{3}\right)^{1/2} t - \frac{2}{3}\alpha_0 \sqrt{\alpha_0}. \qquad (10.19)$$

The unknown $A(t)$ can be obtained by solving for the positive real root of the corresponding cubic equation for $\sqrt{2A + \alpha_0}$.

The age of the universe t_0, determined by $A(t_0) = 1$ in our model, is

$$t_0 = \frac{\alpha_0 \sqrt{\alpha_0} - (\alpha_0 - 1)\sqrt{\alpha_0 + 2}}{\sqrt{12\pi G \rho_0}}. \qquad (10.20)$$

It is interesting to note that the numerator of this expression for t_0 is a positive and monotonically decreasing function of α_0; in fact, it starts from $\sqrt{2}$ at $\alpha_0 = 0$ and vanishes asymptotically as $\alpha_0 \to \infty$. Moreover, the present value of the Hubble parameter is given by eqn (10.17) in this model with $\bar{E} = 0$, namely,

$$H_0^2 = \frac{4\pi G \rho_0}{3} (\alpha_0 + 2), \qquad (10.21)$$

so that

$$3 H_0 t_0 = \alpha_0 \sqrt{\alpha_0 (\alpha_0 + 2)} - (\alpha_0 - 1)(\alpha_0 + 2). \qquad (10.22)$$

For $t/t_0 \to \infty$, it follows from eqn (10.19) that asymptotically $A^3 \sim 6\pi G \rho_0 t^2$, so that this model approaches the Newtonian analog of the Einstein–de Sitter model as $t/t_0 \to \infty$.

The uniformly expanding model of nonlocal Newtonian cosmology under consideration here reduces to the Newtonian version of the spatially flat Einstein–de Sitter model if we formally let $\alpha_0 \to 0$. Indeed, with $\alpha_0 = 0$, $A(t) = (t/t_0)^{2/3}$ and the work throughout this chapter reduces to the standard *local* treatment of gravitational instability in Newtonian cosmology (Mukhanov 2005).

With a Hubble constant of $H_0 \approx 70$ km s^{-1} Mpc^{-1}, we find that, according to the nonlocal model with $\alpha_0 \approx 11$, the present density of baryons in the universe is $\rho_0 \approx 1.4 \times 10^{-30}$ g cm^{-3} and the age of the universe is $t_0 \approx 7.21 \times 10^9$ yr. Thus our model gives a value for the age of the universe that is about half of the age of the currently accepted model; moreover, the currently accepted density of baryons is about $0.3\, \rho_0$ (Gurbatov, Saichev and Shandarin 2012). These cosmological parameters are somewhat reminiscent of the corresponding parameters for the spatially flat Einstein–de Sitter model of standard cosmology.

We emphasize that we are dealing here with a simplified *toy* model of nonlocal Newtonian cosmology, which we employ to explore the possibility that large-scale structure formation in cosmology may be possible without invoking actual dark matter. The next section contains an analysis of the *linear* stability of our model. We then discuss the *nonlinear* instability of the model using the Zel'dovich approach.

It is important to note that, due to the nature of the subject matter, the notation employed in the following sections is partly independent of the previous chapters of this book.

10.3   Jeans Instability

The instability of self-gravitating static configurations was first studied by Jeans within the framework of Newtonian theory of gravitation (Jeans 1902, 1929). Bonnor (1957) extended the study of Newtonian gravitational instability to expanding homogeneous media. In the context of modern cosmology, gravitational instability has been treated extensively by a number of authors (Peebles 1993; Zel'dovich and Novikov 1983; Mukhanov 2005). To investigate Jeans instability in our nonlocal model, this section is devoted to a *linear adiabatic* perturbation analysis of our toy model away from the homogeneous and isotropic solution given in eqn (10.12). The perturbed solution has a baryonic density given by $\bar{\rho} + \delta\rho$, where $\bar{\rho} = \rho_0/A^3$ is the background baryonic density and $\delta\rho/\bar{\rho}$ can be expressed as a sum of Fourier modes each with wave vector $\mathbf{k}$,

$$\frac{\delta\rho}{\bar{\rho}} = \varepsilon_0 \sum_{\mathbf{k}} D_{\mathbf{k}}(t)\, e^{i\mathbf{k}\cdot\mathbf{x}/A(t)}, \tag{10.23}$$

where ε_0, $0 < \varepsilon_0 \ll 1$, is a constant perturbation parameter and we have used the fact that the spatial scale of the perturbation expands with the universe. The perturbed flow velocity is given by $\bar{\mathbf{v}} + \delta\mathbf{v}$, where $\bar{\mathbf{v}} = (\dot{A}/A)\,\mathbf{x}$ and

$$\delta\mathbf{v} = \varepsilon_0 \sum_{\mathbf{k}} \mathbf{V}_{\mathbf{k}}(t)\, e^{i\mathbf{k}\cdot\mathbf{x}/A(t)}. \tag{10.24}$$

For the fluid pressure, $p = p(\rho)$, we find that the net perturbed pressure is

$$p(\bar{\rho} + \delta\rho) = p(\bar{\rho}) + \left.\frac{dp}{d\rho}\right|_{\rho=\bar{\rho}} \delta\rho \tag{10.25}$$

by Taylor expansion. We recall that the speed of sound in the medium is given by

$$c_s(\rho) = \sqrt{\frac{dp}{d\rho}}. \tag{10.26}$$

We are interested here in the *linear* adiabatic perturbation of our perfect fluid model; therefore, in the substitution of perturbed values of the fluid parameters in the model eqns (10.1)–(10.5), we keep terms only up to first order in ε_0. As a direct result of this linearity, it is possible to use complex perturbation amplitudes for the sake of simplicity, with the proviso that only the real parts of the perturbed equations have physical significance.

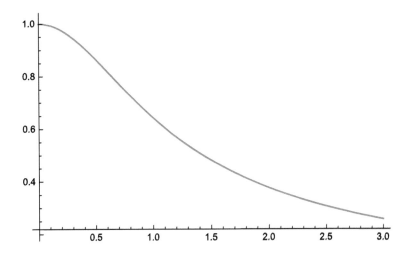

Fig. 10.1 Graph of $\frac{1}{2}[\Theta^{-1}\arctan\Theta + (1+\Theta^2)^{-1}]$ versus Θ. Reproduced from Chicone, C. and Mashhoon, B., 2016, "Nonlocal Newtonian Cosmology", *J. Math. Phys.* **57**, 072501 (27 pages), with the permission of AIP Publishing. DOI: 10.1063/1.4958902

It follows from the perturbed continuity equation that

$$A\,\dot{D}_{\mathbf{k}} + i\,\mathbf{k}\cdot\mathbf{V}_{\mathbf{k}} = 0. \tag{10.27}$$

Moreover, the perturbed Euler equation can be written as

$$-\nabla\Phi = \frac{\ddot{A}}{A}\mathbf{x} + \varepsilon_0\sum_{\mathbf{k}}\left(\dot{\mathbf{V}}_{\mathbf{k}} + \frac{\dot{A}}{A}\mathbf{V}_{\mathbf{k}} + i\,\frac{\bar{c}_s^2}{A}\mathbf{k}\,D_{\mathbf{k}}\right)e^{i\mathbf{k}\cdot\mathbf{x}/A(t)}, \tag{10.28}$$

where $\bar{c}_s := c_s(\bar{\rho})$. The divergence of eqn (10.28) is given by

$$-\nabla^2\Phi = 3\frac{\ddot{A}}{A} + \varepsilon_0\sum_{\mathbf{k}}\left[\frac{1}{A}(i\,\mathbf{k}\cdot\dot{\mathbf{V}}_{\mathbf{k}}) + \frac{\dot{A}}{A^2}(i\,\mathbf{k}\cdot\mathbf{V}_{\mathbf{k}}) - \frac{\bar{c}_s^2}{A^2}k^2\,D_{\mathbf{k}}\right]e^{i\mathbf{k}\cdot\mathbf{x}/A(t)}, \tag{10.29}$$

where $k = |\mathbf{k}|$. To simplify matters, let us note that eqn (10.27) implies, upon differentiation with respect to time, that

$$\dot{A}\,\dot{D}_{\mathbf{k}} + A\,\ddot{D}_{\mathbf{k}} + i\,\mathbf{k}\cdot\dot{\mathbf{V}}_{\mathbf{k}} = 0. \tag{10.30}$$

Thus substituting for $i\,\mathbf{k}\cdot\mathbf{V}_{\mathbf{k}}$ and $i\,\mathbf{k}\cdot\dot{\mathbf{V}}_{\mathbf{k}}$ from eqns (10.27) and (10.30), respectively, in eqn (10.29), we find

$$\nabla^2\Phi = -3\frac{\ddot{A}}{A} + \varepsilon_0\sum_{\mathbf{k}}\left(\ddot{D}_{\mathbf{k}} + 2\frac{\dot{A}}{A}\dot{D}_{\mathbf{k}} + \frac{\bar{c}_s^2}{A^2}k^2\,D_{\mathbf{k}}\right)e^{i\mathbf{k}\cdot\mathbf{x}/A(t)}. \tag{10.31}$$

Finally, inserting this expression in the modified Poisson eqn (10.4) and utilizing eqn (10.13), we obtain

$$\frac{d^2 D_{\mathbf{k}}}{dt^2} + 2\frac{\dot{A}}{A}\frac{dD_{\mathbf{k}}}{dt} + \mathcal{J}\,D_{\mathbf{k}} = 0, \tag{10.32}$$

where $\mathcal{J}$ is given by

$$\mathcal{J} = \frac{k^2\,\bar{c}_s^2}{A^2} - 4\pi G\bar{\rho}\,[1 + Q_k(t)]. \tag{10.33}$$

Here Q_k is essentially the spatial Fourier integral transform of $q(t,\mathbf{x})$, namely,

$$Q_k(t) := \int q(t,\mathbf{x})\,e^{-i\mathbf{k}\cdot\mathbf{x}/A(t)}\,d^3x, \tag{10.34}$$

where the reciprocal kernel is

$$q(t,\mathbf{x}) = \frac{\alpha_0\,\mu_0}{8\pi\,A(t)}\frac{1 + \mu_0\,|\mathbf{x}|}{|\mathbf{x}|^2}\,e^{-\mu_0\,|\mathbf{x}|}. \tag{10.35}$$

Indeed, it is possible to show that

$$Q_k(t) := \frac{\alpha_0}{2\,A(t)}\left[\frac{1}{\Theta}\arctan\Theta + \frac{1}{1+\Theta^2}\right], \tag{10.36}$$

where

$$\Theta := \frac{k}{\mu_0\,A(t)}. \tag{10.37}$$

Thus $Q_k(t)$ depends only on the wave number k and cosmic time t. As $k/\mu_0 \to 0$, we find $Q_0(t) = \alpha_0/A(t) = \alpha(t)$, while for $k/\mu_0 \to \infty$, we have $Q_\infty = 0$; see Fig. 10.1.

Our linear perturbation analysis makes it possible to consider $\delta\rho$ and $\delta\mathbf{v}$ as superpositions of different Fourier modes characterized by the wave vector $\mathbf{k}$. Suppose, for instance, that for a mode $\mathbf{k}$, $\mathcal{J}$ is non-zero and the corresponding peculiar velocity is orthogonal to $\mathbf{k}$; that is, $\mathbf{k}\cdot\mathbf{V}_{\mathbf{k}} = 0$. Then, it follows from the perturbation equations that there is no change in density for this *vector* mode, since $D_{\mathbf{k}} = 0$. On the other hand, suppose that $\mathbf{V}_{\mathbf{k}}$ is parallel to $\mathbf{k}$, then $D_{\mathbf{k}}$ is in general non-zero and is given by eqn (10.32).

The solutions of eqn (10.32) are oscillatory sound waves for $\mathcal{J} > 0$, while for $\mathcal{J} < 0$, we have a linear combination of growing and decaying modes. The transition between the two regimes is characterized by $\mathcal{J} = 0$, which can be solved for wave number k. The result, k_J, can be written as

$$k_J = \frac{2\pi\,A(t)}{\lambda_J}, \tag{10.38}$$

which defines the *Jeans length* λ_J given by

$$\frac{\lambda_J}{\bar{c}_s} = \left[\frac{\pi}{G\bar{\rho}(1+Q_k)}\right]^{1/2}. \tag{10.39}$$

The determination of k_J via $\mathcal{J} = 0$ is illustrated in Fig. 10.2.

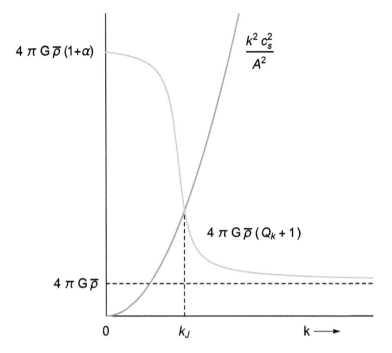

Fig. 10.2 Schematic plot that illustrates the determination of the Jeans length, $\lambda_J = 2\pi A(t)/k_J$, in our model. Here k, $k : 0 \to \infty$, is the wave number. Reproduced from Chicone, C. and Mashhoon, B., 2016, "Nonlocal Newtonian Cosmology", *J. Math. Phys.* **57**, 072501 (27 pages), with the permission of AIP Publishing. DOI: 10.1063/1.4958902

The attraction of gravity tends to produce clumps; however, pressure forces work against this tendency. In eqn (10.39), which expresses the Jeans criterion, these forces balance each other such that $\lambda_J/\bar{c}_s$, which is the time that it would take for a pressure wave to move across a Jeans length, is comparable to the gravitational response time of a self-gravitating fluid of density $\bar{\rho}$. For $\lambda \ll \lambda_J$, or $k \gg k_J$, the density contrast oscillates as a sound wave, while for $\lambda \gg \lambda_J$, or $k \ll k_J$, the pressure term can be neglected and gravitational instability takes over. This is due to the fact that the gravitational response time is very short in comparison with the period of the pressure wave.

We are interested in the study of *nonlinear* gravitational instability in our model. Henceforth, we neglect the pressure term in the Euler eqn (10.2). Thus the structures of interest in the post-recombination era have extensions that are much larger than the corresponding Jeans length. At the recombination era, corresponding to a cosmological redshift of $z \approx 10^3$, when structures are expected to start forming, it is possible to estimate λ_J and the corresponding Jeans mass $M_J := \bar{\rho}\lambda_J^3$. To get a rough idea of the orders of magnitude involved here, we recall that if we formally set $\alpha_0 = 0$ in our nonlocal scheme, we recover the Newtonian analog of the Einstein–de Sitter model. For the standard Einstein–de Sitter model, the result is $\lambda_J \sim 20\,\mathrm{pc}$ and $M_J \sim 10^5 M_\odot$, comparable to the parameters of a globular star cluster (Peebles 1993; Zel'dovich and Novikov 1983).

10.4   Nonlocal Analog of the Zel'dovich Solution

To study nonlinear gravitational instability in the context of Newtonian cosmology, Zel'dovich introduced a method that is based on Lagrangian coordinates (Mukhanov 2005). To follow the Zel'dovich approach in the treatment of the growing mode of gravitational instability, we need to neglect pressure and transform eqns (10.1), (10.2) and (10.4) of our nonlocal fluid model to Lagrangian coordinates.

Before introducing Lagrangian coordinates, let us digress here briefly and note that the continuity eqn (10.1) can be written as

$$[\partial_t + (\mathbf{v} \cdot \nabla)]\,\rho + \rho\,\nabla \cdot \mathbf{v} = 0. \tag{10.40}$$

Moreover, we can combine eqns (10.2) and (10.4) by taking the divergence of both sides of eqn (10.2). The result is

$$\nabla \cdot [\partial_t \mathbf{v} + (\mathbf{v} \cdot \nabla)\,\mathbf{v}] = -4\pi G\,(\rho + \rho_D), \tag{10.41}$$

where pressure has been neglected in accordance with our discussion in the previous section. Henceforth, our nonlocal fluid model will consist of eqns (10.40) and (10.41).

This system can be converted to Lagrangian coordinates $\boldsymbol{\xi}$ by assuming that the position of a fluid particle at a time t is uniquely identified via $(t, \boldsymbol{\xi})$; that is, $\mathbf{x} = \mathbf{x}(t, \boldsymbol{\xi})$, where $\boldsymbol{\xi}$ is a time-independent vector field that is constant along the path of each fluid particle. For instance, $\boldsymbol{\xi}$ could specify the spatial positions of fluid particles, $\mathbf{x}(t_{in}) = \boldsymbol{\xi}$, at some unspecified initial epoch $t_{in} > 0$ after the Big Bang. The Lagrangian coordinates of a fluid particle in spacetime are constants along its path and uniquely identify the particle. For time-dependent velocity fields, these spatial variables are defined only up to the first time when two trajectories with different Lagrangian markers meet in space. At such a time, the Lagrangian coordinate system breaks down and singularities called caustics are formed.

In Lagrangian variables, the fluid velocity can be expressed as

$$\mathbf{v} = \frac{d\mathbf{x}}{dt} = \frac{\partial \mathbf{x}}{\partial t}(t, \boldsymbol{\xi})\bigg|_{\boldsymbol{\xi}} \tag{10.42}$$

and, more generally, we have

$$\frac{\partial}{\partial t}\bigg|_{\mathbf{x}} + \mathbf{v} \cdot \nabla_{\mathbf{x}} = \frac{\partial}{\partial t}\bigg|_{\boldsymbol{\xi}}. \tag{10.43}$$

Let us first consider the transformation of the continuity eqn (10.40) to Lagrangian variables. We note that

$$\nabla_j v^i = \frac{\partial \xi^k}{\partial x^j} \frac{\partial v^i(t, \boldsymbol{\xi})}{\partial \xi^k} = \frac{\partial \xi^k}{\partial x^j} \frac{\partial}{\partial t} \left(\frac{\partial x^i(t, \boldsymbol{\xi})}{\partial \xi^k} \right). \tag{10.44}$$

Hence, it is useful to introduce a matrix Ξ and its inverse Ξ^{-1} with components

$$\Xi^i{}_j := \frac{\partial x^i}{\partial \xi^j}, \qquad (\Xi^{-1})^i{}_j := \frac{\partial \xi^i}{\partial x^j}, \tag{10.45}$$

so that the continuity eqn (10.40) can be written in Lagrangian coordinates as

$$\frac{\partial \varrho(t, \boldsymbol{\xi})}{\partial t} + \varrho \operatorname{tr} \left(\frac{\partial \Xi}{\partial t} \Xi^{-1} \right) = 0, \tag{10.46}$$

where ϱ is the density of baryons in Lagrangian coordinates

$$\rho(t, \mathbf{x}(t, \boldsymbol{\xi})) := \varrho(t, \boldsymbol{\xi}). \tag{10.47}$$

Let us now define

$$\mathbb{J} := \det \Xi, \tag{10.48}$$

and recall that

$$\delta \mathbb{J} = \mathbb{J} \, (\delta \Xi)^i{}_j \, (\Xi^{-1})^j{}_i, \tag{10.49}$$

cf. Section 5.2. Therefore,

$$\frac{\partial \mathbb{J}}{\partial t} = \mathbb{J} \operatorname{tr} \left(\frac{\partial \Xi}{\partial t} \Xi^{-1} \right) \tag{10.50}$$

and the continuity eqn (10.46) finally takes the form

$$\frac{\partial (\varrho \, \mathbb{J})}{\partial t} = 0. \tag{10.51}$$

Let us next consider eqn (10.41) and write the left-hand side of this equation as

$$\nabla \cdot [\partial_t \mathbf{v} + (\mathbf{v} \cdot \nabla) \mathbf{v}] = \left(\frac{\partial}{\partial t} + \mathbf{v} \cdot \nabla \right) \nabla \cdot \mathbf{v} + (\nabla_i v^j)(\nabla_j v^i). \tag{10.52}$$

We recall from eqns (10.44), (10.45) and (10.50) that the divergence of the fluid velocity in Lagrangian variables is $\partial \ln \mathbb{J}/\partial t$. This fact together with eqns (10.43) and (10.44) can be used in eqn (10.52) to transform eqn (10.41) to Lagrangian coordinates. The result is

$$\frac{\partial^2}{\partial t^2} \ln \mathbb{J} + \operatorname{tr} \left[\left(\frac{\partial \Xi}{\partial t} \Xi^{-1} \right)^2 \right] = -4\pi G(\varrho + \varrho_D), \tag{10.53}$$

where ϱ_D is the density of the effective dark matter in Lagrangian coordinates. Henceforth, we consider the problem of solving the Lagrangian system of eqns (10.51) and (10.53) instead of the Eulerian system of eqns (10.40) and (10.41).

To illustrate this approach, let us consider the solution of the spatially homogeneous and isotropic case (10.12) in terms of Lagrangian coordinates. The Hubble flow in this case is in effect a simple scaling with a time-dependent scale factor; therefore, we seek a Lagrangian fluid flow of the form

$$\mathbf{x}(t, \boldsymbol{\xi}) = a(t)\, \boldsymbol{\xi}, \tag{10.54}$$

where $a(t_{in}) = 1$. In this case, $\Xi = a(t)\operatorname{diag}(1,1,1)$ and $\mathbb{J} = a^3(t)$. It follows from eqn (10.51) that $\varrho\mathbb{J} = \varrho(t_{in}, \boldsymbol{\xi})$. Moreover, eqn (10.53) can be expressed as $3\ddot{a}/a = -4\pi G(1 + \varrho_D/\varrho)\,\varrho$. In the homogeneous case, ϱ is independent of $\boldsymbol{\xi}$, $\varrho_D/\varrho = \alpha_0/A(t)$ and $\varrho(t) = \varrho(t_{in})/a^3(t)$. We thus recover eqns (10.12) and (10.16) with $a(t) = A(t)/A(t_{in})$, $\varrho = \bar{\rho}$ and $\rho_0 = \varrho(t_{in})\,A^3(t_{in})$.

10.4.1   Zel'dovich ansatz

To find a spatially inhomogeneous solution of eqns (10.51) and (10.53), we follow Zel'dovich and assume a Lagrangian fluid flow of the form

$$\mathbf{x}(t, \boldsymbol{\xi}) = \mathbb{A}(t)\,[\boldsymbol{\xi} - \mathbf{F}(t, \boldsymbol{\xi})], \tag{10.55}$$

where the general Lagrangian position vector $\boldsymbol{\xi} = (\xi^1, \xi^2, \xi^3)$ is constant along the path of a fluid particle and

$$\mathbf{F}(t, \boldsymbol{\xi}) := (F^1(t, \xi^1), 0, 0). \tag{10.56}$$

It follows from eqns (10.55) and (10.56) that

$$\Xi = \mathbb{A}(t)\operatorname{diag}(1 - \Psi, 1, 1), \qquad \mathbb{J} = \mathbb{A}^3\,(1 - \Psi), \qquad \Psi := \frac{\partial F^1}{\partial \xi^1}. \tag{10.57}$$

The Lagrangian conservation of mass eqn (10.51) implies that $\varrho\mathbb{J}$ is constant in time; therefore,

$$\varrho = \frac{\varrho_0(\boldsymbol{\xi})}{\mathbb{J}}, \tag{10.58}$$

where $\varrho_0(\boldsymbol{\xi}) > 0$ is simply a function of the Lagrangian coordinates. Moreover, the substitution of the Zel'dovich ansatz into the Lagrangian form of the conservation of momentum eqn (10.53) yields, after some algebra,

$$-3\frac{\ddot{\mathbb{A}}}{\mathbb{A}} + 2\frac{\dot{\mathbb{A}}}{\mathbb{A}}\frac{\Psi_t}{1 - \Psi} + \frac{\Psi_{tt}}{1 - \Psi} = 4\pi G\left(\frac{\varrho_0(\boldsymbol{\xi})}{\mathbb{A}^3(1 - \Psi)} + \varrho_D\right). \tag{10.59}$$

We now assume, in conformity with the original Zel'dovich solution (Mukhanov 2005), that

$$\mathbb{A} = A, \qquad \varrho_0(\boldsymbol{\xi}) = \rho_0, \tag{10.60}$$

where $A(t)$, given by eqn (10.16), is the scale factor of the spatially homogeneous and isotropic background and ρ_0 is the uniform background baryonic density at the present epoch. Thus, using eqns (10.16) and (10.60), eqn (10.59) can be written as

$$\frac{\Psi_{tt}}{1 - \Psi} + 2\frac{\dot{A}}{A}\frac{\Psi_t}{1 - \Psi} + \frac{4\pi G\rho_0}{A^3}\left(1 + \frac{\alpha_0}{A}\right) = 4\pi G\left(\frac{\rho_0}{A^3(1 - \Psi)} + \varrho_D\right). \tag{10.61}$$

The quantity $\Psi/(1-\Psi)$ in this context is related to the *density contrast*,

$$\frac{\Psi}{1-\Psi} = \frac{\varrho - \bar{\rho}(t)}{\bar{\rho}(t)}, \tag{10.62}$$

of baryonic matter given as a solution in the Lagrangian formulation of the toy model relative to the background density $\bar{\rho}(t) = \rho_0/A^3$ of the exact homogeneous solution (10.12). Here ϱ is the density of baryonic matter, $\varrho = \rho_0/[A^3(1-\Psi)]$, in our model. We assume that Ψ is a positive function that is less than unity and the background density is bounded. The matter density in Lagrangian coordinates approaches infinity as Ψ approaches unity. This would indicate the breakdown of the Lagrangian coordinate system. On the physical side, this circumstance can be interpreted to mean that the presence of *effective dark matter* ϱ_D in the model produces cosmic structure as time approaches this epoch.

Let us now write eqn (10.61) in the form

$$\Psi_{tt} + 2\frac{\dot{A}}{A}\Psi_t - \frac{4\pi G\rho_0}{A^3}\left(1 + \frac{\alpha_0}{A}\right)\Psi = \mathcal{N}, \tag{10.63}$$

where $\mathcal{N}$ is defined by

$$\mathcal{N} := \frac{4\pi G\rho_0\,\alpha_0}{A^4}\left[\frac{A^4}{\rho_0\,\alpha_0}(1-\Psi)\,\varrho_D - 1\right]. \tag{10.64}$$

The *nonlinear* part of this second-order equation for Ψ is contained in $\mathcal{N}$. It is interesting to note that this nonlinear term vanishes for $\varrho_D = (\alpha_0/A)\,\varrho$. In this case, Ψ can be determined from the linear and homogeneous form of eqn (10.63) with $\mathcal{N} = 0$, which coincides with the $k \to 0$ limit of the linear perturbation eqn (10.32) for the density contrast $D_{\mathbf{k}}(t)$. This limiting case is quite significant and will be discussed in detail later in this section.

The nonlocal Zel'dovich model under consideration here therefore deals with the *nonlinear* perturbation Ψ in the baryonic density of the spatially homogeneous and isotropic background model of nonlocal Newtonian cosmology. The background expands with scale factor $A(t)$ and Ψ satisfies eqn (10.63), where $\mathcal{N}$, the nonlinear part of eqn (10.63), is given by eqn (10.64).

The density of the effective dark matter in the Lagrangian formulation is

$$\varrho_D(t,\boldsymbol{\xi}) = \rho_D(t,\mathbf{x}(t,\boldsymbol{\xi})) = \int_{\mathbb{R}^3} q(t,\mathbf{x}(t,\boldsymbol{\xi}) - \mathbf{y})\,\rho(t,\mathbf{y})\,d^3y, \tag{10.65}$$

where the reciprocal kernel q is given by eqn (10.35). It is natural to complete the transformation of this relation to Lagrangian variables. To this end, we let

$$\mathbf{y} = \mathbf{y}(t,\boldsymbol{\zeta}) \tag{10.66}$$

and note that

$$\rho(t,\mathbf{y}(t,\boldsymbol{\zeta})) := \varrho(t,\boldsymbol{\zeta}) = \frac{\rho_0}{\mathbb{J}(t,\boldsymbol{\zeta})}. \tag{10.67}$$

Moreover, $d^3y = \mathbb{J}(t,\boldsymbol{\zeta})\,d^3\zeta$, since $\mathbb{J}$ is the Jacobian of the transformation from the Eulerian to Lagrangian coordinates. Therefore, the change of variables $\mathbf{y} = \mathbf{y}(t,\boldsymbol{\zeta})$ yields the more useful form of ϱ_D,

$$\varrho_D(t,\boldsymbol{\xi}) = \rho_0 \int_{\mathbb{R}^3} q(t,\mathbf{x}(t,\boldsymbol{\xi}) - \mathbf{y}(t,\boldsymbol{\zeta}))\,d^3\zeta. \tag{10.68}$$

The Lagrangian flow map is here given by the Zel'dovich ansatz, namely,

$$\mathbf{x}(t,\boldsymbol{\xi}) = A(t)\left(\xi^1 - F^1(t,\xi^1),\xi^2,\xi^3\right). \tag{10.69}$$

Using eqn (10.35), the Lagrangian formula for the effective density of dark matter can be written as

$$\varrho_D = \frac{\rho_0\,\alpha_0\,\mu_0^3}{8\pi\,A(t)} \int_{\mathbb{R}^3} \frac{1+\chi}{\chi^2}\,e^{-\chi}\,d^3\zeta, \tag{10.70}$$

where

$$\chi := \mu_0\,|\mathbf{x}(t,\boldsymbol{\xi}) - \mathbf{y}(t,\boldsymbol{\zeta})|. \tag{10.71}$$

From the Zel'dovich ansatz for the Lagrangian fluid flow, we have

$$\mathbf{y}(t,\boldsymbol{\zeta}) - \mathbf{x}(t,\boldsymbol{\xi}) = A(t)\left(\zeta^1 - \xi^1 + F^1(t,\xi^1) - F^1(t,\zeta^1),\zeta^2 - \xi^2,\zeta^3 - \xi^3\right). \tag{10.72}$$

It is therefore useful to define a new integration variable $\boldsymbol{\sigma}$ in eqn (10.70), namely,

$$\boldsymbol{\zeta} - \boldsymbol{\xi} := \frac{\boldsymbol{\sigma}}{\mu_0\,A(t)}. \tag{10.73}$$

Then,

$$\mu_0\left[\mathbf{y}(t,\boldsymbol{\zeta}) - \mathbf{x}(t,\boldsymbol{\xi})\right] = (\mathcal{S},\sigma^2,\sigma^3), \tag{10.74}$$

where

$$\mathcal{S} := \sigma^1 + \mu_0\,A(t)\left[F^1(t,\xi^1) - F^1(t,\xi^1 + \frac{\sigma^1}{\mu_0\,A(t)})\right]. \tag{10.75}$$

To work with dimensionless quantities, we define

$$\gamma := \mu_0\,\xi^1, \qquad \tilde{\Sigma}(t,\gamma) := \mu_0\,F^1(t,\frac{\gamma}{\mu_0}), \tag{10.76}$$

so that

$$\mathcal{S} = \sigma^1 + A(t)\left[\tilde{\Sigma}(t,\gamma) - \tilde{\Sigma}(t,\gamma + \frac{\sigma^1}{A(t)})\right]. \tag{10.77}$$

In this way, we find

$$\varrho_D = \frac{\rho_0\,\alpha_0}{8\pi\,A^4(t)} \int_{\mathbb{R}^3} \frac{1+\chi}{\chi^2}\,e^{-\chi}\,d^3\sigma, \tag{10.78}$$

where

$$\chi = \sqrt{\mathcal{S}^2 + (\sigma^2)^2 + (\sigma^3)^2} \tag{10.79}$$

and $\mathcal{S}$ is given by eqn (10.77).

It is convenient at this point to introduce spherical polar coordinates

$$\sigma^1 = \ell \cos\vartheta, \quad \sigma^2 = \ell \sin\vartheta \cos\varphi, \quad \sigma^3 = \ell \sin\vartheta \sin\varphi \tag{10.80}$$

with corresponding element of volume $\ell^2 \sin\vartheta \, d\ell \, d\vartheta \, d\varphi$. In these coordinates, χ given in eqn (10.79) does not depend on φ, since $(\sigma^2)^2 + (\sigma^3)^2 = \ell^2 \sin^2\vartheta$. Using this fact, integration over φ results in multiplication of the remaining two-dimensional integral over ℓ and ϑ by 2π. With

$$\eta := \cos\vartheta, \tag{10.81}$$

we have that

$$\varrho_D = \frac{\rho_0 \alpha_0}{A^4(t)} \mathcal{I}, \quad \mathcal{I} = \frac{1}{4} \int_0^\infty \int_{-1}^1 \frac{1+\chi}{\chi^2} e^{-\chi} \ell^2 \, d\ell \, d\eta. \tag{10.82}$$

Here,

$$\chi = \sqrt{S^2 + \ell^2 (1 - \eta^2)} \tag{10.83}$$

and

$$S = \ell\eta + A(t)\Big[\tilde{\Sigma}(t, \gamma) - \tilde{\Sigma}\big(t, \gamma + \frac{\ell\eta}{A(t)}\big)\Big]. \tag{10.84}$$

In terms of the new dimensionless quantities defined in eqn (10.76),

$$\Psi = \frac{\partial \tilde{\Sigma}(t, \gamma)}{\partial \gamma} := \tilde{\Sigma}_\gamma(t, \gamma) \tag{10.85}$$

and the integro-differential eqn (10.63) of our model now takes the form

$$\tilde{\Sigma}_{\gamma tt} + 2\frac{\dot{A}}{A}\tilde{\Sigma}_{\gamma t} - \frac{4\pi G\rho_0}{A^3}\big(1 + \frac{\alpha_0}{A}\big)\tilde{\Sigma}_\gamma = \mathcal{N}, \tag{10.86}$$

where the nonlinear part is given by

$$\mathcal{N} = \frac{4\pi G\rho_0 \alpha_0}{A^4}\Big[(1 - \tilde{\Sigma}_\gamma)\mathcal{I} - 1\Big]. \tag{10.87}$$

Finally, it proves convenient to use $s = A(t)$ as the new temporal variable instead of the cosmic time t; that is, we define

$$\Sigma(s, \gamma) := \tilde{\Sigma}(A^{-1}(s), \gamma), \tag{10.88}$$

so that

$$\tilde{\Sigma}_{\gamma t}(A^{-1}(s), \gamma) = \dot{A}(A^{-1}(s))\Sigma_{\gamma s}(s, \gamma),$$
$$\tilde{\Sigma}_{\gamma tt}(A^{-1}(s), \gamma) = \dot{A}^2(A^{-1}(s))\Sigma_{\gamma ss}(s, \gamma) + \ddot{A}(A^{-1}(s))\Sigma_{\gamma s}(s, \gamma). \tag{10.89}$$

We now use eqns (10.16)–(10.17), with $\bar{E} = 0$, to transform the integro-differential eqn (10.86) to

$$s^2(2s + \alpha_0)\Sigma_{\gamma ss} + s(3s + \alpha_0)\Sigma_{\gamma s} - 3(s + \alpha_0)\Sigma_\gamma = 3\alpha_0\Big[(1 - \Sigma_\gamma)\mathcal{I}(\Sigma) - 1\Big], \tag{10.90}$$

where $\alpha_0 = 2/(\lambda_0\,\mu_0) \approx 11$, $\mathcal{I}$ is the integral defined in display (10.82) with χ given by eqn (10.83) and

$$\mathcal{S} = \ell\,\eta + s\left[\Sigma(s,\gamma) - \Sigma(s,\gamma + \frac{\ell\,\eta}{s})\right]. \tag{10.91}$$

Let us note that the dependence of $\mathcal{I}$ on Σ, through $\mathcal{S}$, is such that $\mathcal{I}(\Sigma)$ is unchanged if Σ is replaced by $\Sigma + g(s)$, where $g(s)$ is an arbitrary function of s. Moreover, if Σ is a solution of the integro-differential eqn (10.90), then so is $\Sigma + g(s)$. In particular, it is simple to show that for $\Sigma = g(s)$

$$\mathcal{I}(g) = 1, \tag{10.92}$$

so that

$$\Sigma(s,\gamma) = g(s) \tag{10.93}$$

with $\Sigma_\gamma = 0$ is an *exact* solution of eqn (10.90).

An important feature of this nonlocal Zel'dovich model is that the original work of Zel'dovich within the framework of standard Newtonian cosmology can be simply recovered if we formally set $\alpha_0 = 0$.

10.4.2 Zel'dovich solution

The Zel'dovich solution (Mukhanov 2005) regarding the nonlinear behavior of density perturbation in one spatial dimension can be obtained with $\alpha_0 = 0$ in eqn (10.90). For $s > 0$, eqn (10.90) then reduces to

$$2\,s^2\,\Sigma_{\gamma ss} + 3\,s\,\Sigma_{\gamma s} - 3\,\Sigma_\gamma = 0. \tag{10.94}$$

The solution of this differential equation for Σ_γ is a linear combination of s and $s^{-3/2}$ with coefficients that are arbitrary functions of the spatial variable γ.

We recall that $\Psi = \Sigma_\gamma$ is such that $\Psi/(1 - \Psi)$ is the density contrast in the nonlinear regime. On the other hand, in the linear regime, the perturbation equation for the density contrast is given by eqn (10.32) with $\alpha_0 = 0$ and $\lambda \gg \lambda_J$, namely,

$$\frac{d^2 D_{\mathbf{k}}}{dt^2} + 2\,\frac{\dot{A}}{A}\,\frac{dD_{\mathbf{k}}}{dt} - \frac{4\,\pi G\rho_0}{A^3}\,D_{\mathbf{k}} = 0, \tag{10.95}$$

where $A(t) = (t/t_0)^{2/3}$ in this case, $Q_k = 0$ and the pressure term has been neglected for wavelengths much longer than the Jeans length λ_J. Equation (10.95) has solutions for $D_{\mathbf{k}}$ that are linear combinations of $t^{2/3}$ and t^{-1}. It follows that eqns (10.94) and (10.95) have identical solutions for density perturbations. Thus, so long as the spatial scale of the perturbation is much longer than the Jeans length, *the validity of the linear perturbation scheme extends into the nonlinear regime of the Zel'dovich solution.*

Let us now consider the growing mode, where $\Sigma_\gamma \propto s = (1+z)^{-1}$. This mode grows by a factor of about 10^3 from the epoch of decoupling ($z \approx 10^3$) to the present epoch ($z \approx 0$). Thus baryonic inhomogeneities of amplitude $\delta \approx 10^{-5}$ at the decoupling epoch will only grow to $\delta \approx 10^{-2}$ at the present epoch, far below what is needed to explain the observed large-scale structure of the universe. This simple calculation indicates the necessity of invoking the existence of dark matter in the standard models of cosmology.

The rest of this chapter is about the nature of solutions of the *nonlocal* Zel'dovich model.

10.4.3 $\mathcal{N} \approx 0$

In our nonlocal Zel'dovich model, the density of baryons is given by $\rho_0/[s^3(1 - \Sigma_\gamma)]$, where $s = A(t)$ is the scale factor and Σ is a solution of our integro-differential eqn (10.90). To illustrate certain properties of our nonlocal model, it proves useful at this point to define a limiting form of the model that is in some sense the analog of the original Zel'dovich solution in the present context. The Zel'dovich ansatz for our nonlocal toy model has led to eqn (10.90), where, in contrast to the original Zel'dovich solution, the nonlinear term that appears on the right-hand side of this equation does not in general vanish. However, this nonlinear part of the main differential equation vanishes for a certain limiting form of our toy model if we assume that

$$\rho_D(t, \mathbf{x}) = \frac{\alpha_0}{A} \rho(t, \mathbf{x}), \tag{10.96}$$

where the scale factor $A(t)$ is given by eqn (10.19) and $\rho(t, \mathbf{x})$ is the density of baryonic matter in the universe. In eqn (10.90), the temporal variable s is in fact $A(t)$, so that $s = 0$ at the Big Bang and $s = 1$ at the present epoch. The Zel'dovich ansatz in this limiting case ($\mathcal{N} = 0$) leads to the linear and homogeneous differential equation

$$s^2(2s + \alpha_0)\Sigma_{\gamma ss} + s(3s + \alpha_0)\Sigma_{\gamma s} - 3(s + \alpha_0)\Sigma_\gamma = 0, \tag{10.97}$$

which is essentially the same as the linear perturbation eqn (10.32) for the density contrast $D_{\mathbf{k}}(t)$ in the limit where $k \to 0$.

With the change of variable $s = -\alpha_0 \nu/2$ and some rearrangement, this differential equation for $\Sigma_\gamma(s, \gamma) := \Delta(\nu, \gamma)$ becomes

$$\nu^2(1 - \nu)\Delta_{\nu\nu} + \nu(1 - \frac{3}{2}\nu)\Delta_\nu - 3(1 - \frac{1}{2}\nu)\Delta = 0. \tag{10.98}$$

The solution of this equation is the product of an arbitrary function of the spatial variable γ and a function of ν, which can be expressed in terms of the hypergeometric function (Abramowitz and Stegun 1964). This latter function, expressed in terms of $s = -\alpha_0 \nu/2$, can be written as

$$S_p(s) = s^p F(p + \frac{3}{2}, p - 1; 2p + 1; -\frac{2s}{\alpha_0}). \tag{10.99}$$

Here, $p = \pm\sqrt{3}$, F is the hypergeometric function (Abramowitz and Stegun 1964), $\nu = -2s/\alpha_0$ and $|\nu| < 1$. We note that $S_{\sqrt{3}}$ and $S_{-\sqrt{3}}$ form a fundamental set of solutions of the linear second-order differential eqn (10.98). The general solution of eqn (10.98) is qualitatively similar to the classical Zel'dovich solution; in particular, the given fundamental set of solutions consists of one growing and one decaying mode.

Under what conditions would $\rho_D(t, \mathbf{x})$, given by eqn (10.5), reduce to eqn (10.96)? That is, we wish to determine the conditions under which the limiting form of the

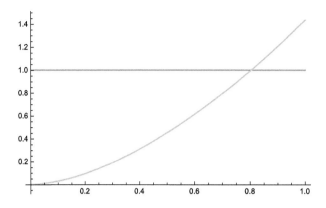

Fig. 10.3 Plot of $s \mapsto 10^{-5} S_{\sqrt{3}}(s)/S_{\sqrt{3}}(10^{-3})$. Reproduced from Chicone, C. and Mashhoon, B., 2016, "Nonlocal Newtonian Cosmology", *J. Math. Phys.* **57**, 072501 (27 pages), with the permission of AIP Publishing. DOI: 10.1063/1.4958902

model under consideration here approximates our basic nonlocal model. Let us note in this connection that kernel q in eqn (10.35) is radial and decays exponentially for $|\mathbf{x}| \gg \mu_0^{-1}$. It follows that if $\rho(t, \mathbf{x})$ always varies in space only over scales much larger than $1/\mu_0$, then its spatial Fourier integral transform $\widehat{\rho}(t, \mathbf{k})$,

$$\widehat{\rho}(t, \mathbf{k}) = \int \rho(t, \mathbf{x})\, e^{-i\mathbf{k}\cdot\mathbf{x}/A(t)}\, d^3x, \tag{10.100}$$

is essentially confined to a region in Fourier space such that $k \ll \mu_0 A(t)$. For instance, if we imagine that $\rho(t, \mathbf{x})$ always has the form of a Gaussian distribution centered around $\mathbf{x} = 0$ with a root-mean-square deviation from the mean of $\sigma_0 \gg \mu_0^{-1}$, then in the Fourier domain, this corresponds to a Gaussian distribution centered around $k = 0$ with a root-mean-square deviation from the mean of $\widehat{\sigma}_0 \ll \mu_0 A(t)$. We can use this fact in conjunction with the convolution theorem for Fourier transforms to show that the solution of the nonlocal model approaches the exact solution of the $\mathcal{N} = 0$ model for large-scale deviations from spatial homogeneity. That is, eqn (10.5) can be written in the Fourier domain, in the sense defined in Section 10.3, as

$$\widehat{\rho}_D(t, \mathbf{k}) = Q_k(t)\, \widehat{\rho}(t, \mathbf{k}), \tag{10.101}$$

where $\widehat{\rho}_D(t, \mathbf{k})$ is the Fourier integral transform of ρ_D and $Q_k(t)$ is given by eqn (10.34). It is clear from Fig. 10.1 that if $k \ll \mu_0 A(t)$, then $Q_k(t) \approx \alpha(t)$, where $\alpha(t) = \alpha_0/A(t)$, so that

$$\widehat{\rho}_D(t, \mathbf{k}) \approx \alpha(t)\, \widehat{\rho}(t, \mathbf{k}). \tag{10.102}$$

The inverse spatial Fourier integral transform of this relation amounts essentially to eqn (10.96). Therefore, on large scales that persist over time, the solution of the nonlocal model corresponding to the growing mode should approach $S_{\sqrt{3}}(s)$. In Fig. 10.3, we plot

$$10^{-5}\, \frac{S_{\sqrt{3}}(s)}{S_{\sqrt{3}}(10^{-3})}, \tag{10.103}$$

where it is assumed that at the epoch of recombination corresponding to cosmological redshift $z \approx 10^3$ and $s = 1/(1 + z)$, the density contrast is $\sim 10^{-5}$. It is demonstrated in Fig. 10.3 that Σ_γ for the growing mode approaches unity well before the present era $s = 1$.

Let us recall here that as $\Sigma_\gamma \to 1$, the density of baryons $\rho_0/[A^3 (1 - \Sigma_\gamma)]$ approaches infinity. At some stage in this process, the density contrast is so large that the perturbation separates from the background and collapses under its own gravitational attraction, thus leading to the formation of structure in the universe (Peebles 1993; Zel'dovich and Novikov 1983; Mukhanov 2005; Gurbatov, Saichev and Shandarin 2012). The result presented in Fig. 10.3 indicates that the nonlocal model under consideration here is such that for sufficiently large-scale perturbations after recombination, structure formation is theoretically possible in this toy model.

10.4.4 $\mathcal{N} \neq 0$

Chicone has reformulated the nonlocal Zel'dovich model in terms of a nonlinear nonautonomous first-order system of ordinary differential equations in some function space (of functions of γ). The analysis of the resulting system seems to be beyond current understanding of infinite-dimensional ordinary differential equations (Chicone 2006). Therefore, a numerical algorithm has been developed by Chicone for dealing with this system. He has verified numerically that the solution of the full nonlocal nonlinear model indeed approaches that of the linear homogeneous $\mathcal{N} = 0$ model as the spatial scale of the perturbation increases; see Chicone and Mashhoon (2016b) for a detailed treatment.

The positive result of our nonlocal Zel'dovich model indicates that the nonlocal gravity approach to the problem of structure formation in cosmology without invoking the existence of dark matter deserves further investigation.

10.5   An Exact Solution

It is possible to find an exact nontrivial solution of our nonlocal model. To this end, let

$$\Sigma(s, \gamma) = g(s) + \beta(s) \, \gamma, \tag{10.104}$$

where g and β are only functions of the temporal variable s. It proves interesting to start by computing $\mathcal{I}(\Sigma)(s, \gamma)$ given in eqn (10.82). We find that $\mathcal{S} = \ell \eta (1 - \beta)$ and $\chi = \ell \chi_0$, where χ_0 is independent of ℓ and is given by

$$\chi_0 := [\eta^2 (1 - \beta)^2 + 1 - \eta^2]^{1/2}. \tag{10.105}$$

The integration in $\mathcal{I}$ over $\ell : 0 \to \infty$ can be carried out first using the relation

$$\int_0^\infty \ell^n \, e^{-\ell \chi_0} \, d\ell = \frac{n!}{\chi_0^{n+1}} \tag{10.106}$$

for $n = 0, 1, 2, \ldots$. For the integration over $\eta : -1 \to 1$, we note that

$$\int \frac{x^{2n}}{(1 + c x^2)^{n+3/2}} \, dx = \frac{1}{2n + 1} \frac{x^{2n+1}}{(1 + c x^2)^{n+1/2}}. \tag{10.107}$$

Therefore, for $c > -1$,

$$\int_{-1}^{1} \frac{\eta^{2n}}{(1+c\,\eta^2)^{n+3/2}} \, d\eta = \frac{2}{2n+1} \frac{1}{(1+c)^{n+1/2}}, \tag{10.108}$$

for $n = 0, 1, 2, \ldots$. For $n = 0$, we find that

$$\mathcal{I}(\Sigma) = \int_{0}^{1} \frac{d\eta}{\{1 + [(1-\beta)^2 - 1]\,\eta^2\}^{3/2}} = \frac{1}{|1-\beta|}. \tag{10.109}$$

It follows from this result that for $\beta < 1$, the right-hand side of eqn (10.90) vanishes, so that $\Sigma_\gamma = \beta(s)$ must satisfy

$$s^2(2s + \alpha_0)\beta_{ss} + s(3s + \alpha_0)\beta_s - 3(s + \alpha_0)\beta = 0. \tag{10.110}$$

Thus β can be written as

$$\beta(s) = C_+ \, S_{\sqrt{3}}(s) + C_- \, S_{-\sqrt{3}}(s), \tag{10.111}$$

where $S_{\pm\sqrt{3}}(s)$ are given in eqn (10.99) and $C_\pm$ are constants such that $\beta(s) < 1$.

Let $\tilde{B}(t) := \beta(A(t))$, then the exact solution under consideration corresponds to the Zel'dovich ansatz (10.55) such that

$$x^1 = A(t)\,[1 - \tilde{B}(t)]\,\xi^1, \qquad x^2 = A(t)\,\xi^2, \qquad x^3 = A(t)\,\xi^3, \tag{10.112}$$

which represents an expanding homogeneous but *anisotropic* cosmological model. It is straightforward to check that with the spatially uniform matter density $\rho_0/[A^3\,(1-\tilde{B})]$, our original model eqns (10.40)–(10.41) are satisfied in this case.

If in eqn (10.111), $C_+ > 0$, then in time β will increase monotonically and eventually approach unity, in which case the oblate spheroidal model universe collapses to form an expanding circular disk of matter.

It was argued in Section 10.4 that the solution of the nonlocal model should approach the solution of the $\mathcal{N} = 0$ model if the density $\rho(t, \mathbf{x})$ always varies in space over distances that are much larger than μ_0^{-1}. The limiting situation, where $\rho(t, \mathbf{x})$ loses all dependence upon space and depends only upon time corresponds to $\Sigma_\gamma(s, \gamma) = \beta(s)$, from which the exact solution under consideration would necessarily follow. These remarks then provide the physical interpretation for the exact solution of the nonlocal Zel'dovich model. Furthermore, this exact solution turns out to be unstable, as demonstrated via a linear perturbation method in Section 10.7.

10.6   Appendix A: Distention of Massive Early-Type Galaxies with Decreasing z

The nonlocal model of Newtonian cosmology employed in this chapter has been based on the assumption that memory fades over time and hence the strength of the gravitational interaction as well as the dark matter fraction monotonically decreases with cosmic time. As before, this temporal dependence can be expressed, for instance, as

$1/A^{\varpi}$, where A is the scale factor and $\varpi > 0$. For the nonlocal Zel'dovich model developed in this chapter, we have set $\varpi = 1$ for the sake of simplicity. Writing $A^{-1} = 1 + z$, we note that the *cosmological memory drag* can be incorporated in the attractive gravitational force. That is, eqn (8.1), which expresses the force of gravity on a point mass m due to a point mass m' in the Newtonian regime of nonlocal gravity, takes the form

$$\mathbf{F}_z(\mathbf{r}) = -Gmm'\frac{\hat{\mathbf{r}}}{r^2}\left\{1 + (1+z)^{\varpi}\left[\alpha_0 - \alpha_0\left(1 + \frac{1}{2}\mu_0\,r\right)e^{-\mu_0\,r} - \mathcal{E}(r)\right]\right\} \quad (10.113)$$

in the cosmological context under consideration here. Equation (10.113) consists of two parts: a standard Newtonian part and a dominant effective dark matter part whose strength monotonically decreases with decreasing cosmological redshift z.

Let us now imagine that, as a consequence of gravitational instability, inhomogeneities grow after recombination to produce the observed large-scale structure of the universe. That is, after recombination, clumps of matter form that separate from the expanding background and collapse under their own gravitational attraction. Subsequent interaction and merger of such clumps eventually leads to the formation of galaxies. It follows from eqn (10.113) that the process of galaxy formation slows down with decreasing z. Moreover, the evolution of the internal gravitational dynamics of galaxies, from the time of formation to the present era ($z = 0$), would be governed by eqn (10.113). For an isolated galaxy, if the strength of the internal gravitational attraction slowly decreases with time, the size of the self-gravitating system slowly increases. The precise manner in which this *distention* comes about depends upon the nature of the attractive force. For the Newtonian inverse square law of attraction, for instance, the adiabatic invariants of the Keplerian two-body system indicate that if the strength of the force slowly decreases, the eccentricity of the orbit remains unchanged, while its dimensions increase in inverse proportion to the strength of the force (Landau and Lifshitz 1988). For the nonlocal gravity force (10.113), let us tentatively assume that the distention factor is more generally of the form $(1 + z)^{\varpi'}$ with $\varpi' > 0$.

It is possible that the distention mechanism we have described here is at work in the significant size evolution of high-mass quiescent (i.e. with no recent star formation) early-type galaxies. This swelling has been observationally established from $z \approx 2$–3 to $z \approx 0$; see for example van Dokkum *et al.* (2008, 2010), Damjanov *et al.* (2009) and the references therein. To illustrate this point, let us adopt the measurements of nine massive quiescent early-type galaxies at average $z \approx 2.3$ contained in van Dokkum *et al.* (2008). The average effective radius—r_e, following the usual de Vaucouleurs light profile for early-type galaxies—of these galaxies at $z \approx 2.3$ is 0.9 kpc, while galaxies with similar masses in the nearby universe have sizes of ≈ 5 kpc (van Dokkum *et al.* 2008). According to van Dokkum *et al.* (2008), several known astrophysical mechanisms (such as mergers) *might* be able to account for a distention factor of only ≈ 1.5–2, unless some of these known mechanisms are all at work here at the same time. While this is not impossible, it is worth noting that the nonlocal gravity model actually offers an alternative explanation, because the distention in size would occur naturally in this model. For simplicity, let us attribute the observed size distention solely to nonlocal gravity. This implies $5.5 \approx (1+z)^{\varpi'}$, and hence $\varpi' \approx 1.4$. Further consideration of the distention mechanism due to cosmological memory drag is beyond the scope of this work.

10.7 Appendix B: Instability of the Exact Solution

Consider the linear operator

$$\mathcal{L} := s^2(2s + \alpha_0)\frac{\partial^2}{\partial s^2} + s(3s + \alpha_0)\frac{\partial}{\partial s} - 3(s + \alpha_0); \qquad (10.114)$$

then, our integro-differential eqn (10.90) can be written as

$$\mathcal{L}\,\Sigma_\gamma = 3\,\alpha_0\left[(1 - \Sigma_\gamma)\mathcal{I} - 1\right]. \qquad (10.115)$$

As discussed in Section 10.5, eqn (10.115) has an exact solution

$$\Sigma^0(s,\gamma) = g(s) + \beta(s)\,\gamma, \qquad (10.116)$$

where $g(s)$ is arbitrary and $\beta(s) < 1$ is a solution of the homogeneous linear equation $\mathcal{L}\,\beta(s) = 0$.

10.7.1 Linear perturbation of the exact solution

We would now like to look for a solution of eqn (10.115) of the form

$$\Sigma = \Sigma^0 + \epsilon\,\Pi(s,\gamma), \qquad (10.117)$$

where Π is the perturbing function and ϵ, $0 < \epsilon \ll 1$, is the perturbation parameter such that only terms linear in ϵ will be considered. To compute $\mathcal{I}(\Sigma)(s,\gamma)$ in this case, we note that

$$\mathcal{S} = \ell\,\eta\,(1 - \beta) - \epsilon\,\mathcal{T}, \qquad (10.118)$$

where

$$\mathcal{T} := s\left[\Pi(s,\gamma + \frac{\ell\,\eta}{s}) - \Pi(s,\gamma)\right]. \qquad (10.119)$$

Hence,

$$\chi = \ell\,\chi_0 - \epsilon\,\eta\,(1 - \beta)\,\chi_o^{-1}\,\mathcal{T} \qquad (10.120)$$

and

$$\frac{1 + \chi}{\chi^2}e^{-\chi} = \frac{1 + \ell\,\chi_0}{\ell^2\,\chi_0^2}e^{-\ell\,\chi_0} + \epsilon\,\eta\,(1 - \beta)\,\mathcal{T}\,\frac{2 + 2\,\ell\,\chi_0 + \ell^2\,\chi_0^2}{\ell^3\,\chi_0^4}e^{-\ell\,\chi_0}. \qquad (10.121)$$

To proceed, we assume that perturbation Π is analytic in the spatial variable so that $\mathcal{T}$ is represented by its Taylor series,

$$\mathcal{T} := \sum_{m=1}^{\infty}\frac{1}{m!}\Pi^{(m)}(s,\gamma)\frac{\ell^m\,\eta^m}{s^{m-1}}, \qquad (10.122)$$

where $\Pi^{(m)} := \partial^m\,\Pi/\partial\gamma^m$. Moreover, we can write

$$\mathcal{I}(\Sigma) = \mathcal{I}(\Sigma^0) + \epsilon\,\mathcal{K}, \tag{10.123}$$

where for $\beta < 1$, we have from eqn (10.109) that

$$\mathcal{I}(\Sigma^0) = \frac{1}{1-\beta}. \tag{10.124}$$

In evaluating $\mathcal{K}$, the integration over ℓ is straightforward using eqn (10.106) and we find

$$\int_0^\infty \ell^{m-1}\left[\ell^2\,\chi_0^2 + 2\,\ell\,\chi_0 + 2\right]e^{-\ell\,\chi_0}\,d\ell = \frac{(m+1)(m+2)(m-1)!}{\chi_0^m}. \tag{10.125}$$

The subsequent integration over η vanishes by symmetry unless the integrand is even in η, which means that m must be odd, namely, $m = 2n+1$, where $n = 0,1,2,\ldots$. Then, using eqn (10.108), we find

$$\mathcal{K} = \sum_{n=0}^\infty \frac{n+1}{2n+1}\frac{\Pi^{(2n+1)}(s,\gamma)}{s^{2n}\,(1-\beta)^{2n+2}}. \tag{10.126}$$

It follows from eqn (10.115) that to first order in ϵ, we have the equation for the linear perturbation of the exact solution, namely,

$$\mathcal{L}\,\Pi_\gamma(s,\gamma) = 3\,\alpha_0\left[(1-\beta)\mathcal{K} - \frac{\Pi_\gamma}{1-\beta}\right]. \tag{10.127}$$

Combining eqns (10.126) and (10.127), we finally get

$$\mathcal{L}\,\Pi_\gamma(s,\gamma) = 3\,\alpha_0\sum_{n=1}^\infty \frac{n+1}{2n+1}\frac{\Pi^{(2n+1)}(s,\gamma)}{s^{2n}\,(1-\beta)^{2n+1}}. \tag{10.128}$$

10.7.2 Solutions of the perturbation equation

Inspection of eqn (10.128) for the linear perturbation Π reveals that

$$\Pi = \frac{1}{2}\,\theta(s)\,\gamma^2, \qquad \mathcal{L}\,\theta(s) = 0 \tag{10.129}$$

is a solution, since the right-hand side of eqn (10.128) vanishes identically in this case. Another possibility involves a solution of eqn (10.128) of the form

$$\Pi = e^{b\gamma}\,\mathcal{E}(s), \tag{10.130}$$

where b is a constant. Substituting eqn (10.130) into eqn (10.128) results in

$$\mathcal{L}\,\mathcal{E}(s) = \frac{3\,\alpha_0}{b}\,s\,\mathcal{E}(s)\sum_{n=1}^\infty \frac{n+1}{2n+1}\,\mathbb{B}^{2n+1}, \tag{10.131}$$

where

$$\mathbb{B}(s) := \frac{b}{s\,(1-\beta)}. \tag{10.132}$$

We recall that s is the scale factor and for an expanding universe model, $s : 0 \to \infty$. Moreover, $\beta(s) < 1$. If we assume that the constant $|b|$ is such that $|\mathbb{B}(s)| < 1$, then the series in eqn (10.131) converges uniformly and we have

$$\sum_{n=1}^{\infty} \frac{n+1}{2n+1} \mathbb{B}^{2n+1} = \frac{1}{2} \left[\frac{\mathbb{B}^3}{1 - \mathbb{B}^2} - \mathbb{B} - \frac{1}{2} \ln \left(\frac{1 - \mathbb{B}}{1 + \mathbb{B}} \right) \right] := \mathcal{B}(s). \tag{10.133}$$

The function $\mathcal{E}(s)$ can now be determined from the linear differential equation

$$\mathcal{L}\mathcal{E}(s) = \frac{3\,\alpha_0}{b} \, s \, \mathcal{B}(s)\, \mathcal{E}(s). \tag{10.134}$$

These possible solutions of the linear perturbation equation indicate that the exact solution of our nonlocal model is *unstable*. A complete analysis of the solutions of eqn (10.128) is beyond the scope of this work.

10.7.3 $\Sigma^0 = g(s)$

Finally, it is important to point out a seemingly trivial special case, namely, when $\beta(s) = 0$. In this case, $\Sigma_\gamma^0 = 0$ and we start with a small density perturbation of $\epsilon\, \Pi_\gamma$. The linear perturbation away from this exact zero solution can be unstable due to solution (10.129) leading to $\Pi_\gamma = \theta(s)\,\gamma$ as well as solution (10.130), which diverges exponentially in the spatial variable γ, where b, $|b| < s$, can be positive or negative.

Bibliography

Abbott, B.P., *et al.* (LIGO Scientific Collaboration and Virgo Collaboration), 2016a, *Phys. Rev. Lett.* **116**, 061102.

Abbott, B.P., *et al.* (LIGO Scientific Collaboration and Virgo Collaboration), 2016b, *Phys. Rev. Lett.* **116**, 241103.

Abramowitz, M. and Stegun, I.A., 1964, *Handbook of Mathematical Functions*, National Bureau of Standards, Washington, DC.

Adelberger, E.G., Heckel, B.R. and Nelson, A.E., 2003, *Ann. Rev. Nucl. Part. Sci.* **53**, 77. [arXiv: hep-ph/0307284]

Adelberger, E.G., *et al.*, 2007, *Phys. Rev. Lett.* **98**, 131104. [arXiv:hep-ph/0611223]

Agnese, R., *et al.*, (SuperCDMS), 2014, *Phys. Rev. Lett.* **112**, 241302. [arXiv:1402.7137 [hep-ex]]

Akerib, D.S., *et al.*, (LUX), 2014, *Phys. Rev. Lett.* **112**, 091303. [arXiv:1310.8214 [astro-ph.CO]]

Aldrovandi, R. and Pereira, J.G., 2013, *Teleparallel Gravity: An Introduction*, Springer, New York.

Allen, P.J., 1966, *Am. J. Phys.* **34**, 1185.

Aprile, E., *et al.*, (XENON100), 2012, *Phys. Rev. Lett.* **109**, 181301. [arXiv:1207.5988 [astro-ph.CO]]

Ashby, N., 2003, *Living Rev. Relativity* **6**, 1.

ATLAS Collaboration, 2012, *Phys. Lett. B* **716**, 1.

Auchmann, B. and Kurz, S., 2014, *J. Phys. A: Math. Theor.* **47**, 435202.

Barmby, P., *et al.*, 2006, *Astrophys. J.* **650**, L45.

Barmby, P., *et al.*, 2007, *Astrophys. J.* **655**, L61.

Barvinsky, A.O., 2003, *Phys. Lett. B* **572**, 109.

Baudis, L., 2016, *Ann. Phys. (Berlin)* **528**, 74.

Becker, D. and Reuter, M., 2014, *J. High Energy Phys.* **12**, 025. [arXiv:1407.5848 [hep-th]]

Beem, J.K., Ehrlich, P.E. and Easley, K.L., 1996, *Global Lorentzian Geometry*, 2nd edn, Marcel Dekker, New York.

Bekenstein, J.D., 1988, in *Second Canadian Conference on General Relativity and Relativistic Astrophysics*, edited by A. Coley, C. Dyer and T. Tupper, World Scientific, Singapore.

Bel, Ll., 2008, arXiv:0805.0846 [gr-qc].

Bel, Ll., 2016, arXiv:1607.03797 [gr-qc].

Bell, J.S., 1987, *Speakable and Unspeakable in Quantum Mechanics*, Cambridge University Press, Cambridge, UK.

Bertotti, G., 1998, *Hysteresis in Magnetism*, Academic Press, San Diego.

Bialynicki-Birula, I. and Bialynicka-Birula, Z., 1997, *Phys. Rev. Lett.* **78**, 2539.

Bini, D., Chicone, C. and Mashhoon, B., 2012, *Phys. Rev. D* **85**, 104020. [arXiv: 1203.3454 [gr-qc]]

Bini, D. and Lusanna, L., 2008, *Gen. Relativ. Gravit.* **40**, 1145.

Bini, D., Lusanna, L. and Mashhoon, B., 2005, *Int. J. Mod. Phys. D* **14**, 1413. [arXiv: gr-qc/0409052]

Bini, D. and Mashhoon, B., 2015, *Phys. Rev. D* **91**, 084026. [arXiv: 1502.04183 [gr-qc]]   DOI: 10.1103/PhysRevD.91.084026

Bini, D. and Mashhoon, B., 2016, *Int. J. Geom. Methods Mod. Phys.* **13**, 1650081. [arXiv: 1603.09477 [gr-qc]]   DOI: 10.1142/S021988781650081X

Birkhoff, G. and MacLane, S., 1953, *A survey of Modern Algebra*, Macmillan, New York.

Biswas, T., Mazumdar, A. and Siegel, W., 2006, *J. Cosmol. Astropart. Phys.* **0603**, 009. [arXiv: hep-th/0508194]

Biswas, T., *et al.*, 2012, *Phys. Rev. Lett.* **108**, 031101. [arXiv: 1110.5249 [gr-qc]]

Blagojević, M. and Hehl, F.W., editors, 2013, *Gauge Theories of Gravitation*, Imperial College Press, London, UK.

Blanchet, L., 2014, *Living Rev. Relativity* **17**, 2.

Bleistein, N. and Handelsman, R.A., 1986, *Asymptotic Expansions of Integrals*, Dover, New York.

Bliokh, K.Y., *et al.*, 2008, *Phys. Rev. Lett.* **101**, 030404.

Bliokh, K.Y., 2009, *J. Opt. A: Pure Appl. Opt.* **11**, 094009.

Bliokh, K.Y. and Aiello, A., 2013, *J. Opt.* **15**, 014001.

Blome, H.-J., *et al.*, 2010, *Phys. Rev. D* **81**, 065020. [arXiv:1002.1425 [gr-qc]]

Bogachev, V.I., 2007, *Measure Theory*, vol. I, Springer-Verlag, Berlin.

Bohr, N. and Rosenfeld, L., 1933, *K. Dan. Vidensk. Selsk. Mat. Fys. Medd.* **12**, 8; translated in *Quantum Theory and Measurement*, edited by J.A. Wheeler and W.H. Zurek, 1983, Princeton University Press, Princeton, NJ.

Bohr, N. and Rosenfeld, L., 1950, *Phys. Rev.* **78**, 794.

Bonnor, W.B., 1957, *Mon. Not. Roy. Astron. Soc.* **117**, 104.

Born, M., 1909, *Ann. Phys. (Berlin)* **335**, 1.

Botermann, B., *et al.*, 2014, *Phys. Rev. Lett.* **113**, 120405.

Bremm, G.N. and Falciano, F.T., 2015, *Ann. Phys. (Berlin)* **527**, 265.

Briscese, F., *et al.*, 2013, *Phys. Rev. D* **87**, 083507. [arXiv:1212.3611 [hep-th]]

Buchholz, D., Mund, J. and Summers, S.J., 2002, *Class. Quantum Grav.* **19**, 6417.

Buscaino, B., *et al.*, 2015, *Phys. Rev. D* **92**, 104048. [arXiv: 1508.06273[gr-qc]]

Canovan, C.E.S. and Tucker, R.W., 2010, *Am. J. Phys.* **78**, 1181. [arXiv: 1104.0574 [math-ph]]

Chicone, C., 2006, *Ordinary Differential Equations with Applications*, 2nd edn, Springer-Verlag, New York.

Chicone, C. and Mashhoon, B., 2002a, *Ann. Phys. (Berlin)* **11**, 309. [arXiv: gr-qc/0110109]

Chicone, C. and Mashhoon, B., 2002b, *Phys. Lett. A* **298**, 229. [arXiv: gr-qc/0202054]

Chicone, C. and Mashhoon, B., 2002c, *Class. Quantum Grav.* **19**, 4231. [arXiv:gr-qc/0203073]

Chicone, C. and Mashhoon, B., 2005, *Class. Quantum Grav.* **22**, 195. [arXiv: gr-qc/0409017]

Chicone, C. and Mashhoon, B., 2006, *Phys. Rev. D* **74**, 064019. [arXiv: gr-qc/0511129]

Chicone, C. and Mashhoon, B., 2007, *Ann. Phys. (Berlin)* **16**, 811. [arXiv: 0708.2744 [hep-th]]

Chicone, C. and Mashhoon, B., 2012, *J. Math. Phys.* **53**, 042501. [arXiv:1111.4702 [gr-qc]] DOI: 10.1063/1.3702449

Chicone, C. and Mashhoon, B., 2013, *Phys. Rev. D* **87**, 064015. [arXiv:1210.3860 [gr-qc]] DOI: 10.1103/PhysRevD.87.064015

Chicone, C. and Mashhoon, B., 2016a, *Class. Quantum Grav.* **33**, 075005. [arXiv:1508.01508 [gr-qc]] DOI: 10.1088/0264-9381/33/7/075005

Chicone, C. and Mashhoon, B., 2016b, *J. Math. Phys.* **57**, 072501. [arXiv:1510.07316 [gr-qc]] DOI: 10.1063/1.4958902

Cho, Y.M., 1976, *Phys. Rev. D* **14**, 3335.

Chowdhury, D. and Basu, B., 2013, *Ann. Phys.* **339**, 358.

Clowe, D., *et al.*, 2006, *Astrophys. J. Lett.* **648**, L109.

Clowe, D., Randall, S.W. and Markevitch, M., 2007, *Nucl. Phys. B, Proc. Suppl.* **173**, 28.

CMS Collaboration, 2012, *Phys. Lett. B* **716**, 30.

Cohen, I.B., 1960, *The Birth of a New Physics*, Doubleday Anchor, Garden City, NY.

Conroy, A., *et al.*, 2015, *Class. Quantum Grav.* **32**, 015024. [arXiv:1406.4998 [hep-th]]

Corbelli, E., 2003, *Mon. Not. R. Astron. Soc.* **342**, 199.

Dahia, F. and Felix da Silva, P.J., 2015, *Class. Quantum Grav.* **32**, 177001.

Damião Soares, I. and Tiomno, J., 1996, *Phys. Rev. D* **54**, 2808.

Damjanov, I., *et al.*, 2009, *Astrophys. J.* **695**, 101.

Davis, H.T., 1930, *The Theory of the Volterra Integral Equation of Second Kind*, Indiana University Studies, vol. 17.

Demirel, B., Sponar, S. and Hasegawa, Y., 2015, *New J. Phys.* **17**, 023065.

Deng, X.M. and Xie, Y., 2015, *Ann. Phys.* **361**, 62.

Dewan, E. and Beran, M., 1959, *Am. J. Phys.* **27**, 517.

Einstein, A., 1930, *Math. Ann.* **102**, 685.

Einstein, A., 1949, in *Albert Einstein: Philosopher-Scientist*, edited by P. A. Schilpp, Library of Living Philosophers, Evanston, IL.

Einstein, A., 1950, *The Meaning of Relativity*, Princeton University Press, Princeton, NJ.

Fabris, J.C. and Pereira Campos, J., 2009, *Gen. Relativ. Gravit.* **41**, 93.

Faraut, J. and Viano, G.A., 1986, *J. Math. Phys.* **27**, 840.

Fermi, E., 1936, *Ric. Sci.* **VII-11**, 13.

Fienga, A., *et al.*, 2015, *Celest. Mech. Dyn. Astron.* **123**, 325.

Fleck, J.-J. and Kuhn, J.R., 2003, *Astrophys. J.* **592**, 147.

Friedman, Y. and Scarr, T., 2013, *Phys. Scr.* **87**, 055004.

Friedman, Y. and Scarr, T., 2015, *Gen. Relativ. Gravit.* **47**, 121.

Frolov, V.P. and Shoom, A.A., 2011, *Phys. Rev. D* **84**, 044026.

Frolov, V.P. and Shoom, A.A., 2012, *Phys. Rev. D* **86**, 024010.

Galley, C.R., 2013, *Phys. Rev. Lett.* **110**, 174301.

Garetz, B.A., 1981, *J. Opt. Soc. Am.* **71**, 609.

Garetz, B.A. and Arnold, S., 1979, *Opt. Commun.* **31**, 1.

Gödel, K., 1949, in *Albert Einstein: Philosopher-Scientist*, edited by P. A. Schilpp, Library of Living Philosophers, Evanston, IL.

Goldberger, M.L. and Watson, K.M., 1964, *Collision Theory*, Wiley, New York.

Gottfried, K., 1966, *Quantum Mechanics*, Benjamin, New York.

Gradshteyn, I.S. and Ryzhik, I.M., 1980, *Table of Integrals, Series and Products*, Academic Press, New York.

Gurbatov, S.N., Saichev, A.I. and Shandarin, S.F., 2012, *Physics-Uspekhi* **55**, 223.

Hadamard, J., 1952, *Lectures on Cauchy's Problem in Linear Partial Differential Equations*, Dover, New York.

Hamada, M., Yokoyama, T. and Murakami, S., 2015, *Phys. Rev. B* **92**, 060409(R).

Hardy, G.H., Littlewood, J.E. and Pólya, G., 1988, *Inequalities*, 2nd edn, Cambridge University Press, Cambridge, UK.

Harrison, E., 2000, *Cosmology*, 2nd edn, Cambridge University Press, Cambridge, UK.

Harvey, D., *et al.*, 2015, *Science* **347**, 1462.

Hauck, J.C. and Mashhoon, B., 2003, *Ann. Phys. (Berlin)* **12**, 275. [arXiv: gr-qc/0304069]

Hayashi, K. and Shirafuji, T., 1979, *Phys. Rev. D* **19**, 3524.

Hees, A., *et al.*, 2015, in *SF2A-2015: Proceedings of the Annual Meeting of the French Society of Astronomy and Astrophysics*, edited by F. Martins, S. Boissier, V. Buat, L. Cambrésy and P. Petit, pp. 125–131. [arXiv: 1509.06868 [gr-qc]]

Hehl, F.W., 1971, *Phys. Lett. A* **36**, 225.

Hehl, F.W., 2008, *Ann. Phys. (Berlin)* **17**, 691.

Hehl, F.W. and Mashhoon, B., 2009a, *Phys. Lett. B* **673**, 279. [arXiv: 0812.1059 [gr-qc]]

Hehl, F.W. and Mashhoon, B., 2009b, *Phys. Rev. D* **79**, 064028. [arXiv: 0902.0560 [gr-qc]]

Hehl, F.W. and Ni, W.-T., 1990, *Phys. Rev. D* **42**, 2045.

Hehl, F.W., Nitsch, J. and Von der Heyde, P., 1980, in *General Relativity and Gravitation*, edited by A. Held, vol. 1, Plenum, New York.

Hehl, F.W. and Obukhov, Yu.N., 2003, *Foundations of Classical Electrodynamics: Charge, Flux, and Metric*, Birkhäuser, Boston, MA.

Hehl, F.W., Obukhov, Yu.N. and Puetzfeld, D., 2013, *Phys. Lett. A* **377**, 1775. [arXiv:1304.2769 [gr-qc]]

Hicks, N.J., 1965, *Notes on Differential Geometry*, D. Van Nostrand, Princeton, NJ.

Hopkinson, J., 1877, *Phil. Trans. Roy. Soc. London* **167**, 599.

Hoyle, C.D., *et al.*, 2004, *Phys. Rev. D* **70**, 042004. [arXiv: hep-ph/0405262]

Ieda, J., Matsuo, M. and Maekawa, S., 2014, *Solid State Communications* **198**, 52.

Iorio, L., 2015, *Int. J. Mod. Phys. D* **24**, 1530015. [arXiv: 1412.7673 [gr-qc]]

Iorio, L. and Giudice, G., 2006, *New Astron.* **11**, 600.

Jackson, J.D., 1999, *Classical Electrodynamics*, 3rd edn, Wiley, Hoboken, NJ.

Jantzen, R.T., Carini, P. and Bini, D., 1992, *Ann. Phys. (N.Y.)* **215**, 1.

Jeans, J.H., 1902, *Phil. Trans. Roy. Soc. London A* **199**, 1.

Jeans, J.H., 1929, *Astronomy and Cosmology*, Cambridge University Press, Cambridge, UK.

Jentschura, U.D. and Noble, J.H., 2014, *J. Phys. A: Math. Theor.* **47**, 045402.

Kapner, D.J., *et al.*, 2007, *Phys. Rev. Lett.* **98**, 021101. [arXiv:hep-ph/0611184]

Kennard, E.H., 1917, *Phil. Mag.* **33**, 179.

Kerr, A.W., Hauck, J.C. and Mashhoon, B., 2003, *Class. Quantum Grav.* **20**, 2727. [arXiv: gr-qc/0301057]

Kiefer, C. and Weber, C., 2005, *Ann. Phys. (Berlin)* **14**, 253.

Konno, K. and Takahashi, R., 2012, *Phys. Rev. D* **85**, 061502(R).

Kuhn, J.R., Burns, C.A. and Schorr, A.J., 1986, "Numerical Coincidences, Fictional Forces, and the Galactic Dark Matter Distribution", Princeton preprint, unpublished.

Kuhn, J.R. and Kruglyak, L., 1987, *Astrophys. J.* **313**, 1.

Kuhn, J.R. and Miller, R.H., 1989, *Astrophys. J. Lett.* **341**, L41.

Lämmerzahl, C., 1997, *Phys. Lett. A* **228**, 223.

Landau, L.D. and Lifshitz, E.M., 1988, *Mechanics*, Pergamon, Oxford.

Landau, L.D. and Lifshitz, E.M., 1960, *Electrodynamics of Continuous Media*, Pergamon, Oxford.

Landau, L.D. and Lifshitz, E.M., 1971, *The Classical Theory of Fields*, Pergamon, Oxford.

Landau, L.D. and Lifshitz, E.M., 1977, *Quantum Mechanics (Non-Relativistic Theory)*, Pergamon, Oxford.

Lendinez, S., Chudnovsky, E.M. and Tejada, J., 2010, *Phys. Rev. B* **82**, 174418.

Levi-Civita, T., 1926, *Math. Ann.* **97**, 291.

Lima, J.R.F. and Moraes, F., 2015, *Eur. Phys. J. B* **88**, 63.

Liouville, J., 1837, *J. Math.* **2**, 439.

Little, S. and Little, M., 2014, *Class. Quantum Grav.* **31**, 195008.

Lorentz, H.A., 1952, *The Theory of Electrons*, Dover, New York.

Lovitt, W.V., 1950, *Linear Integral Equations*, Dover, New York.

Lu, C., *et al.*, 2014, *Research in Astronomy and Astrophysics* **14**, 1301.

Mainwaring, S.R. and Stedman, G.E., 1993, *Phys. Rev. A* **47**, 3611.

Maluf, J.W., 2013, *Ann. Phys. (Berlin)* **525**, 339.

Maluf, J.W., Ulhoa, S.C. and Faria, F.F., 2009, *Phys. Rev. D* **80**, 044036.

Mashhoon, B., 1974, *Nature* **250**, 316.

Mashhoon, B., 1977, *Astrophys. J.* **216**, 591.

Mashhoon, B., 1978, *Astrophys. J.* **223**, 285.

Mashhoon, B., 1986, *Found. Phys.* **16** (Wheeler Festschrift), 619.

Mashhoon, B., 1987, *Phys. Lett. A* **122**, 299.

Mashhoon, B., 1988, *Phys. Rev. Lett.* **61**, 2639.

Mashhoon, B., 1989, *Phys. Lett. A* **139**, 103.

Mashhoon, B., 1990a, *Phys. Lett. A* **143**, 176.

Mashhoon, B., 1990b, *Phys. Lett. A* **145**, 147.

Mashhoon, B., 1992, *Phys. Rev. Lett.* **68**, 3812.

Mashhoon, B., 1993a, *Phys. Rev. A* **47**, 4498.

Mashhoon, B., 1993b, *Phys. Lett. A* **173**, 347.

Mashhoon, B., 1995, *Phys. Lett. A* **198**, 9.

Mashhoon, B., 1999, *Gen. Relativ. Gravit.* **31**, 681. [arXiv: gr-qc/9803017]

Mashhoon, B., 2000, *Class. Quantum Grav.* **17**, 2399. [arXiv: gr-qc/0003022]

Mashhoon, B., 2002, *Phys. Lett. A* **306**, 66. [arXiv: gr-qc/0209079]

Mashhoon, B., 2003a, in *Advances in General Relativity and Cosmology*, edited by G. Ferrarese, Pitagora, Bologna. [arXiv: gr-qc/0301065]

Mashhoon, B., 2003b, *Ann. Phys. (Berlin)* **12**, 586. [arXiv: hep-th/0309124] DOI: 10.1002/andp.200310028

Mashhoon, B., 2004a, in *Relativity in Rotating Frames*, edited by G. Rizzi and M.L. Ruggiero, Kluwer, Dordrecht. [arXiv: gr-qc/0303029]

Mashhoon, B., 2004b, *Phys. Rev. A* **70**, 062103. [arXiv: hep-th/0407278]

Mashhoon, B., 2005, *Phys. Rev. A* **72**, 052105. [arXiv: hep-th/0503205]

Mashhoon, B., 2006, *Lect. Notes Phys.* **702**, 112. [arXiv: hep-th/0507157]

Mashhoon, B., 2007a, *Ann. Phys. (Berlin)* **16**, 57. [arXiv: hep-th/0608010] DOI: 10.1002/andp.200610221

Mashhoon, B., 2007b, *Phys. Rev. A* **75**, 042112. [arXiv: hep-th/0611319]

Mashhoon, B., 2007c, *Phys. Lett. A* **366**, 545. [arXiv: hep-th/0702074]

Mashhoon, B., 2007d, in *The Measurement of Gravitomagnetism: A Challenging Enterprise*, edited by L. Iorio, NOVA Science, Hauppage, NY. [arXiv: gr-qc/0311030]

Mashhoon, B., 2008, *Ann. Phys. (Berlin)* **17**, 705. [arXiv: 0805.2926 [gr-qc]] DOI: 10.1002/andp.200810308

Mashhoon, B., 2009, *Phys. Rev. A* **79**, 062111. [arXiv: 0903.1315 [gr-qc]] DOI: 10.1103/PhysRevA.79.062111

Mashhoon, B., 2011a, *Ann. Phys. (Berlin)* **523**, 226. [arXiv: 1006.4150 [gr-qc]] DOI: 10.1002/andp.201010464

Mashhoon, B., 2011b, in *Cosmology and Gravitation*, edited by M. Novello and S. E. Perez Begliaffa, Cambridge Scientific Publishers, Cambridge, UK. [arXiv:1101.3752 [gr-qc]]

Mashhoon, B., 2012, in *Proceedings of Mario Novello's 70th Anniversary Symposium*, edited by N. Pinto Neto and S. E. Perez Bergliaffa, Editora Livraria da Fisica, Sao Paulo. [arXiv: 1204.6069 [gr-qc]]

Mashhoon, B., 2013a, *Ann. Phys. (Berlin)* **525**, 235. [arXiv: 1211.1077 [gr-qc]] DOI: 10.1002/andp.201200208

Mashhoon, B., 2013b, *Class. Quantum Grav.* **30**, 155008. [arXiv:1304.1769 [gr-qc]] DOI: 10.1088/0264-9381/30/15/155008

Mashhoon, B., 2014, *Phys. Rev. D* **90**, 124031. [arXiv:1409.4472 [gr-qc]] DOI: 10.1103/PhysRevD.90.124031

Mashhoon, B., 2015, *Galaxies* **3**, 1. [arXiv:1411.5411 [gr-qc]]

Mashhoon, B., 2016, *Universe* **2**, 9. [arXiv:1512.01193 [astro-ph.GA]]

Mashhoon, B., *et al.*, 1998, *Phys. Lett. A* **249**, 161. [arXiv: gr-qc/9808077]

Mashhoon, B. and Kaiser, H., 2006, *Physica B* **385–386**, 1381. [arXiv: quantum-ph/0508182]

Mashhoon, B. and Muench, U., 2002, *Ann. Phys. (Berlin)* **11**, 532. [arXiv: gr-qc/0206082]

Mashhoon, B. and Obukhov, Yu.N., 2013, *Phys. Rev. D* **88**, 064037. [arXiv: 1307.5470 [gr-qc]]

Mashhoon, B., Paik, H.J. and Will, C.M., 1989, *Phys. Rev. D* **39**, 2825.

Mashhoon, B. and Theiss, D.S., 1982, *Phys. Rev. Lett.* **49**, 1542.

Matsuo, M., *et al.*, 2011a, *Phys. Rev. Lett.* **106**, 076601.

Matsuo, M., *et al.*, 2011b, *Phys. Rev. B* **84**, 104410.

Matsuo, M., *et al.*, 2013, *Phys. Rev. B* **87**, 180402(R).

Matsuo, M., Ieda, J. and Maekawa, S., 2013, *Phys. Rev. B* **87**, 115301.

Matte, A., 1953, *Canadian J. Math.* **5**, 1.

Maxwell, J.C., 1880, *Nature* **21**, 314.

McCrea, W.H. and Milne, E.A., 1934, *Quart. J. Math. (Oxford)* **5**, 73.

Meyer, H., *et al.*, 2012, *Gen. Relativ. Gravit.* **44**, 2537.

Milne, E.A., 1934, *Quart. J. Math. (Oxford)* **5**, 64.

Minkowski, H., 1952, in *The Principle of Relativity*, by H. A. Lorentz, A. Einstein, H. Minkowski and H. Weyl, Dover, New York.

Misner, C.W., Thorne, K.S. and Wheeler, J.A., 1973, *Gravitation*, Freeman, San Francisco.

Møller, C., 1961, *K. Dan. Vidensk. Selsk. Mat. Fys. Skr.* **1**, 10.

Moorhead, G.F. and Opat, G.I., 1996, *Class. Quantum Grav.* **13**, 3129.

Morganti, L., *et al.*, 2013, *Mon. Not. R. Astron. Soc.* **431**, 3570.

Muench, U., Gronwald, F. and Hehl, F.W., 1998, *Gen. Relativ. Gravit.* **30**, 933.

Mukhanov, V., 2005, *Physical Foundations of Cosmology*, Cambridge University Press, Cambridge, UK.

Muñoz, R.R., *et al.*, 2005, *Astrophys. J. Lett.* **631**, L137.

Newton, R.G., 1982, *Scattering Theory of Waves and Particles*, 2nd edn, Springer-Verlag, New York.

Ni, W.-T. and Zimmermann, M., 1978, *Phys. Rev. D* **17**, 1473.

Obukhov, Yu.N., Silenko, A.J. and Teryaev, O.V., 2009, *Phys. Rev. D* **80**, 064044.

Obukhov, Yu.N., Silenko, A.J. and Teryaev, O.V., 2011, *Phys. Rev. D* **84**, 024025.

Obukhov, Yu.N., Silenko, A.J. and Teryaev, O.V., 2013, *Phys. Rev. D* **88**, 084014.

Obukhov, Yu.N., Silenko, A.J. and Teryaev, O.V., 2014, *Phys. Rev. D* **90**, 124068.

Obukhov, Yu.N., Silenko, A.J. and Teryaev, O.V., 2016, *Phys. Rev. D* **94**, 044019.

Oh, S.-H., *et al.*, 2015, *Astron. J.* **149**, 180.

O'Neill, B., 1966, *Elementary Differential Geometry*, Academic Press, New York.

Paik, H.J., 2008, *Gen. Relativ. Gravit.* **40**, 907.

Papini, G., 2002, *Phys. Rev. D* **65**, 077901.

Papini, G., 2013, *Phys. Lett. A* **377**, 960.

Peebles, P.J.E., 1993, *Principles of Physical Cosmology*, Princeton University Press, Princeton, NJ.

Pegram, G.B., 1917, *Phys. Rev.* **10**, 591.

Pellegrini, C. and Plebański, J., 1963, *K. Dan. Vidensk. Selsk. Mat. Fys. Skr.* **2**, 4.

Peralta de Arriba, L., *et al.*, 2014, *Mon. Not. R. Astron. Soc.* **440**, 1634.

Perlick, V., 2004, *Phys. Rev. D* **69**, 064017.

Petters, A.O., Levine, H. and Wambsganss, J., 2001, *Singularity Theory and Gravitational Lensing*, Birkhäuser, Boston, MA.

Pitjeva, E.V. and Pitjev, N.P., 2013, *Mon. Not. R. Astron. Soc.* **432**, 3431.

Poisson, S., 1823, *Mém. Acad. Roy. Sci. Inst. France* **6**, 441.

Poisson, E., Pound, A. and Vega, I., 2011, *Living Rev. Relativity* **14**, 7.

Pota, V., *et al.*, 2015, *Mon. Not. R. Astron. Soc.* **450**, 3345.

Puetzfeld, D. and Obukhov, Yu.N., 2014, *Int. J. Mod. Phys. D* **23**, 1442004.

Rahvar, S. and Mashhoon, B., 2014, *Phys. Rev. D* **89**, 104011. [arXiv:1401.4819 [gr-qc]]

Ramos, J. and Mashhoon, B., 2006, *Phys. Rev. D* **73**, 084003. [arXiv: gr-qc/0601054]

Randono, A., 2010, *Phys. Rev. D* **81**, 024027.

Rauch, H. and Werner, S.A., 2015, *Neutron Interferometry*, 2nd edn, Oxford University Press, Oxford, UK.

Riesz, M., 1949, *Acta Math.* **81**, 1.

Riles, K., 2013, *Prog. Part. Nucl. Phys.* **68**, 1. [arXiv: 1209.0667 [hep-ex]]

Roberts, M.D., 1993, *Astrophys. Lett. Commun.* **28**, 349. [arXiv: gr-qc/ 9905005]

Roberts, M.S. and Whitehurst, R.N., 1975, *Astrophys. J.* **201**, 327.

Robertson, H.P., 1949, *Rev. Mod. Phys.* **21**, 378.

Rubin, V.C. and Ford, W.K., 1970, *Astrophys. J.* **159**, 379.

Rudin, W., 1966, *Real and Complex Analysis*, McGraw-Hill, New York.

Ruse, H.S., 1930, *Quart. J. Math. (Oxford)* **1**, 146.

Ruse, H.S., 1931, *Proc. London Math. Soc.* **32**, 87.

Ryder, L., 1998, *J. Phys. A* **31**, 2465.

Ryder, L., 2009, *Introduction to General Relativity*, Cambridge University Press, Cambridge, UK.

Sanders, R.H., 2006, *Mon. Not. R. Astron. Soc.* **370**, 1519.

Sarazin, C.L., 1988, *X-Ray Emission from Clusters of Galaxies*, Cambridge University Press, Cambridge, UK.

Sauerbrey, R., 1996, *Phys. Plasmas* **3**, 4712.

Scarr,T. and Friedman, Y., 2016, *Gen. Relativ. Gravit.* **48**, 65.

Schneider, P., Ehlers, J. and Falco, E.E., 1999, *Gravitational Lenses*, Springer-Verlag, Berlin.

Schweizer, M. and Straumann, N., 1979, *Phys. Lett. A* **71**, 493.

Schweizer, M., Straumann, N. and Wipf, A., 1980, *Gen. Relativ. Gravit.* **12**, 951.

Seigar, M.S., 2015, *Dark Matter in the Universe*, Morgan and Claypool, San Rafael, CA.

Sereno, M. and Jetzer, Ph., 2006, *Mon. Not. R. Astron. Soc.* **371**, 626.

Shapiro, I.I., 1980, in *General Relativity and Gravitation*, edited by A. Held, Plenum, New York.

Shen, J.-Q. and He, S.-L., 2003, *Phys. Rev. B* **68**, 195421.

Silenko, A.J. and Teryaev, O.V., 2007, *Phys. Rev. D* **76**, 061101(R).

Singh, D. and Papini, G., 2000, *Nuovo Cimento B* **115**, 223.

Sofue, Y. and Rubin, V., 2001, *Annu. Rev. Astron. Astrophys.* **39**, 137.

Sollima, A., Bellazzini, M. and Lee, J.W., 2012, *Astrophys. J.* **755**, 156.

Soussa, M.E. and Woodard, R.P., 2003, *Class. Quantum Grav.* **20**, 2737.

Stairs, I.H., 2003, *Living Rev. Relativity* **6**, 5.

Stedman, G.E., 1997, *Rep. Prog. Phys.* **60**, 615.

Stephani, H., *et al.*, 2003, *Exact Solutions of Einstein's Field Equations*, 2nd edn, Cambridge University Press, Cambridge, UK.

Streater, R.F. and Wightman, A.S., 1964, *PCT, Spin and Statistics, and All That*, Benjamin, New York.

Swann, W.F.G., 1920, *Phys. Rev.* **15**, 365.

Synge, J.L., 1931, *Proc. London Math. Soc.* **32**, 241.

Synge, J.L., 1965, *Relativity: The Special Theory*, 2nd edn, North-Holland, Amsterdam.

Synge, J.L., 1971, *Relativity: The General Theory*, North-Holland, Amsterdam.

Tamm, A., *et al.*, 2012, *Astron. Astrophys.* **546**, A4.

Tejada, J., *et al.*, 2010, *Phys. Rev. Lett.* **104**, 027202.

Teyssandier, P., 1977, *Phys. Rev. D* **16**, 946.

Teyssandier, P., 1978, *Phys. Rev. D* **18**, 1037.

Tohline, J.E., 1983, in *IAU Symposium 100, Internal Kinematics and Dynamics of Galaxies*, edited by E. Athanassoula, Reidel, Dordrecht.

Tohline, J.E., 1984, *Ann. N.Y. Acad. Sci.* **422**, 390.

Tricomi, F.G., 1957, *Integral Equations*, Interscience, New York.

Tsamis, N.C. and Woodard, R.P., 2014, *J. Cosmol. Astropart. Phys.* **09**, 008. [arXiv:1405.4470 [astro-ph.CO]]

Tully, R.B. and Fisher, J.R., 1977, *Astron. Astrophys.* **54**, 661.

Umstadter, D., 2001, *Phys. Plasmas* **8**, 1774.

Van Bladel, J., 1976, *Proc. IEEE* **64**, 301.

Van Bladel, J., 1984, *Relativity and Engineering*, Springer, Berlin.

van Dokkum, P.G., *et al.*, 2008, *Astrophys. J. Lett.* **677**, L5.

van Dokkum, P.G., *et al.*, 2010, *Astrophys. J.* **709**, 1018.

Volterra, V., 1959, *Theory of Functionals and of Integral and Integro-Differential Equations*, Dover, New York.

Weinberg, S., 1972, *Gravitation and Cosmology*, Wiley, New York.

Weitzenböck, R., 1923, *Invariantentheorie*, Noordhoff, Groningen.

Werner, S., 2008, *Gen. Relativ. Gravit.* **40**, 921.

Werner, S.A., Staudenmann, J.-L. and Colella, R., 1979, *Phys. Rev. Lett.* **42**, 1103.

Weyl, H., 1952, *Space–Time–Matter*, Dover, New York.

Wigner, E.P., 1939, *Ann. Math.* **40**, 149.

Wucknitz, O. and Sperhake, U., 2004, *Phys. Rev. D* **69**, 063001.

Zel'dovich, Ya.B. and Novikov, I.D., 1983, *Relativistic Astrophysics*, vol. 2, *The Structure and Evolution of the Universe*, University of Chicago Press, Chicago, IL.

Zhang, Y.Z., 1997, *Special Relativity and its Experimental Foundations*, World Scientific, Singapore.

Zwicky, F., 1933, *Helv. Phys. Acta* **6**, 110.

Zwicky, F., 1937, *Astrophys. J.* **86**, 217.

Index